Detection and Analysis of Microorganisms by Mass Spectrometry

New Developments in Mass Spectrometry

Editor-in-chief:
Frank Sobott, *University of Leeds, UK*

Series editors:
Marek Domin, *Boston College, USA*
Zhongping Yao, *The Hong Kong Polytechnic University, China*

Titles in the Series:

1: Quantitative Proteomics
2: Ambient Ionization Mass Spectrometry
3: Sector Field Mass Spectrometry for Elemental and Isotopic Analysis
4: Tandem Mass Spectrometry of Lipids: Molecular Analysis of Complex Lipids
5: Proteome Informatics
6: Capillary Electrophoresis–Mass Spectrometry for Metabolomics
7: Lipidomics: Current and Emerging Techniques
8: Processing Metabolomics and Proteomics Data with Open Software: A Practical Guide
9: Advanced Fragmentation Methods in Biomolecular Mass Spectrometry: Probing Primary and Higher-order Structure with Electrons, Photons and Surfaces
10: Advanced Mass Spectrometry-based Analytical Separation Techniques for Probing the Polar Metabolome
11: Ion Mobility–Mass Spectrometry: Fundamentals and Applications
12: MALDI Mass Spectrometry Imaging: From Fundamentals to Spatial Omics
13: Detection and Analysis of Microorganisms by Mass Spectrometry

How to obtain future titles on publication:
A standing order plan is available for this series. A standing order will bring delivery of each new volume immediately on publication.

For further information please contact:
Book Sales Department, Royal Society of Chemistry, Thomas Graham House, Science Park, Milton Road, Cambridge, CB4 0WF, UK
Telephone: +44 (0)1223 420066, Fax: +44 (0)1223 420247
Email: booksales@rsc.org
Visit our website at www.rsc.org/books

Detection and Analysis of Microorganisms by Mass Spectrometry

Edited by

Liang Qiao
Fudan University, China
Email: liang_qiao@fudan.edu.cn

and

Jia Yi
Fudan University, China
Email: jia_yi@fudan.edu.cn

ROYAL SOCIETY OF **CHEMISTRY**

New Developments in Mass Spectrometry No. 13

Print ISBN: 978-1-83916-763-8
PDF ISBN: 978-1-83767-033-8
EPUB ISBN: 978-1-83767-034-5
Print ISSN: 2045-7545
Electronic ISSN: 2045-7553

A catalogue record for this book is available from the British Library

The Royal Society of Chemistry is a charity, registered in England and Wales, Number 207890, and a company incorporated in England by Royal Charter (Registered No. RC000524), registered office: Burlington House, Piccadilly, London W1J 0BA, UK, Telephone: +44 (0)20 7437 8656.

For further information see our web site at www.rsc.org

Printed in the United Kingdom by CPI Group (UK) Ltd, Croydon, CR0 4YY, UK

Preface

Microorganisms, including bacteria, viruses, fungi and other eukaryotic microbes, are basic components of the earth's ecosystem, which are involved in photosynthesis, biodegradation, biomass production, *etc.* In the human body, there are also millions of living microorganisms, involved in protecting the body from invaders, helping in digestion, regulating moods, *etc.* Microorganisms can also be harmful pathogens to humans, animals and plants, causing infectious diseases. For instance, the coronavirus disease 2019 (COVID-19) caused by severe acute respiratory syndrome coronavirus 2 (SARS-CoV-2) has caused a considerable loss of life and property since its outbreak. Comprehensive analysis and characterization of microbes is of significant importance to understand the function and role of microorganisms, as well as to control the spread and development of pathogens.

Mass spectrometry is a technique used to ionize molecules and detect the mass-to-charge ratio of the generated ions. The technique was developed originally for elemental analysis in the early 20th century. Since the 1980s, thanks to the development of soft ionization techniques, mass spectrometry has been extensively applied to bioanalysis, to characterize proteins, metabolites, lipids, nucleic acids, *etc.* In the 1990s, matrix-assisted laser desorption/ionization time-of-flight mass spectrometry (MALDI-TOF MS) was used to identify bacteria. Whole cells can be directly analyzed by MALDI-TOF MS to collect mass spectra with dozens of peaks as the molecular mass fingerprints of the microbes. Based on the characteristic mass fingerprint, bacteria can be identified at the species level. Nowadays, the technique is widely used in hospitals for pathogenic bacterial identification, as well as in environmental science and food science for biosafety control. Mass spectrometry is also combined with gene amplification techniques to detect the characteristic genes of

New Developments in Mass Spectrometry No. 13
Detection and Analysis of Microorganisms by Mass Spectrometry
Edited by Liang Qiao and Jia Yi

Published by the Royal Society of Chemistry, www.rsc.org

microbes for microbial identification. Mass spectrometry-based proteomics and metabolomics are used for microbial identification, microbial metabolism studies and microbiota research. By quantifying proteins and metabolites, the specific function regulation in drug-resistant bacteria, as well as the function regulation in microflora, can be revealed.

This book summarizes the most recent developments of mass spectrometry techniques in microbial analysis, including mass spectrometry-based microbial identification, bacterial antimicrobial resistance studies, data mining algorithm development, omics for microbial research, and applications in clinical diagnosis, environmental science and food science. This book aims at being a reference for researchers in the field and for those who are about to enter the field.

Liang Qiao
Jia Yi

Contents

New Developments in Mass Spectrometry No. 13
Detection and Analysis of Microorganisms by Mass Spectrometry
Edited by Liang Qiao and Jia Yi

Published by the Royal Society of Chemistry, www.rsc.org

CHAPTER 1

Introduction of Mass Spectrometry-based Microorganism Detection

JIA YI, DONGXUE ZHANG AND LIANG QIAO*

Chemistry Department, Fudan University, Shanghai, 200000, China
*Email: liang_qiao@fudan.edu.cn

1.1 Introduction

The intractability of the epidemic caused by SARS-CoV-2 has brought our attention to pathogenic microorganisms to a new level. In addition to COVID-19, infectious diseases such as Ebola, Legionnaires' disease, and monkeypox pose a huge threat to public health.[1,2] These facts demonstrate the deleterious effects of microorganisms on higher organisms. Furthermore, microorganisms are abundant and have versatile metabolism and physiology, playing an important role in the earth's system processes, such as photosynthesis, biodegradation, and biomass production. In the human body, millions of living microorganisms are involved in protecting the body from invaders,[3] facilitating digestion,[4] regulating moods,[5] *etc.*

To better cope with problems caused by microorganisms and utilize microorganism sources, it is important to understand the role and function of microorganisms through comprehensive analysis and characterization. Confirming the microorganism's identity is the first step, providing clues for further investigations and applications. With the development of detection techniques and the increasing number of discovered microorganisms, more and more shared characteristics between different strains are being

New Developments in Mass Spectrometry No. 13
Detection and Analysis of Microorganisms by Mass Spectrometry
Edited by Liang Qiao and Jia Yi

Published by the Royal Society of Chemistry, www.rsc.org

detected. This, in turn, raises the need for analytical techniques with higher sensitivity and resolution.

Microorganism detection has been developed from simple biochemical reaction-based conventional methods to molecular technique-based and mass spectrometry (MS)-based modern methods. It has been improved a lot in throughput and accuracy. MS has its merits in measuring various biological molecules, such as peptides, proteins, metabolites, lipids, and nucleic acids, making it a fundamental tool for characterizing microorganisms. Compared with conventional methods, it shortens the turnover time and provides more powerful discrimination. Due to their brilliant ability to analyze microbial community composition, molecular biology techniques are regarded as the "gold standard" in microbial analysis, and these techniques will continue to have an important role in the field.

MS-based microorganism detection has been developed for nearly 50 years, involving many aspects. In general, it can be divided into the development of application-oriented rapid microorganism detection methods, such as microbial identification and classification, and the development of discovery-based microbial analysis methods, such as different "omics" and MS imaging. This chapter is intended to provide a general overview of rapid MS-based microorganism detection methods. They include the main MS techniques used for rapid microorganism detection, such as matrix-assisted laser ionization/desorption time-to-flight MS (MALDI-TOF-MS) and electrospray ionization MS (ESI-MS). Due to the popularity of MS-based methods in clinical applications, the development of microorganism detection in clinical laboratories is the fastest growing field of all application fields. In recent years, there has been growing interest in its application in environmental science and the food industry. The chapter aims to help researchers in these fields or other related fields to have a preliminary understanding of the availability of MS for their research, especially for researchers who have no experience in using MS.

1.2 Development of Microorganism Detection

There are various approaches available for the detection of microorganisms, which can be briefly classified into four categories: conventional methods, MS-based methods, nucleic acid-based methods, and other modern techniques (Figure 1.1). The advancement of microorganism detection methods is closely related to instrument technology development, and modern methods for microorganism identification have recently been centered around MS.

Prior to the advent of molecular techniques in microbiology, phenotypic methods were the only means of microorganism characterization, and hence, the most conventional methods for microorganism detection are phenotypic methods. Direct observation with the naked eye or microscopy remains a valuable step in microorganism detection and is based on differences in morphology and appearance. It is often used alone or in

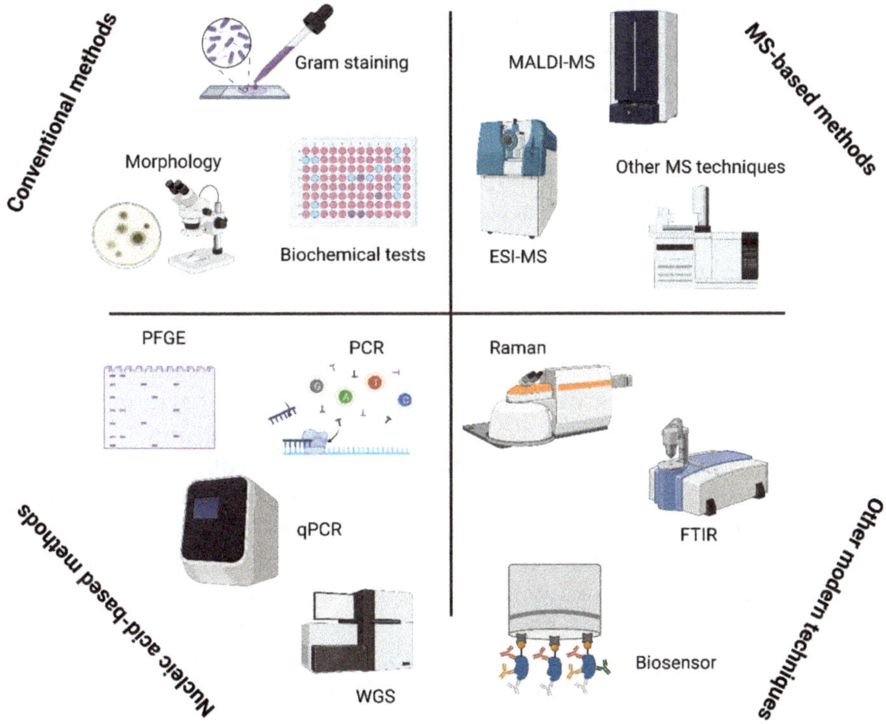

Figure 1.1 Methods used for microorganism detection. ESI: electrospray ionization. PFGE: pulsed-field gel electrophoresis. PCR: polymerase chain reaction. qPCR: quantitative polymerase chain reaction. WGS: whole-genome sequencing. FTIR: Fourier transform infrared spectroscopy.

conjunction with other complementary techniques. Combining Gram staining and microscopy is a rapid screening technique for bacteria in clinical laboratories and the food industry.[6] With the invention of electron microscopy, the resolution of distinguishing microorganisms has been significantly improved, even for viruses.[7] However, when it comes to identification at the species and subspecies levels, the principle of microscopy limits its discriminating power. Microscopy alone is insufficient to identify microorganisms because most microorganisms are small and vary considerably in size, and it cannot show the phylogenetic diversity of the microorganisms under study.

Various system reactivity assays, such as pH-based reactions, enzyme assays, carbon source utilization, acid detection, and analysis of carbohydrate utilization, preformed enzymes, organic products, and cellular fatty acids, can be systematically combined to identify microorganisms.[8] Commercial automatic systems and kits have been developed for microbial biochemical tests, which allow microbial identification in as little as 4 to 48 hours.

In the 1970s, molecular tools and MS began to be increasingly applied to life science, thereby improving the ability to understand complex biological systems.[9,10] In 1985, the polymerase chain reaction was developed for bacterial identification. Subsequently, different molecular techniques, such as PFGE, qPCR, and WGS, were developed and applied to microbial characterization and identification.[11,12] These techniques have revolutionized microorganism detection.

MS is a powerful tool for analyzing biomolecules such as metabolites, lipids, proteins, and nucleic acids.[13] It has gained attention for microorganism detection since the 1970s. In 1975, MS was first tried for identifying bacteria.[14] Gently heated bacterial samples were subjected to electron impact ionization to produce ions of volatile metabolites like diglycerides and ubiquinones, which could help distinguish bacterial species. Subsequently, many methods combining pyrolysis of bacteria and MS to produce lipid biomarkers were developed to distinguish bacteria.[15,16] Fast atom bombardment was also applied to detect bacteria, and the spectra quantitatively revealed differences in the phospholipid compositions among different bacteria, indicating its aid in differentiating bacteria.[17] The advent of soft ionization techniques made the analysis of proteins and nucleic acids easier, producing fewer hard-to-discriminate ion fragments compared to hard ionization techniques. MALDI and ESI are two representative soft ionization approaches that have greatly promoted the development and application of biomolecule analysis by MS. As current methods for microorganism detection are mainly based on MALDI and ESI, we will focus on introducing them. Additionally, we will briefly introduce other novel microorganism detection techniques based on MS.

1.3 MS-based Microorganism Detection

1.3.1 MALDI-TOF-MS

When proteins are exposed to laser radiation directly, they tend to become fragmented. To address this issue, small molecules can be used as a matrix to protect the proteins from fragmentation. Although the potential of MALDI-TOF-MS to identify microorganisms was proven, it took more than a decade before a commercialized MALDI-TOF-MS-based identification system was established. Prior to commercialization, several challenges needed to be addressed, including the composition of the protein extraction solvent, the choice of matrix, and the establishment of a reference database. These efforts aimed to provide reproducible mass spectra of microorganisms and accurate identification results.

In 1987, Gain *et al.* used MALDI-TOF-MS to analyze water-soluble proteins of lyzed bacteria to differentiate several bacterial species.[18] Later, bacteria were differentiated based on the MALDI-TOF-MS analysis of intact bacterial cells.[19] As a result of its simple sample preparation and high-throughput capabilities, MALDI-TOF-MS-based microorganism detection methods have become increasingly popular. The mass range (m/z) typically used for MALDI-TOF-MS-based microbial identification is 2000 to 20 000 Da. Most of the peaks detected are from ribosomal proteins, which are conserved within the species. Within the mass range, up to 50 ribosomal proteins can be identified, resulting in reproducible spectra even when the cultivation conditions change.[20–22] In addition to the ribosomal proteins, the typical whole-cell mass spectrum also contains other identifiable proteins, most of which are structural or regulatory in nature, such as ribosome modulation factors, carbon storage regulators, cold-shock proteins, and DNA-binding proteins.[21] The stable spectral patterns of conserved ribosomal and house-keeping proteins support the fact that the MS signals of house-keeping proteins can be used as phylogenetic markers, similar to multilocus sequence analysis of house-keeping genes. The similarity of patterns indicates the closeness of microorganisms' phylogenetic relationships.[20–22]

Commercialized MALDI-TOF-MS-based microbial identification systems, such as Bruker Biotyper and Shimadzu VITEK MS systems, have been approved for clinical use by the U.S. FDA.[23] Compared to traditional identification systems, these commercial systems provide consistent or better findings, making MALDI-TOF-MS technology the preferred method for microorganism detection.[24,25] With the standardization of protocols, expansion of spectral libraries and databases, and creation of more user-friendly computer programs, more diagnostic methods based on MALDI-TOF-MS are being developed. The two main types of approaches to characterizing microorganisms are the library-based approach and the bioinformatics-based approach. Given the popularity of library-based methods, this chapter will focus on the introduction of the library-based technique and its application.

MALDI-TOF-MS-based microbial identification offers numerous advantages. However, there are ongoing needs to decrease the turnaround time, broaden the coverage of identifiable species, increase the accuracy of identification, enhance the discrimination power of closely related species and subspecies, and expand its application in microbial detection beyond identification.[25,26]

Although the high throughput and simple sample preparation of MALDI-TOF-MS help speed up the identification process, the total turnaround time remains longer due to the culture required to collect sufficient cells before analysis. To shorten or avoid microbial culture, microbial enrichment techniques have been developed, such as immunoaffinity enrichment[27] and microfluidics with different enrichment mechanisms.[28–31] Zhu *et al.* demonstrated a method that combines antibody-modified magnetic beads and MALDI-TOF-MS to enrich and identify bacteria in blood samples, enabling identification within hours.[32] Shen *et al.* developed a MALDI-compatible method to enrich bacteria by combining microfluidics and vancomycin-modified magnetic beads (Figure 1.2A).[33] Condina *et al.* described an approach that combined inertial microfluidics and secondary flows in a spiral microchannel to isolate yeasts (*Saccharomyces pastorianus* and *Saccharomyces cerevisiae*) from beer spoilage microorganisms (*Lactobacillus brevis* and *Pediococcus damnosus*), which were then analyzed using MALDI-TOF-MS (Figure 1.2B).[34] This method has the potential to be applied to the brewing industry for biological control.

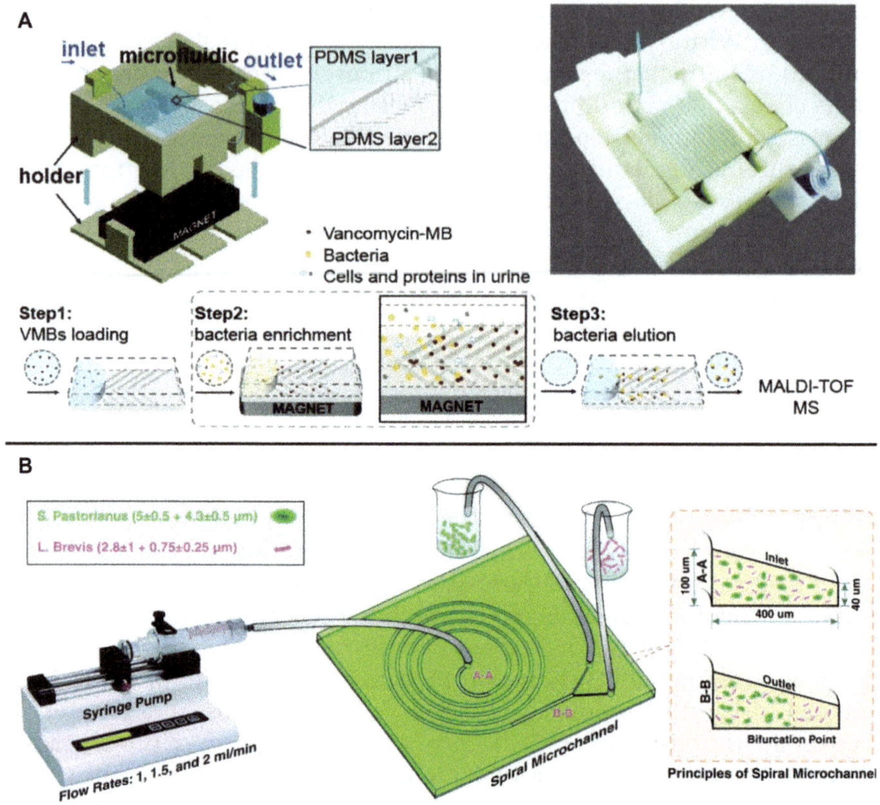

Figure 1.2 (A) The assembly of the integrated device, and workflow of enrichment and MALDI-TOF analysis of bacteria that cause urinary tract infections. Reproduced from ref. 33 with permission from the Royal Society of Chemistry. (B) Schematic illustration of the experimental setup for *Saccharomyces pastorianus* and *Lactobacillus brevis* separation. Reproduced from ref. 34 with permission from the Royal Society of Chemistry.

MALDI-TOF-MS has provided brilliant identification and discrimination power for plenty of microorganisms; differentiating some closely related species/subspecies, particularly Gram-positive bacteria like *Streptococcus pneumoniae* and members of the *Streptococcus oralis*/*Streptococcus mitis* group,[35,36] and *Listeria* spp.,[37] and some Gram-negative bacteria like *Shigella* spp. with low rates of differences in their ribosomal protein sequences with *Escherichia coli*,[38] some *Enterobacter* spp.,[39] and some *Stenotrophomonas maltophilia* strains, remains a challenge.[40] The ability of MALDI-TOF-MS to distinguish between *E. coli* serotypes, *Brucella suis* biovars, and clinically significant multidrug-resistant *Klebsiella pneumoniae* clones below the species level has been shown to be limited.[41] Additionally, there are difficulties in identifying some anaerobes at the species level.[42] Related species are usually genotypically regarded as the same species, indicating the difficulty of differentiation based on protein profiling. Therefore, characterization at the subspecies/strain level requires more information content, repeatability, and mass accuracy than characterization at the species level, as proposed.[43] To broaden the coverage of identifiable species and enhance the classification and identification of closely related species/subspecies, expanding the strain mass spectral database and improving algorithms are two key issues. The success of species matching grows as the number of strains per species in the database is increased.[44]

There have been attempts to improve commercial databases, open-source databases, and laboratories' customized databases in terms of their coverage and matching algorithms.[45] The microbial coverages of commercialized databases (*i.e.*, MALDI Biotyper and VITEK MS) are continuously increased. These databases often allow customers build in-house libraries of mass spectra to identify strains not included in the original databases. Although the commercialized databases are available and applicable to common strains, it is important to establish custom-made reference databases, which will promote identification or classification of those localized strains and be applied to those specific situations. To improve the identification of anaerobic bacteria, the European Network for the Rapid Identification of Anaerobes project was conducted and it expanded the anaerobic bacterial mass spectra in the Biotyper database.[46] A reference database was built to identify the microorganism yeast forms and early mycelial cultures, and it can be applied to identify the fungus *Histoplasma capsulatum*.[47] The traditional identification of *H. capsulatum* is based on intracellular visualization in tissue after culture, which is time-consuming and needs to be performed in a biosafety level 3 facility. Paauw *et al.* proposed a MALDI-MS-based method for classifying the *E. coli–Shigella* phylogroup by constructing a custom reference library involving the genetic diversity of *E. coli* and *Shigella* strains. 94.4% of 180 tested isolates were correctly classified as *E. coli* or *Shigellae*.[48] However, there are only a few MALDI-MS spectral databases that are publicly available for microbial identification, such as FoodBIMS[49] and SpectraBank.[50]

In 1999, Holland *et al.* were one of the first groups to state the necessity for robust and reliable software techniques to compare spectra.[19] Currently, it is possible to compare the spectra of various microorganisms and quantify how similar their spectra are by using open-source and commercial software tools. The improvements in algorithms can increase the precision of identification and classification. Those software packages supporting comparison of large datasets will improve the precision of identification. For example, ClinProTools can distinguish *S. pneumonia*, *S. oralis*, and *S. mitis*.[51] It can also distinguish methicillin-resistant *Staphylococcus aureus* (MRSA) and methicillin-sensitive *Staphylococcus aureus* (MSSA).[52] Karger *et al.* categorized three serovars of *E. coli* that produce Shiga toxins.[53] *Yersinia enterocolitica* strains were classified as pathogenic or non-pathogenic by MALDI-TOF-MS.[54] Similar to this, *Francisella tularensis* is the highly infectious causative agent of tularemia and a potential bioweapon. *Francisella tularensis* strains have been categorized into one of four subspecies.[55]

In addition to these methods, other techniques have been developed to enhance the discrimination power of MALDI-TOF-MS-based microorganism characterization, such as the high-intensity focused ultrasound-assisted proteomics approach developed by Gekenidis *et al.*[56] This approach allows for improved taxonomic differentiation by increasing the mass range and variety of possible biomarker peptides. Other studies have shown that a wider mass range (from 15 to 75 kDa) can also improve the discrimination power for protein biomarkers.[43,57,58] Additional tools for high-resolution typing are still needed, and statistical approaches can be used in some circumstances to increase the discriminative power of MALDI typing methods and uncover reliable biomarkers.[59] For example, Zautner *et al.*[60] described a method that combines the analysis of variable masses seen during whole-cell MALDI-TOF-MS with ribosomal and whole-genome multilocus sequence typing (MLST) database-deduced isoform lists to improve the reliability and reproducibility of MALDI typing results. Different from conventional whole-spectrum clustering techniques, their method takes only variations in mass associated with one particular set of allelic isoforms for the same protein into account. The effective identification of important biomarkers utilizing a proteogenomics technique also serves as a demonstration that more reliable methods depend on peak shifts and the integration of genomic/proteomic data. In proteogenomics research, after sequence-based prediction, potential biomarkers are identified utilizing a shotgun proteomics technique before their discriminatory power is confirmed using linear MALDI-TOF-MS profiles.[22] Zhang *et al.* developed a method based on TiO_2-assisted laser desorption/ionization MS to discriminate different strains and characterize antimicrobial resistance by detecting metabolites from intact cells.[61] Sorensen *et al.* reported a method combining on-target lipid extraction and MALDI-TOF-MS analysis to realize the identification of bacterial and fungal species.[62] Alternative MS techniques, particularly those that enable more thorough structural biomarker characterization (such as ESI-MS), may be necessary to achieve the required level of taxonomic resolution for microorganisms that are unable to provide strain-specific MALDI-TOF-MS profiles.

Despite the recent advancements in monomicrobial identification, a significant challenge in microbiology is the accurate identification of microorganisms in polymicrobial samples.[65,66] While MALDI-TOF-MS was used to characterize simple mixtures without purification procedures in some early studies, these models were limited to equal mixing of two or three bacterial species. When spectra of such mixtures are manually compared to reference spectra of pure cultures, only a few biomarker peaks of each component are observed, and the identification results are highly influenced by the composition ratio of each kind of bacteria in the mixture. In clinical samples containing two species, a study found that only when the concentration of a bacterial species is greater than 5% in the mixture, it can be detected by MALDI.[66] A biomarker-based approach was proposed by Mahé *et al.* to analyze polymicrobial samples, which showed ~85% correct or partially correct identification of mixtures involving two to five species (Figure 1.3A and B).[63] It involves a step-by-step subtraction of individual peak profiles and comparison between the remaining peaks in the mixture spectra and

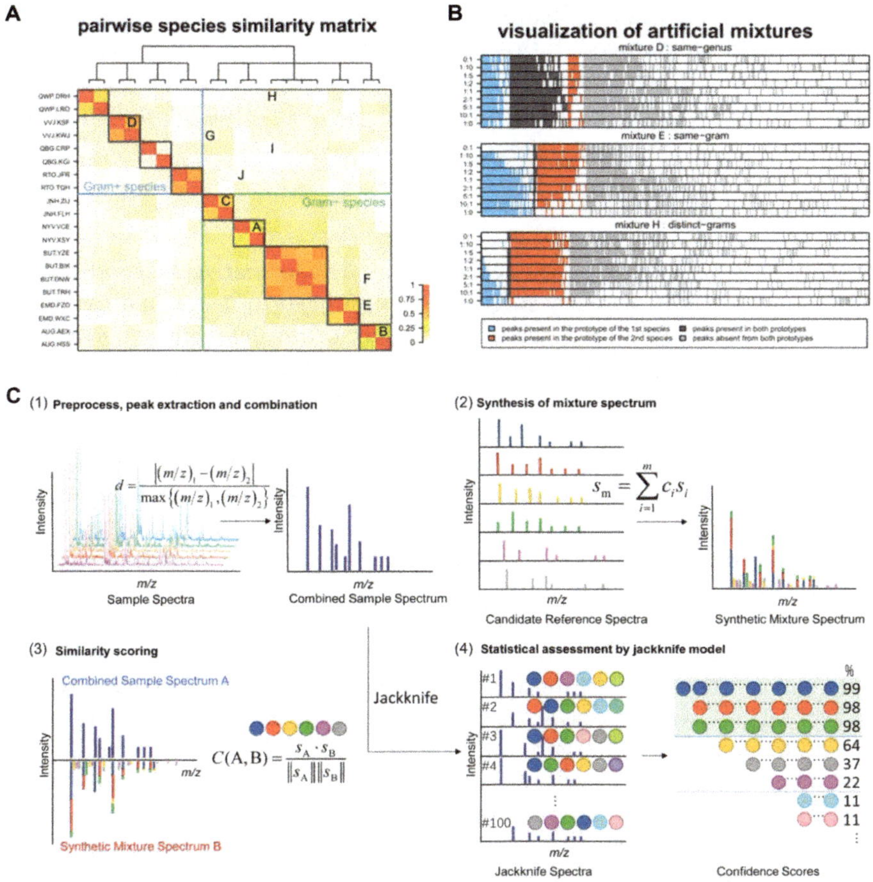

Figure 1.3 Visualization of the two datasets in Mahé's study: (A) reference database and (B) mixture dataset. Reproduced from ref. 63 with permission from Oxford University Press, Copyright 2014. (C) Overview of the framework for bacterial mixture identification by MALDI-TOF. Reproduced from ref. 64 with permission from American Chemical Society, Copyright 2018.

the databases. Commercial software like Bruker's Biotyper can be applied to both monomicrobial and polymicrobial identification, but its performance greatly depends on the species, their phylogenetic distance, and the ratio of each species.[67] Yang *et al.* proposed a framework to analyze MALDI-TOF-MS spectra of bacterial mixtures, which consists of a synthetic mixture model based on a non-negative linear combination of candidate reference spectra and a statistical assessment by *in silico* generated spectra *via* jackknife resampling, realizing characterization of each component without purification procedures. The sensitivity of this approach was found to be >80% for balanced binary and ternary mixtures, >60% for balanced quaternary and pentabasic mixtures, and 48–71% for asymmetric mixtures (Figure 1.3C).[64]

However, these results indicate that identification of real polymicrobial samples with unbalanced ratios remains challenging. Therefore, further research is needed, for example, investigating more complex models with a broader range of species, to improve the identification of polymicrobial samples.

1.3.2 ESI-MS

ESI-MS, along with MALDI, has emerged as an alternative method for biomolecule analysis. In its early stages, ESI-MS proved to be a valuable tool for characterizing bacteria by detecting lipopolysaccharides,[68] lipooligosaccharides,[69] phospholipids,[70] muramic acids,[71] and proteins.[72] Additionally, ESI-MS is frequently coupled with liquid chromatographic separation, that is, tandem MS combined with liquid chromatography (LC–MS/MS). The first report of using ESI-MS to produce information-rich spectra from bacterial suspensions in both positive and negative ion modes dates back to 1999.[73]

While whole-genome sequencing is considered to be the gold standard for identifying microbial phylogeny and taxonomy, most microbial systematic investigations still rely on biomarker-based approaches. Proteomics provide an approach to study the expression of genes by analyzing the resulting proteins, offering indirect measurements of genome sequences. The amino acid sequences of the related proteins will vary between various microbial species just as their gene sequences do. In this way, studying a microbe's proteins can be viewed as studying its genome's coding regions in a roundabout way. Proteomics can hypothetically be used to analyze microbial traits, with applications for microbial characterization, classification, and identification, provided that the changes in target molecules can be detected. For closely related strains with only minor differences in their genomes, it is necessary to recognize that the differences are also subtle in their proteomes when conducting proteomics. Informative differences can be obtained by analyzing most parts of the entire proteome of the microorganism. Proteomics can be applied to the characterization of microorganisms at various taxonomic levels.

Proteotyping of microorganisms, which refers to proteomics-based microorganism identification and classification,[74] is classified into two main categories: those that use confidently identified taxon-specific peptides[75–77] and those that use all confidently identified peptides.[78–80] Compared to analysis of intact proteins, bottom-up proteomics has the advantages of higher sensitivity and a higher dynamic range. Initially named Mass Spectrometry-based PhyloProteomics (MSPP), this approach has successfully differentiated taxonomically similar species, such as *S. pseudopneumoniae*, *S. mitis* of the *Mitis* Group of the genus *Streptococcus* and commensal species of *S. pneumoniae*.[74,81,82] Proteotyping can elucidate not only features for identification but also features crucial for clinical responses, such as biomarkers of antibiotic resistance and virulence, or biological and environmental applications, such as biomarkers of catabolic and anabolic pathways in cell metabolism.

To perform reliable proteomics-based characterization, it is necessary to develop gene-encoding databases, which is governed by the availability of genome sequences. The expanding databases of microbial genome sequences will determine the effectiveness of proteomics-based methods for microbial typing. The expanding databases of information on microbial genome sequences determine the successful application of systematic proteomics-based approaches for microbial typing. Identifying peptides from genome sequence data in turn validates putative genes and helps reduce erroneous annotations.[83] To realize higher discriminatory levels, it is necessary to increase the coverage of detected proteins in proteomics by optimizing the process of microorganisms for MS, such as changing the culture conditions.[84] Previous studies have demonstrated that fractionation and selective separation of microbial cells can improve proteomic characterization of microorganisms.[85] For example, gel-based electrophoresis of the outer membrane protein fractions was helpful to the proteomics-based differentiation of *Haemophilus influenzae* strains.[86] Furthermore, investigations using bottom-up proteomics have shown that the outer membrane fractions of bacteria have a greater discriminating power than whole-cell lysates.[87]

To determine which peptides in a sample are useful for differentiating between strains in proteotyping applications, finding discriminative peptide fragments is a crucial step in this procedure. Regarding the dependence of the identification of discriminative peptides on the database, researchers suggest prioritizing the developments of bioinformatics tools with advanced and robust algorithms.[74] Several bioinformatics pipelines, including the Microorganism Classification and Identification (MiCId), have been developed to obtain peptides specific to different taxonomic levels.[81,89–91] Chen *et al.* developed a pipeline that integrates bottom-up proteomics and filtering on a BLAST search for the classification of *Salmonella* at the serovar level, as BLAST search facilitates the determination of the specificity of those peptides as classification biomarkers (Figure 1.4).[88] As many proteotyping methods require pure culture of strains, Boulund *et al.* developed a new computational method, Typing and Characterization of bacteria Using bottom-up tandem MS Proteomics (TCUP), to determine the compositions of bacterial taxonomy and detect antibiotic-resistant proteins (Figure 1.5).[89] It identifies discriminative peptides by comparing the generated peptides to reference databases. Therefore, it is important to develop databases with larger coverages of references, high-quality annotations of genomes and proteomes, and bioinformatics with advanced algorithms for classification of informative peptides.

1.3.3 MS of Nucleic Acids

The accurate identification, characterization, and classification of microorganisms have been accomplished using MS technologies that determine microbial nucleic acids. ESI-MS and MALDI-TOF-MS are the two most popular MS techniques of them.

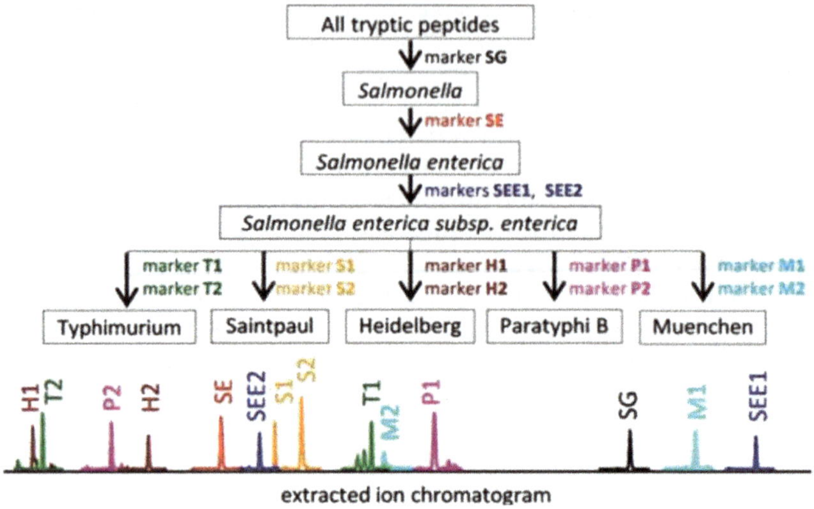

Figure 1.4 Pipeline combining bottom-up proteomics and filtering based on a BLAST search. Reproduced from ref. 88 with permission from American Chemical Society, Copyright 2019. Subset of markers for the identification of *Salmonella* from the genus to serovar levels – SG: *Salmonella* genus marker, SE: *S. enterica* marker, SEE: *S. enterica* subsp. *enterica* markers, and T: *S.* Typhimurium, S: *S.* Saintpaul, H: *S.* Heidelberg; P: *S.* Paratyphi B, and M: *S.* Muenchen markers.

Due to the improvement in nucleic acid analysis by MS, the ionization of intact PCR products by ESI was realized, which in turn promotes the development of MS-based nucleic acid analysis for microbial identification.[92,93] After amplification of targeted and conserved regions, amplicons are produced and analyzed by ESI-MS. The obtained amplicon fingerprint is then used to compare to a database consisting of spectral fingerprints of known microbes. In a multilocus PCR approach, broad-range primers are utilized to amplify informative genomic loci, followed by detection of PCR products and analysis of their base composition using ESI-MS. By using universal primers to target conserved sequences, this approach enables the identification and quantification of a wide variety of microorganisms, including bacteria and viruses, as well as the detection of resistance and/or virulence factors.

In 2005, Hofstadler *et al.* developed Triangulation Identification of Genetic Evaluation of Risks (TIGER), which combines PCR with ESI-MS to identify different microorganisms, such as bacteria, fungi, and viruses.[94] The workflow of this method is depicted in Figure 1.6. Later, the authors demonstrated an advanced and automatic platform based on TIGER, named Ibis T5000.[95] The further development of Ibis led to the product PLEX-ID, which obtained a European CE Mark and was redesigned into the IRIDICA system.[96,97] This system has many advantages and applications in clinical diagnosis, particularly for non-culturable or low-concentration microorganisms.[98,99] However, economic and logistical considerations should be taken into account when

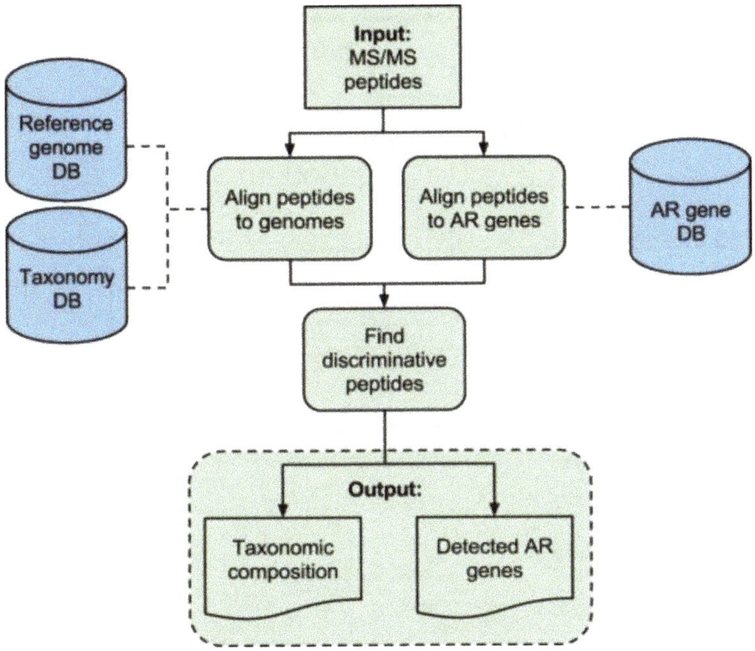

Figure 1.5 TCUP workflow. Reproduced from ref. 89 with permission from Elsevier, Copyright 2017. The figure illustrates the two tracks of the analytical workflow, namely taxonomic composition estimation and antibiotic resistance (AR) protein detection. In the left track, peptides from the sample are aligned to reference genome sequences, and the lowest common ancestor algorithm is applied to identify organisms uniquely. In the right track, peptides are aligned to a database of reference AR gene sequences, and the method estimates the relative abundances of all taxonomic entities detected in the sample and lists the expressed AR proteins detected.

implementing modern methods in clinical practice, as there are numerous challenges. Notably, since 2017, PCR-ESI-MS has been unavailable in the commercial market for unknown reasons.[100] In the future, we believe that other alternative methods like PCR-ESI-MS might be available in clinical practice.

In 1990, MALDI-TOF-MS was first employed for the analysis of oligonucleotides.[102] The improvement of matrices greatly enhanced the ionization of nucleic acids, leading to the development of MALDI-TOF-MS-based nucleic acid characterization. This technique has been successfully applied to the identification of viruses, bacteria, and fungi. Moreover, the MALDI-TOF-MS platform has been combined with standard molecular techniques, including spoligotyping, genotyping, 16S rDNA-based typing, MLST, restriction fragment length polymorphism analysis, and comparative sequencing of nucleic acids. In these methods, traditional molecular techniques are either substituted or refined by using particular or broad-range PCR primers in combination with various post-PCR biochemistries.

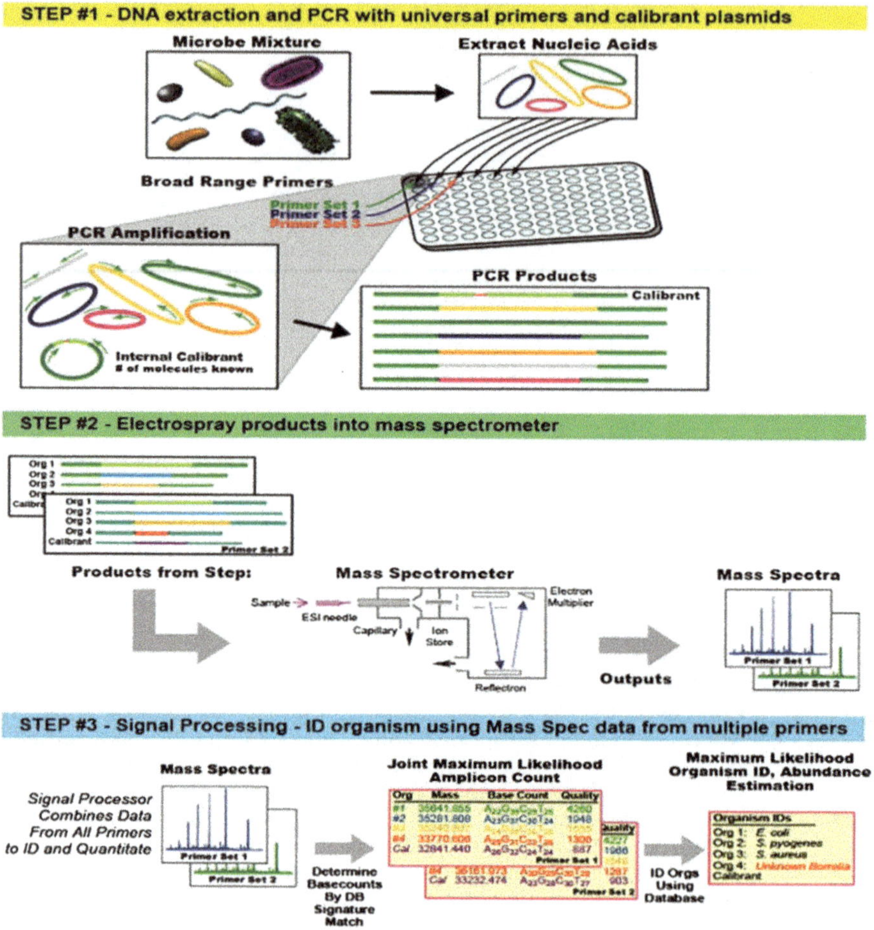

Figure 1.6 The workflow of TIGER. Reproduced from ref. 94 with permission from Elsevier, Copyright 2005. Nucleic acid extracts from the sample of interest are amplified using broad-range PCR primers in Step #1. MS-based analysis of PCR-derived amplicons is performed in Step #2. Step #3 involves signal processing to generate unambiguous base composition signatures from multiple genomic regions, which are used to identify the microbe(s) present in the sample.

In a general workflow, the genomic target regions of interest are amplified *via* PCR, followed by the detection of SNPs *via* primer extension reactions or comparative sequence analysis of mass peak profiles generated by *in vitro* transcription and base-specific cleavage. Benefiting from the commercialized Sequenom "MassARRAY iPLEX" system (Figure 1.7) and the classical pin-point assay, primer extension is the most popular allelic discrimination method in large-scale SNP genotyping research with MALDI-MS detection.[103,104] MALDI-TOF-MS-based SNP genotyping has been utilized to screen different viruses, such as the human papillomavirus,[105] hepatitis B

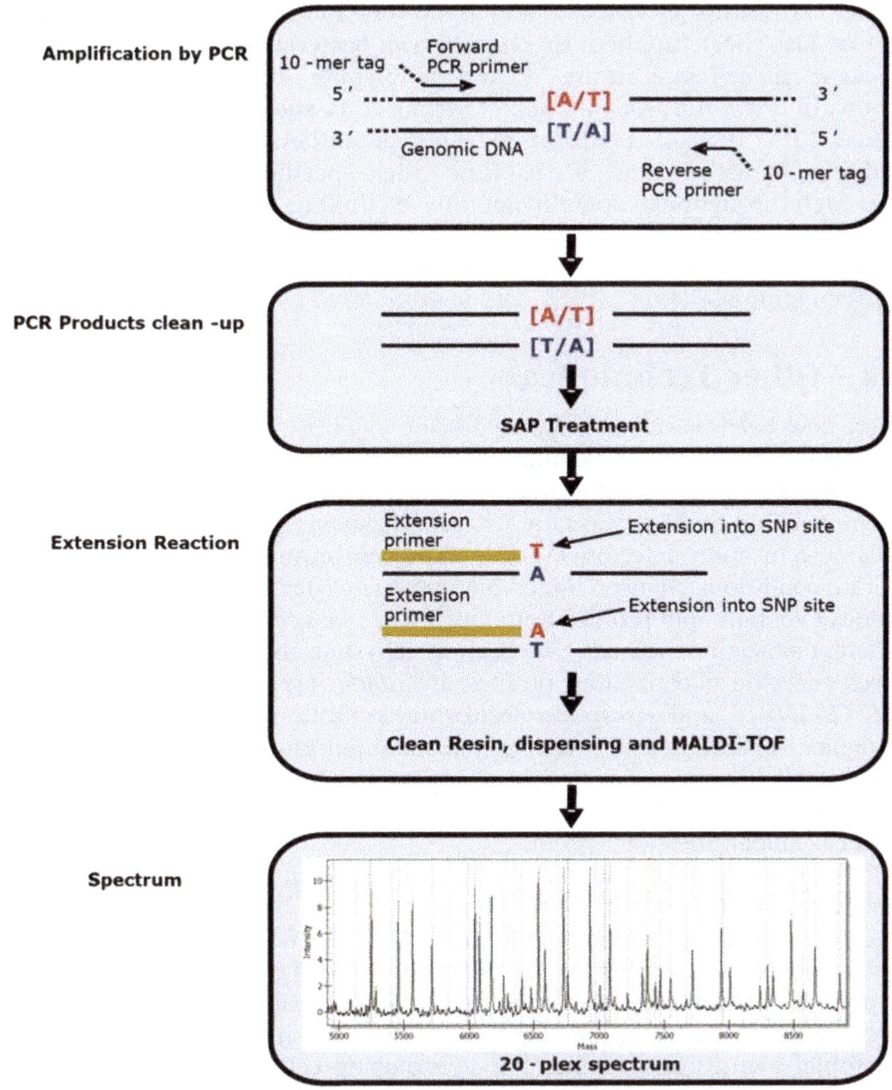

Figure 1.7 The workflow of iPLEX. Reproduced from ref. 101 with permission from Springer Nature, Copyright 2017. The iPLEX chemistry involves several steps in the generation of single-nucleotide polymorphism (SNP) genotypes. Initially, regions targeted by the multiplex assay are amplified using PCR. Next, the PCR products are treated with shrimp alkaline phosphatase to neutralize unincorporated nucleotides. An extension reaction is then performed to elongate the PCR fragments by one base into the SNP site. Subsequently, the mass of the extended fragments is measured using MALDI-TOF, which produces a spectrum of distinct mass peaks for the multiplex reaction.

virus,[106] hepatitis C virus,[107] and hand-foot-and-mouth disease virus.[108] It has also been applied to characterize bacteria, such as genotyping *Neisseria gonorrhoeae* strains,[104] and genotyping *Pseudomonas aeruginosa* strains in cystic fibrosis patients.[109] Syrmis *et al.* successfully employed the "MassARRAY iPLEX" platform to genotype MRSA, which is more cost- and time-effective than a real-time allele-specific PCR technology.[110] The high-throughput capability of this technique makes it a promising biomolecular approach. However, the diversity of microorganisms and the occurrence of mutations limit single-nucleotide-based discrimination between strains.

1.4 Other Techniques

There have been several attempts to develop identification methods for microorganisms based on their metabolite and lipid profiles. One of the fastest ways to generate lipid profiles is through ambient ionization techniques. The first attempt at lipid profiling was fatty acid profiling using gas chromatography and a flame ionization detector, but this approach proved to be unstable when culture conditions changed. Since the first use of desorption ionization techniques to obtain lipid profiles from intact bacterial cells in 1987,[113] numerous different ionization methods have been used to show that various bacteria have species-specific phospholipid profiles, including fast atom bombardment,[17] ESI,[73] MALDI,[114] and desorption electrospray ionization (DESI)[111] (Figure 1.8A). Compared with fatty acid profiling, phospholipid profiling was proved to be more stable. However, none of these approaches have been demonstrated to have the precision and robustness needed to serve as the cornerstone of a broad lipid-based identification system.

By the use of ambient ionization techniques, such DESI, nanospray DESI, and other electrospray setups, methods for *in vivo* analysis of bacterial metabolites and phospholipids were further developed.[115,116] A more recently developed ambient ionization technique (REIMS; Figure 1.8B and C) is able to produce highly specific phospholipid profiles of various microorganisms, opening up new possibilities for the creation of a lipid-based, sample-free microbial identification system.[112] By exposing cellular biomass to radiofrequency alternating electric current, species-specific mass spectral fingerprints are produced. REIMS shares a lot of technical similarities with pyrolysis techniques, but unlike pyrolysis, it performs soft ionization that primarily produces molecular ions.[117] The REIMS platform has been designed to realize automated and high-throughput microbial analysis.[118]

1.5 Applications

Although many of the novel MS-based methods for microorganism detection are now still in the proof-of-concept stage, it is anticipated that they will play a significant role in various fields, such as clinical laboratories, environmental science, and the food industry, in the future. While most published

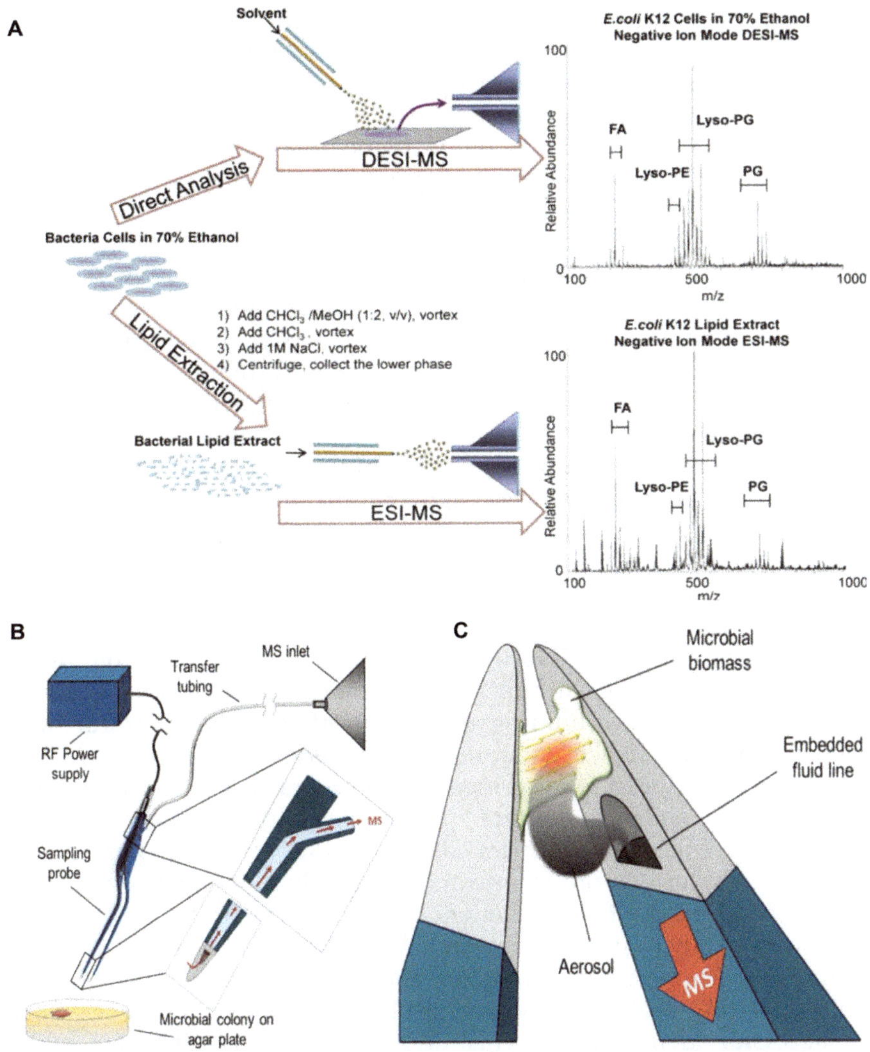

Figure 1.8 Workflow comparison between DESI-MS and ESI-MS. (A) The direct detection of polar lipids in intact bacterial samples through DESI-MS yields spectra that closely resemble those obtained from lipid extracts analyzed by ESI-MS. (B) Setup used for analysis of bacteria by rapid evaporative ionization mass spectrometry (REIMS). Reproduced from ref. 111 with permission from Elsevier, Copyright 2011. (C) Scheme of analysis. Reproduced from ref. 112 with permission from American Chemical Society, Copyright 2014. The irrigated bipolar forceps trap microbial biomass between their two electrodes, which is then subjected to an electric current. The sample is subsequently thermally evaporated, generating an aerosol that is aspirated into the embedded fluid line opening.

articles on the applications of these methods have focused on clinical detection of microorganisms, especially bacteria and fungi, there is a need for expansion to virus detection. The promotion of MS techniques in professional laboratories for database construction and application development has lagged behind that in clinical laboratories, which highlights the lack of communication between MS experts and experts in these two fields. There is a need for more MS promotion in these areas.

The constant need for innovative techniques enabling fast microbial identification and delivery of antibiotic therapy has served as a metaphor for clinical microbiology efforts. Traditional culture and identification methods take at least 36 to 48 hours to produce a clinical report, delaying patient care, particularly in cases of serious infections where prompt treatment has a substantial impact on survival. For a dozen years, clinical microbiology has made extensive use of MALDI-TOF-MS. In clinical laboratories, MS-based techniques have their mature applications in microbial identification and they are undergoing method developments for improvements in speed of analysis, coverage of species, and accuracy of identification and classification, as well as detection of virulence factors and antimicrobial sensitivity.

Environmental microorganisms provide a number of ecological functions. They play a huge role in the biological decay of organic matter in the soil, the biogeochemical cycling of various elements, the detoxification or biodegradation of various environmental pollutants, the provision of nutrients for plant growth, the inhibition of the growth of other harmful microorganisms, and a variety of other natural processes. MS-based microorganism detection has been used to identify and characterize new strains from environmental samples, explore the biodiversity of microorganisms, track the source of microorganism contamination, identify bioremediation-related species, and so on. Due to the incredibly different habitats of the microbes isolated from environmental samples, identification of microorganisms in environmental research is more difficult than in clinical microbiology. However, less is known about environmental microbes than medically significant bacteria, which results in insufficient reference spectra for successful identification. This underlines the necessity for reliable databases for microbial identification in environmental research.

In the food industry, microbiological safety is of significance, which relies on the identity and quantity of microorganisms present, as well as the amount of toxins produced by them. Microorganisms in food influence its taste, appearance, and smell. Certain microorganisms in food are beneficial given that they improve the nutrition and quality of food and biosynthesize food ingredients, preservatives, and biomass by carrying out the fermentation process.[119] On the other hand, some microorganisms in food can cause spoilage, deteriorate its quality, and reduce its nutritional value. Moreover, they can pose a threat to human health and life when consumed. It has been reported that contamination has brought huge economic losses to the food industry. MS-based microorganism detection has been applied to monitor and control food quality. It focuses primarily on the identification and typing of microorganisms such as filamentous yeasts, fungi, and bacteria found in fermented beverages (beer and wine), honey, dairy products (yogurt and pasteurized milk), meat, and seafood.[120]

1.6 Conclusions

MS has emerged as a promising technique for detecting microorganisms, offering comparable results to other methods while reducing time and cost. Recent advances in MS have enabled more accurate and efficient analysis of microorganisms, garnering attention from researchers in various fields. However, limitations such as the size and scope of available databases, as well as the need for improved software to discriminate between closely related species and subspecies, pose challenges to its widespread implementation. Therefore, in addition to continued advancements in instrumentation, database construction and algorithm development are critical areas of focus for improving MS-based microorganism detection. Furthermore, while MS has already shown promise in this area, there are still untapped applications that warrant exploration. Later chapters will provide further insight into the fundamental principles and cutting-edge methods of MS-based microorganism detection.

Abbreviations

AR	Antibiotic resistance
ESI	Electrospray ionization
FTIR	Fourier transform infrared spectroscopy
MS	Mass spectrometry
MSPP	Mass Spectrometry-based PhyloProteomics
MALDI	Matrix-assisted laser desorption/ionization
MRSA	Methicillin-resistant *Staphylococcus aureus*
MSSA	Methicillin-sensitive *Staphylococcus aureus*
MiCId	Microorganism Classification and Identification
MLST	Multilocus sequence typing
PCR	Polymerase chain reaction
PFGE	Pulsed-field gel electrophoresis
qPCR	Quantitative polymerase chain reaction
REIMS	Rapid evaporative ionization mass spectrometry
LC–MS/MS	Tandem mass spectrometry combined with liquid chromatography
TOF	Time-of-flight
TiO_2	Titanium dioxide
TIGER	Triangulation Identification of Genetic Evaluation of Risks
TCUP	Typing and Characterization of bacteria Using bottom-up tandem mass spectrometry Proteomics
SNP	Single-nucleotide polymorphism
WGS	Whole-genome sequencing

References

1. D. J. Weber, W. A. Rutala, W. A. Fischer, H. Kanamori and E. E. Sickbert-Bennett, *Am. J. Infect. Control*, 2016, **44**, E91–E100.
2. J. Giradkar and M. N. Khatib, *Cureus J. Med. Sci.*, 2022, **14**, e31736.
3. R. Khan, F. C. Petersen and S. Shekhar, *Front. Immunol.*, 2019, **10**, 1203.
4. A. Caminero, M. Meisel, B. Jabri and E. F. Verdu, *Nat. Rev. Gastroenterol. Hepatol.*, 2019, **16**, 7–18.
5. A. Agusti, M. P. Garcia-Pardo, I. Lopez-Almela, I. Campillo, M. Maes, M. Romani-Perez and Y. Sanz, *Front. Neurosci.*, 2018, **12**, 155.
6. A. Ahmed, J. V. Rushworth, N. A. Hirst and P. A. Millner, *Clin. Microbiol. Rev.*, 2014, **27**, 631–646.
7. P. Hugon, J. C. Lagier, C. Robert, C. Lepolard, L. Papazian, D. Musso, B. Vialettes and D. Raoult, *J. Clin. Microbiol.*, 2013, **51**, 3286–3293.
8. J. Miller and C. J. M. O. C. M. O'Hara, *Substrate utilization systems for the identification of bacteria and yeasts,* ASM Press, Washington, DC, 9th edn, 2007.
9. J. R. Yates, *Nat. Methods*, 2011, **8**, 633–637.
10. J. C. Lagier, P. Hugon, S. Khelaifia, P. E. Fournier, B. La Scola and D. Raoult, *Clin. Microbiol. Rev.*, 2015, **28**, 237–264.
11. A. Singh, R. V. Goering, S. Simjee, S. L. Foley and M. J. Zervos, *Clin. Microbiol. Rev.*, 2006, **19**, 512–530.
12. F. Balloux, O. B. Brynildsrud, L. van Dorp, L. P. Shaw, H. B. Chen, K. A. Harris, H. Wang and V. Eldholm, *Trends Microbiol.*, 2018, **26**, 1035–1048.
13. K. Chughtai and R. M. A. Heeren, *Chem. Rev.*, 2010, **110**, 3237–3277.
14. J. P. Anhalt and C. Fenselau, *Anal. Chem.*, 1975, **47**, 219–225.
15. F. Basile, M. B. Beverly, K. J. Voorhees and T. L. Hadfield, *TrAC, Trends Anal. Chem.*, 1998, **17**, 95–109.
16. A. P. Snyder, P. B. W. Smith, J. P. Dworzanski and H. L. C. Meuzelaar, in *Mass Spectrometry for the Characterization of Microorganisms*, ed. C. Fenselau, 1994, vol. 541, pp. 62–84.
17. D. N. Heller, R. J. Cotter and C. Fenselau, *Anal. Chem.*, 1987, **59**, 2806–2809.
18. T. C. Cain, D. M. Lubman and W. J. Weber, *Rapid Commun. Mass Spectrom.*, 1994, **8**, 1026–1030.
19. R. D. Holland, J. G. Wilkes, F. Rafii, J. B. Sutherland, C. C. Persons, K. J. Voorhees and J. O. Lay, *Rapid Commun. Mass Spectrom.*, 1996, **10**, 1227–1232.
20. V. Ryzhov and C. Fenselau, *Anal. Chem.*, 2001, **73**, 746–750.
21. L. W. Sun, K. Teramoto, H. Sato, M. Torimura, H. Tao and T. Shintani, *Rapid Commun. Mass Spectrom.*, 2006, **20**, 3789–3798.
22. S. Suarez, A. Ferroni, A. Lotz, K. A. Jolley, P. Guerin, J. Leto, B. Dauphin, A. Jamet, M. C. J. Maiden, X. Nassif and J. Armengaud, *J. Microbiol. Methods*, 2013, **94**, 390–396.
23. M. Sanguinetti and B. Posteraro, *J. Clin. Microbiol.*, 2017, **55**, 369–379.

24. K. E. Tan, B. C. Ellis, R. Lee, P. D. Stamper, S. X. Zhang and K. C. Carroll, *J. Clin. Microbiol.*, 2012, **50**, 3301–3308.
25. A. Bizzini, C. Durussel, J. Bille, G. Greub and G. Prod'hom, *J. Clin. Microbiol.*, 2010, **48**, 1549–1554.
26. F. Alcaide, J. Amlerova, G. Bou, P. J. Ceyssens, P. Coll, D. Corcoran, M. S. Fangous, I. Gonzalez-Alvarez, R. Gorton, G. Greub, G. Hery-Arnaud, J. Hrabak, A. Ingebretsen, B. Lucey, I. Marekovic, C. Mediavilla-Gradolph, M. R. Monte, J. O'Connor, J. O'Mahony, O. Opota, B. O'Reilly, D. Orth-Holler, M. Oviano, J. J. Palacios, B. Palop, A. B. Pranada, L. Quiroga, D. Rodriguez-Temporal, M. J. Ruiz-Serrano, G. Tudo, A. Van den Bossche, J. van Ingen, B. Rodriguez-Sanchez and M. European Study Grp Genomics, *Clin. Microbiol. Infect.*, 2018, **24**, 599–603.
27. J. Yi, Q. Qin, Y. Wang, R. T. Zhang, H. Y. Bi, S. N. Yu, B. H. Liu and L. Qiao, *Microchim. Acta*, 2018, **185**, 542.
28. S. C. Hong, J. S. Kang, J. E. Lee, S. S. Kim and J. H. Jung, *Lab Chip*, 2015, **15**, 1889–1897.
29. Z. G. Wu, B. Willing, J. Bjerketorp, J. K. Jansson and K. Hjort, *Lab Chip*, 2009, **9**, 1193–1199.
30. P. Ohlsson, M. Evander, K. Petersson, L. Mellhammar, A. Lehmusvuori, U. Karhunen, M. Soikkeli, T. Seppa, E. Tuunainen, A. Spangar, P. von Lode, K. Rantakokko-Jalava, G. Otto, S. Scheding, T. Soukka, S. Wittfooth and T. Laurell, *Anal. Chem.*, 2016, **88**, 9403–9411.
31. S. P. Ning, S. C. Liu, Y. J. Xiao, G. Y. Zhang, W. W. Cui and M. Reed, *Lab Chip*, 2021, **21**, 4608–4617.
32. Y. D. Zhu, L. Qiao, M. Prudent, A. Bondarenko, N. Gasilova, S. B. Moller, N. Lion, H. Pick, T. Q. Gong, Z. X. Chen, P. Y. Yang, L. T. Loveyf and H. H. Girault, *Chem. Sci.*, 2016, 7, 2987–2995.
33. Y. Q. Shen, J. Yi, M. H. Song, D. D. Li, Y. Wu, Y. J. Liu, M. C. Yang and L. Qiao, *Analyst*, 2021, **146**, 4146–4153.
34. M. R. Condina, B. A. Dilmetz, S. R. Bazaz, J. Meneses, M. E. Warkiani and P. Hoffmann, *Lab Chip*, 2019, **19**, 1961–1970.
35. M. Marin, E. Cercenado, C. Sanchez-Carrillo, A. Ruiz, A. G. Gonzalez, B. Rodriguez-Sanchez and E. Bouza, *Front. Microbiol.*, 2017, **8**, 698.
36. H. C. Slotved, R. R. Facklam and K. Fuursted, *Sci. Rep.*, 2017, 7, 7167.
37. E. Farfour, J. Leto, M. Barritault, C. Barberis, J. Meyer, B. Dauphin, A. S. Le Guern, A. Lefleche, E. Badell, N. Guiso, A. Leclercq, A. Le Monnier, M. Lecuit, V. Rodriguez-Nava, E. Bergeron, J. Raymond, S. Vimont, E. Bille, E. Carbonnelle, H. Guet-Revillet, H. Lecuyer, J. L. Beretti, C. Vay, P. Berche, A. Ferroni, X. Nassif and O. Join-Lambert, *J. Clin. Microbiol.*, 2012, **50**, 2702–2707.
38. P. D. Khot and M. A. Fisher, *J. Clin. Microbiol.*, 2013, **51**, 3711–3716.
39. A. Elbehiry, E. Marzouk, M. Hamada, M. Al-Dubaib, E. Alyamani, I. M. Moussa, A. AlRowaidhan and H. A. Hemeg, *New Microbiol.*, 2017, **40**, 269–278.

40. H. de Souza, L. M. Dalla-Costa, F. J. Vicenzi, D. C. de Souza, C. A. Riedi, N. A. Rosario and M. Pilonetto, *J. Med. Microbiol.*, 2014, **63**, 1148–1153.
41. N. M. B. Rodrigues, G. F. Bronzato, G. S. Santiago, L. A. B. Botelho, B. M. Moreira, I. D. Coelho, M. M. S. de Souza and S. D. D. Coelho, *Braz. J. Microbiol.*, 2017, **48**, 132–138.
42. B. Rodriguez-Sanchez, L. Alcala, M. Marin, A. Ruiz, E. Alonso and E. Bouza, *Anaerobe*, 2016, **42**, 101–107.
43. R. Dieckmann, R. Helmuth, M. Erhard and B. Malorny, *Appl. Environ. Microbiol.*, 2008, **74**, 7767–7778.
44. A. Freiwald and S. Sauer, *Nat. Protoc.*, 2009, **4**, 732–742.
45. A. Psaroulaki and D. Chochlakis, *Expert Rev. Proteomics*, 2018, **15**, 537–539.
46. A. C. M. Veloo, H. Jean-Pierre, U. S. Justesen, T. Morris, E. Urban, I. Wybo, M. Kostrzewa, A. W. Friedrich and E. Workgrp, *Anaerobe*, 2018, **54**, 224–230.
47. C. Valero, M. J. Buitrago, S. Gago, I. Quiles-Melero and J. Garcia-Rodriguez, *Med. Mycol.*, 2018, **56**, 307–314.
48. A. Paauw, D. Jonker, G. Roeselers, J. M. E. Heng, R. H. Mars-Groenendijk, H. Trip, E. M. Molhoek, H. J. Jansen, J. van der Plas, A. L. de Jong, J. A. Majchrzykiewicz-Koehorsta and A. Speksnijder, *Int. J. Med. Microbiol.*, 2015, **305**, 446–452.
49. M. F. Mazzeo, A. Sorrentino, M. Gaita, G. Cacace, M. Di Stasio, A. Facchiano, G. Comi, A. Malorni and R. A. Siciliano, *Appl. Environ. Microbiol.*, 2006, **72**, 1180–1189.
50. K. Bohme, I. C. Fernandez-No, J. Barros-Velazquez, J. M. Gallardo, B. Canas and P. Calo-Mata, *Electrophoresis*, 2012, **33**, 2138–2142.
51. J. H. K. Chen, K. K. K. She, O. Y. Wong, J. L. L. Teng, W. C. Yam, S. K. P. Lau, P. C. Y. Woo, V. C. C. Cheng and K. Y. Yuen, *J. Clin. Pathol.*, 2015, **68**, 652–656.
52. T. T. Zhang, J. Y. Ding, X. C. Rao, J. B. Yu, M. L. Chu, W. Ren, L. Wang and W. C. Xue, *J. Microbiol. Methods*, 2015, **117**, 122–127.
53. A. Karger, M. Ziller, B. Bettin, B. Mintel, S. Schares and L. Geue, *Appl. Environ. Microbiol.*, 2011, 77, 896–905.
54. R. Stephan, N. Cernela, D. Ziegler, V. Pfluger, M. Tonolla, D. Ravasi, M. Fredriksson-Ahomaa and H. Hachler, *J. Microbiol. Methods*, 2011, **87**, 150–153.
55. M. Lundquist, M. B. Caspersen, P. Wikstrom and M. Forsman, *FEMS Microbiol. Lett.*, 2005, **243**, 303–310.
56. M. T. Gekenidis, P. Studer, S. Wuthrich, R. Brunisholz and D. Drissner, *Appl. Environ. Microbiol.*, 2014, **80**, 4234–4241.
57. A. J. Madonna, F. Basile, I. Ferrer, M. A. Meetani, J. C. Rees and K. J. Voorhees, *Rapid Commun. Mass Spectrom.*, 2000, **14**, 2220–2229.
58. R. Dieckmann and B. Malorny, *Appl. Environ. Microbiol.*, 2011, 77, 4136–4146.
59. A. Karger, *Proteomics Clin. Appl.*, 2016, **10**, 982–993.
60. A. E. Zautner, W. O. Masanta, M. Weig, U. Gross and O. Bader, *Sci. Rep.*, 2015, **5**, 13431.

61. R. T. Zhang, Q. Qin, B. H. Liu and L. Qiao, *Anal. Chem.*, 2018, **90**, 3863–3870.
62. M. Sorensen, C. E. Chandler, F. M. Gardner, S. Ramadan, P. D. Khot, L. M. Leung, C. E. Farrance, D. R. Goodlett, R. K. Ernst and E. Nilsson, *Sci. Rep.*, 2020, **10**, 21536.
63. P. Mahe, M. Arsac, S. Chatellier, V. Monnin, N. Perrot, S. Mailler, V. Girard, M. Ramjeet, J. Surre, B. Lacroix, A. van Belkum and J. B. Veyrieras, *Bioinformatics*, 2014, **30**, 1280–1286.
64. Y. Yang, Y. Lin and L. Qiao, *Anal. Chem.*, 2018, **90**, 10400–10408.
65. K. L. Wahl, S. C. Wunschel, K. H. Jarman, N. B. Valentine, C. E. Petersen, M. T. Kingsley, K. A. Zartolas and A. J. Saenz, *Anal. Chem.*, 2002, **74**, 6191–6199.
66. R. E. Mandrell, L. A. Harden, A. Bates, W. G. Miller, W. F. Haddon and C. K. Fagerquist, *Appl. Environ. Microbiol.*, 2005, **71**, 6292–6307.
67. C. Mortelmaier, S. Panda, I. Robertson, M. Krell, M. Christodoulou, N. Reichardt and I. Mulder, *Anal. Bioanal. Chem.*, 2019, **411**, 7027–7038.
68. J. Li, P. Thibault, A. Martin, J. C. Richards, W. W. Wakarchuk and W. van der Wilp, *J. Chromatogr. A*, 1998, **817**, 325–336.
69. B. W. Gibson, N. J. Phillips, C. M. John and W. Melaugh, in *Mass Spectrometry for the Characterization of Microorganisms*, ed. C. Fenselau, 1994, vol. 541, pp. 185–202.
70. G. E. Black, A. P. Snyder and K. S. Heroux, *J. Microbiol. Methods*, 1997, **28**, 187–199.
71. D. S. Wunschel, K. F. Fox, A. Fox, M. L. Nagpal, K. Kim, G. C. Stewart and M. Shahgholi, *J. Chromatogr. A*, 1997, **776**, 205–219.
72. C. L. Liu, S. A. Hofstadler, J. A. Bresson, H. R. Udseth, T. Tsukuda, R. D. Smith and A. P. Snyder, *Anal. Chem.*, 1998, **70**, 1797–1801.
73. R. Goodacre, J. K. Heald and D. B. Kell, *FEMS Microbiol. Lett.*, 1999, **176**, 17–24.
74. R. Karlsson, L. Gonzales-Siles, F. Boulund, L. Svensson-Stadler, S. Skovbjerg, A. Karlsson, M. Davidson, S. Hulth, E. Kristiansson and E. R. B. Moore, *Syst. Appl. Microbiol.*, 2015, **38**, 246–257.
75. D. H. Huson, S. Beier, I. Flade, A. Gorska, M. El-Hadidi, S. Mitra, H. J. Ruscheweyh and R. Tappu, *PLoS Comput. Biol.*, 2016, **12**, e100495.
76. D. H. Huson, S. Mitra, H. J. Ruscheweyh, N. Weber and S. C. Schuster, *Genome Res.*, 2011, **21**, 1552–1560.
77. K. Rooijers, C. Kolmeder, C. Juste, J. Dore, M. de Been, S. Boeren, P. Galan, C. Beauvallet, W. M. de Vos and P. J. Schaap, *BMC Genomics*, 2011, **12**, 6.
78. G. Alves, G. H. Wang, A. Y. Ogurtsov, S. K. Drake, M. Gucek, A. F. Suffredini, D. B. Sacks and Y. K. Yu, *J. Am. Soc. Mass Spectrom.*, 2016, **27**, 194–210.
79. J. P. Dworzanski, S. V. Deshpande, R. Chen, R. E. Jabbour, A. P. Snyder, C. H. Wick and L. Li, *J. Proteome Res.*, 2006, **5**, 76–87.
80. A. Penzlin, M. S. Lindner, J. Doellinger, P. W. Dabrowski, A. Nitsche and B. Y. Renard, *Bioinformatics*, 2014, **30**, 149–156.

81. L. Grenga, O. Pible and J. Armengaud, *Clin. Mass Spectrom.*, 2019, **14**, 9–17.
82. R. Karlsson, L. Gonzales-Siles, M. Gomila, A. Busquets, F. Salva-Serra, D. Jaen-Luchoro, H. E. Jakobsson, A. Karlsson, F. Boulund, E. Kristiansson and E. R. B. Moore, *PLoS One*, 2018, **13**, e0208804.
83. J. Armengaud, *Curr. Opin. Microbiol.*, 2009, **12**, 292–300.
84. J. A. Christie-Oleza, B. Fernandez, B. Nogales, R. Bosch and J. Armengaud, *ISME J.*, 2012, **6**, 124–135.
85. R. Karlsson, M. Davidson, L. Svensson-Stadler, A. Karlsson, K. Olesen, E. Carlsohn and E. R. B. Moore, *J. Proteome Res.*, 2012, **11**, 2710–2720.
86. D. A. Clements, S. J. Macinnes and G. L. Gilbert, *J. Clin. Microbiol.*, 1992, **30**, 1879–1881.
87. R. E. Jabbour, M. M. Wade, S. V. Deshpande, M. F. Stanford, C. H. Wick, A. W. Zulich and A. P. Snyder, *J. Proteome Res.*, 2010, **9**, 3647–3655.
88. S. H. Chen, C. H. Parker, T. R. Croley and M. A. McFarland, *Anal. Chem.*, 2019, **91**, 4388–4395.
89. F. Boulund, R. Karlsson, L. Gonzales-Siles, A. Johnning, N. Karami, O. Al-Bayati, C. Ahren, E. R. B. Moore and E. Kristiansson, *Mol. Cell. Proteomics*, 2017, **16**, 1052–1063.
90. O. Pible, F. Allain, V. Jouffret, K. Culotta, G. Miotello and J. Armengaud, *Microbiome*, 2020, **8**, 30.
91. G. Alves, G. H. Wang, A. Y. Ogurtsov, S. K. Drake, M. Gucek, D. B. Sacks and Y. K. Yu, *J. Am. Soc. Mass Spectrom.*, 2018, **29**, 1721–1737.
92. D. C. Muddiman, G. A. Anderson, S. A. Hofstadler and R. D. Smith, *Anal. Chem.*, 1997, **69**, 1543–1549.
93. D. C. Muddiman, D. S. Wunschel, C. L. Liu, L. PasaTolic, K. F. Fox, A. Fox, G. A. Anderson and R. D. Smith, *Anal. Chem.*, 1996, **68**, 3705–3712.
94. S. A. Hofstadler, R. Sampath, L. B. Blyn, M. W. Eshoo, T. A. Hall, Y. Jiang, J. J. Drader, J. C. Hannis, K. A. Sannes-Lowery, L. L. Cummins, B. Libby, D. J. Walcott, A. Schink, C. Massire, R. Ranken, J. Gutierrez, S. Manalili, C. Ivy, R. Melton, H. Levene, G. Barrett-Wilt, F. Li, V. Zapp, N. White, V. Samant, J. A. McNeil, D. Knize, D. Robbins, K. Rudnick, A. Desai, E. Moradi and D. J. Ecker, *Int. J. Mass Spectrom.*, 2005, **242**, 23–41.
95. D. J. Ecker, R. Sampath, C. Massire, L. B. Blyn, T. A. Hall, M. W. Eshoo and S. A. Hofstadler, *Nat. Rev. Microbiol.*, 2008, **6**, 553–558.
96. D. Jacob, U. Sauer, R. Housley, C. Washington, K. Sannes-Lowery, D. J. Ecker, R. Sampath and R. Grunow, *PLoS One*, 2012, 7, e39928.
97. V. Ozenci and K. Stralin, *Expert Rev. Mol. Diagn.*, 2019, **19**, 389–395.
98. A. Bacconi, G. S. Richmond, M. A. Baroldi, T. G. Laffler, L. B. Blyn, H. E. Carolan, M. R. Frinder, D. M. Toleno, D. Metzgar, J. R. Gutierrez, C. Massire, M. Rounds, N. J. Kennel, R. E. Rothman, S. Peterson,

K. C. Carroll, T. Wakefield, D. J. Ecker and R. Sampath, *J. Clin. Microbiol.*, 2014, **52**, 3164–3174.
99. D. Metzgar, M. W. Frinder, R. E. Rothman, S. Peterson, K. C. Carroll, S. X. Zhang, G. D. Avornu, M. A. Rounds, H. E. Carolan, D. M. Toleno, D. Moore, T. A. Hall, C. Massire, G. S. Richmond, J. R. Gutierrez, R. Sampath, D. J. Ecker and L. B. Blyn, *PLoS One*, 2016, **11**, e0158186.
100. V. Ozenci, R. Patel, M. Ullberg and K. Stralin, *Clin. Infect. Dis.*, 2018, **66**, 452–455.
101. J. A. Ellis and B. Ong, in *Genotyping*, ed. S. White and S. Cantsilieris, Methods in Molecular Biology, 2017, vol. 1492.
102. B. Spengler, Y. Pan, R. J. Cotter and L. S. Kan, *Rapid Commun. Mass Spectrom.*, 1990, **4**, 99–102.
103. J. Tost and I. G. Gut, *Clin. Biochem.*, 2005, **38**, 335–350.
104. E. Trembizki, H. Smith, M. M. Lahra, M. Chen, B. Donovan, C. K. Fairley, R. Guy, J. Kaldor, D. Regan, J. Ward, M. D. Nissen, T. P. Sloots and D. M. Whiley, *J. Antimicrob. Chemother.*, 2014, **69**, 1526–1532.
105. J. P. Peng, L. Gao, J. H. Guo, T. Wang, L. Wang, Q. Yao, H. J. Zhu and Q. Jin, *J. Clin. Microbiol.*, 2013, **51**, 402–408.
106. M. I. L. Sjoholm, J. Dillner and J. Carlson, *J. Clin. Microbiol.*, 2008, **46**, 540–545.
107. E. N. Ilina, M. V. Malakhova, E. V. Generozov, E. N. Nikolaev and V. M. Govorun, *J. Clin. Microbiol.*, 2005, **43**, 2810–2815.
108. J. P. Peng, F. Yang, Z. H. Xiong, J. H. Guo, J. Du, Y. F. Hu and Q. Jin, *J. Clin. Virol.*, 2013, **56**, 170–174.
109. M. W. Syrmis, T. J. Kidd, R. J. Moser, K. A. Ramsay, K. M. Gibson, S. Anuj, S. C. Bell, C. E. Wainwright, K. Grimwood, M. Nissen, T. P. Sloots and D. M. Whiley, *BMC Infect. Dis.*, 2014, **14**, 307.
110. M. W. Syrmis, R. J. Moser, D. M. Whiley, V. Vaska, G. W. Coombs, M. D. Nissen, T. P. Sloots and G. R. Nimmo, *Clin. Microbiol. Infect.*, 2011, **17**, 1804–1810.
111. J. I. Zhang, N. Talaty, A. B. Costa, Y. Xia, W. A. Tao, R. Bell, J. H. Callahan and R. G. Cooks, *Int. J. Mass Spectrom.*, 2011, **301**, 37–44.
112. N. Strittmatter, M. Rebec, E. A. Jones, O. Golf, A. Abdolrasouli, J. Balog, V. Behrends, K. A. Veselkov and Z. Takats, *Anal. Chem.*, 2014, **86**, 6555–6562.
113. D. N. Heller, C. Fenselau, R. J. Cotter, P. Demirev, J. K. Olthoff, J. Honovich, M. Uy, T. Tanaka and Y. Kishimoto, *Biochem. Biophys. Res. Commun.*, 1987, **142**, 194–199.
114. Y. Ishida, A. J. Madonna, J. C. Rees, M. A. Meetani and K. J. Voorhees, *Rapid Commun. Mass Spectrom.*, 2002, **16**, 1877–1882.
115. Y. Song, N. Talaty, K. Datsenko, B. L. Wanner and R. G. Cooks, *Analyst*, 2009, **134**, 838–841.
116. J. Watrous, P. Roach, T. Alexandrov, B. S. Heath, J. Y. Yang, R. D. Kersten, M. van der Voort, K. Pogliano, H. Gross,

J. M. Raaijmakers, B. S. Moore, J. Laskin, N. Bandeira and P. C. Dorrestein, *Proc. Natl. Acad. Sci. U. S. A.*, 2012, **109**, E1743–E1752.

117. N. Strittmatter, E. A. Jones, K. A. Veselkov, M. Rebec, J. G. Bundy and Z. Takats, *Chem. Commun.*, 2013, **49**, 6188–6190.
118. F. Bolt, S. J. S. Cameron, T. Karancsi, D. Simon, R. Schaffer, T. Rickards, K. Hardirnan, A. Burke, Z. Bodai, A. Perdones-Montero, M. Rebec, J. Balog and Z. Takats, *Anal. Chem.*, 2016, **88**, 9419–9426.
119. E. Caplice and G. F. Fitzgerald, *Int. J. Food Microbiol.*, 1999, **50**, 131–149.
120. M. Akimowicz and J. Bucka-Kolendo, *Acta Biochim. Pol.*, 2020, **67**, 327–332.

CHAPTER 2

MALDI-TOF Mass Fingerprinting for Rapid Identification of Bacteria: Experimental Procedures

YINGDI ZHU*[a,b] AND HUBERT H. GIRAULT*[a]

[a] Department of Chemistry and Chemical Engineering, Laboratory of Analytical and Physical Electrochemistry, École Polytechnique Fédérale de Lausanne, Station 6, 1015 Lausanne, Switzerland; [b] Institute of Basic Medicine and Cancer, Chinese Academy of Sciences, Fucheng Road 150, 310000 Hangzhou, China
*Emails: yingdi.zhu@outlook.com; hubert.girault@epfl.ch

2.1 Instrumental Fundamentals of MALDI-TOF Mass Fingerprinting for Bacterial Identification

2.1.1 A Brief Description of the Instrument

In the past decades MALDI-TOF mass spectrometry (MS) has become one of the most powerful tools for the analysis of biomolecules like metabolites, lipids, peptides, and proteins, as well as biological entities like bacteria, fungi, viruses, and mammalian cells. It has been employed not only in fundamental scientific research but also in practical clinical diagnosis. The importance of this MS technique was recognised by the Nobel Prize in Chemistry in 2002.

New Developments in Mass Spectrometry No. 13
Detection and Analysis of Microorganisms by Mass Spectrometry
Edited by Liang Qiao and Jia Yi

Published by the Royal Society of Chemistry, www.rsc.org

Generally, a mass spectrometer consists of five parts: (i) an inlet to introduce the sample, *e.g.*, a liquid chromatograph or gas chromatograph, an injection pump, or a direct insert target plate; (ii) an ion source to produce ions from the sample, with the ionisation technique varying from hard gas phase ionisation to soft condensed phase ionisation; (iii) one or several mass analysers to separate ions according to their mass-to-charge ratio, sometimes also including ion selection and fragmentation steps; (iv) an ion detector to "receive" or "count" the ions coming from the mass analyser and convert the ions into electrical signals; and (v) a digital data acquisition system to process the signals from the detector and create a mass spectrum.[1] The general structure of a MALDI-TOF mass spectrometer is shown in Figure 2.1. The analyte is introduced by directly inserting an electrically conductive target plate, which bears 48, 96, 192, or 384 sample spots and allows high-throughput analysis. Thereafter, the analyte is ionised by the soft ionisation technique matrix-assisted laser desorption/ionisation (MALDI), which employs a pulsed laser to irradiate the co-crystals of the analyte and a laser energy absorbing matrix to generate protonated or deprotonated ions. The generated ions are then separated using a time-of-flight (TOF) analyser according to their mass-to-charge ratios, and finally detected using a mass detector, like a microchannel plate (MCP) detector. The mass spectrometer is software-controlled for spectrum acquisition and instrumental parameter adjustment. All the mass spectra presented or instrumental parameters mentioned in this chapter were derived from the Bruker microflex™ LRF instrument, which is currently one of the most powerful benchtop MALDI-TOF mass spectrometers.

2.1.2 The Generation of Bacterial Mass Fingerprints

MALDI is a soft condensed phase ionisation process, which is achieved in two steps. In the first step, the analytes are dissolved in an aqueous/organic solvent containing small organic molecules, the so-called matrix. The matrix molecules have a strong ability to absorb energy from a laser. The analyte/matrix mixture is dried on the target plate at room temperature

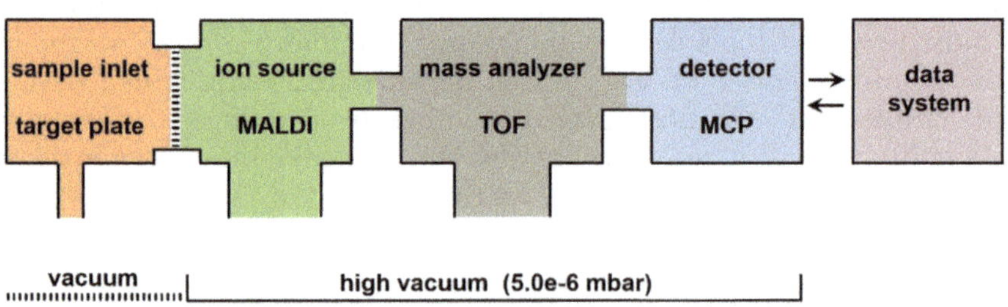

Figure 2.1 General structure of a MALDI-TOF mass spectrometer. The measurements are performed under high vacuum conditions. The exact vacuum level depends on the individual instrument. For instance, the Bruker microflex™ LRF requires that the ion source vacuum be lower than 5×10^{-6} mbar.

to remove any liquid solvent, resulting in the formation of a "solid solution" where analyte molecules are doped into the crystals of matrix molecules. In the cases of bacterial identification, the intact bacterial cells or protein mixtures extracted from the bacterial cells are transferred to the target plate and dried at room temperature, and the matrix solution thereafter is added dropwise to cover the bacterial sample. The bacterial cells or proteins are thus embedded throughout the matrix crystals. The target plate bearing the analyte/matrix mixture is inserted into the sample inlet of the mass spectrometer, and is transferred to the ion source under high vacuum conditions. In the ion source, the analyte/matrix mixture is irradiated with intense laser pulses. Each of the pulses lasts a short time, usually a few nanoseconds to a few hundred nanoseconds. Under the laser irradiation, the analyte/matrix mixture is rapidly heated by the accumulation of laser energy in the condensed phase. The rapid heating excites the matrix crystals, and leads to the localised ablation of the analyte/matrix surface, the expansion of the analyte/matrix mixture into the gas phase, the formation of a bacterial protein/matrix expanding plume, and thus the generation of bacterial protein ions.[2,3] The produced bacterial protein ions in the gas phase are thereafter accelerated using an electrostatic field towards the mass analyser.

The TOF mass analyser separates ions according to their velocities.[1] Figure 2.2 displays the scheme of a linear TOF system for the analysis of ions produced by MALDI. The generated bacterial protein ions are accelerated to the flight tube by a potential difference (U) applied between the target plate and the extraction grid. When leaving the acceleration region, all the bacterial protein ions acquire a certain velocity and fly into a field-free region in the flight tube, where they are separated according to their velocities. Ions with different velocities reach the detector at different time points. The m/z of ions is determined by measuring the time that ions take to fly through the flight tube.

After passing through the TOF analyser, bacterial protein ions are captured using a detector. The detector is able to generate an electric current

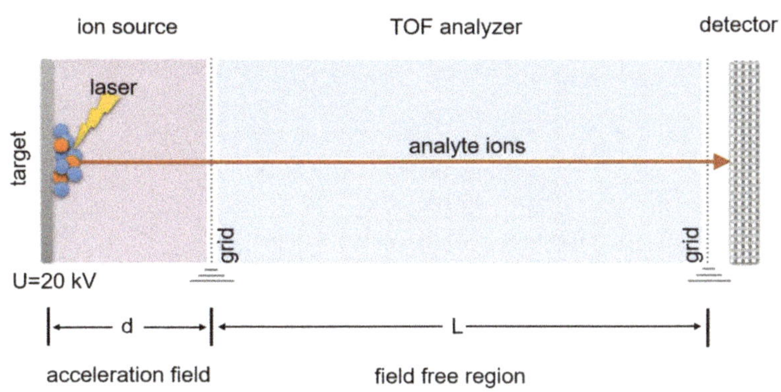

Figure 2.2 Scheme of a linear TOF system for the analysis of ions produced by MALDI under positive mode.

from the incident ions, which is proportional to the ion abundance in the bacterial sample. The detector is often a microchannel plate, which relies on parallel cylindrical channels to multiply incident ions under a strong electric field.[4] Each channel acts as a continuous dynode electron multiplier. The electrical signal from the detector is then recorded using a time-to-digital converter or a fast analog-to-digital converter to generate a mass spectrum (*i.e.*, bacterial mass fingerprint).

2.1.3 Structure of Bacterial Mass Fingerprints

The generated mass spectrum is a two-dimensional representation of signal intensity *versus* the mass-to-charge ratio (m/z), with the observation of a panel of bacterial protein ion peaks, with an example shown in Figure 2.3. The bacterial protein ion peaks are separated according to the protein mass-to-charge ratio, with the peak intensity related to the abundance of the corresponding protein expressed in the bacterial cells. Each bacterial species or strain has its unique protein composition or protein profile, and thus has a unique mass spectral pattern. The spectral pattern is therefore used as a fingerprint for bacterial identification at the species or strain level.

The mass fingerprint is usually normalised to make the most intense peak a 100% relative intensity, as shown in Figure 2.3. The normalisation facilitates the comparison of mass spectra obtained from different bacterial species and from different batches of experiment. MS measurements are "ion counting" processes, with the peak intensities greatly dependent on the

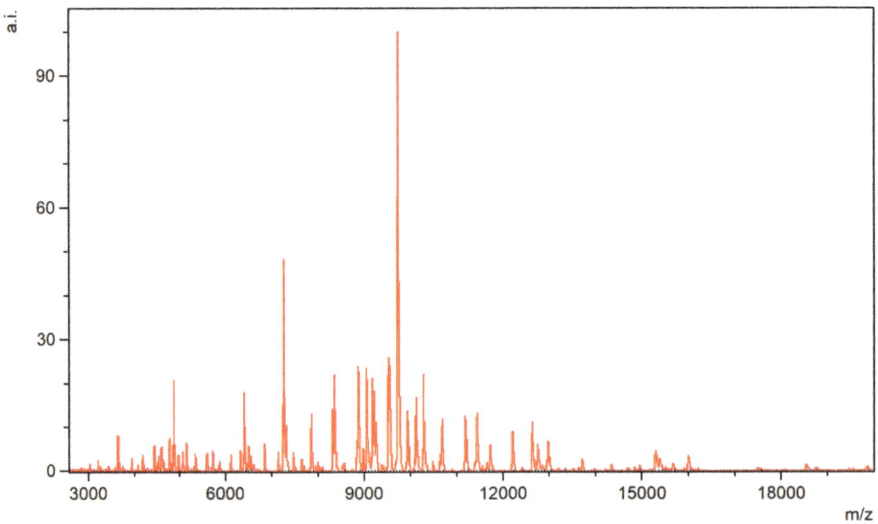

Figure 2.3 An example of a bacterial MALDI-TOF mass fingerprint. The fingerprint was collected from *Escherichia coli* strain DH5α, a Gram-negative bacterial strain (~10^5 intact bacterial cells prepared on the sample spot during the measurement). The mass spectrum is normalised to make the most intense peak a 100% relative intensity.

numbers of ions reaching the detector. Therefore, during the measurement of a mixture analyte, the peak intensities represent, to a large extent, the relative molecular abundance of the analyte components. The performance of a mass spectrometer is usually characterised in terms of sensitivity and detection limit.[5] Sensitivity is the change in peak intensity in response to analyte amount, reported as the electric charge of a particular ion species reaching the detector per mass of the analyte. It is given in the units of $C\,\mu g^{-1}$ for solid analytes. The detection limit is defined as the smallest amount of an analyte required to generate a mass spectral peak that can be distinguished from the background noise. The detection limit can be improved by adjusting the sample preparation protocol. A good quality mass spectrum makes data explanation easy. The spectrum quality is evaluated according to the signal-to-noise ratio and the peak resolution. The signal-to-noise ratio represents the ability to read an analyte from the mass spectrum. The noise mainly comes from the matrix, especially in the mass range lower than 2000 m/z. Peak resolution is defined as the observed mass value (m/z) divided by the smallest mass difference ($\Delta(m/z)$) allowing the separation of two nearby peaks, *i.e.*, $(m/z)/\Delta(m/z)$. Higher resolution refers to the higher possibility to distinguish two adjacent protein peaks, and usually leads to more accurate bacterial identification.

2.2 Preparation of Bacterial Samples

2.2.1 Sampling, Culture, and Isolation of Bacteria

In clinics, bacterial infections are usually diagnosed by identification of bacterial pathogens from patient samples like tissue biopsies, urine, or blood. Typical MALDI-TOF MS identification requires a sufficient number of purified bacterial cells. The initial bacterial abundance in patient samples is usually quite low, and the presence of other biocomponents in the patient samples interferes with the MALDI-TOF mass fingerprinting. Therefore, culture steps to recover and purify the bacterial pathogens are usually required before mass fingerprinting.

As for patient sampling, tissue biopsy is often applied when infections occur on skin surfaces, for instance, on hands, feet, fingernails, toenails, or skin wounds. A small piece of tissue or epidermal cells are collected from a suspicious skin region by using sterile surgical tools like knives, nipples, needles, and adhesive tape. Urine sampling is performed when an infection occurs in the urinary system, like in the kidneys, ureters, bladder, and urethra. Normally, 5–30 mL of urine is required for bacterial identification, with fresh urine collected in sterile plastic cups or tubes. Infections occurring inside the body could be diagnosed from blood samples in most cases, for instance, when patients have fever, chills, leukocytosis, septic shock, or suspected endocarditis. Before blood collection, proper skin antisepsis is required to reduce the risk of contamination.[6] On the basis of published reports, tincture of iodine, chlorine peroxide, and alcoholic chlorhexidine gluconate are good candidates for skin antisepsis. Among them, alcoholic chlorhexidine gluconate is recommended by the Clinical and Laboratory Standards Institute guidelines for blood collection from adults, children, and infants older than 2 months of age.[7] According to several studies, peripheral venepuncture is the preferred method for blood collection. Arterial blood sampling is not necessary, as it will not increase the rate of a successful diagnosis. Blood sampling from intravascular lines should be avoided, because it brings a high possibility of contamination.[8] The volume of collected blood directly affects the blood culture yield. The concentration of bacteria is quite low in most patients, and their recovery rate can be higher if more blood is collected. Data from modelling and clinical studies show that the bacterial concentration is in the range of 0.01–1 colony-forming units (CFU) per mL of blood for around 50% of the infection episodes. And Washington's empirical data published in 1975 showed that at least 30 mL of blood is required for detecting 99% of bloodstream infections in adult patients. Currently, it is recommended to collect 2 or 3 sets of blood from an individual patient. From each set, 10–20 mL of blood should be drawn and afterwards distributed into two bottles for continuous culture (one aerobic bottle and one anaerobic bottle). In each bottle, 5–10 mL of blood is injected for culture.[9] The volume of blood drawn from infants and children is not well prescribed, but should be dependent on the patient age and the estimated total body blood volume of the patient. Basically, the drawn blood should not exceed 1% of the total body blood volume.[10]

Once patient samples are collected, they should not be kept at room temperature for more than a few hours, and should not be refrigerated or frozen, in case of the death or degradation of bacterial pathogens. Instead, the freshly collected samples should be cultured as soon as possible, so as to let the pathogens proliferate in an ideal environment. Delayed incubation could result in delayed or impeded bacterial identification. The culture of patient samples in clinical laboratories is usually performed in instruments which provide an optical environment for the growth of a wide range of bacterial pathogens and enable continuous monitoring of the pathogen growth. Examples of the incubation instruments approved by the U.S. Food and Drug Administration (FDA) include, but not limited to, the BACTEC™ system series from Becton Dickinson and the BACT/ALERT® system series from bioMérieux. The incubation happens in specially designed culture bottles which contain carbohydrate substrates and culture broth. The carbohydrate substrates produce CO_2 during the growth of bacteria, and thus result in a drop of pH value in the culture bottle. The pH value is monitored using a colorimetric or fluorescent sensor installed on the bottom of the bottle (examples are shown in Figure 2.4).[11] The instrument measures the colour or fluorescence change regularly, for instance, every 10 minutes in the BACT/ALERT® 3D system, and analyses the changes using sophisticated algorithms, so as to monitor the bacterial growth dynamically and ensure an early detection of the bacteria. The culture broth is often a mixture of nutritional culture medium to promote bacterial proliferation, an

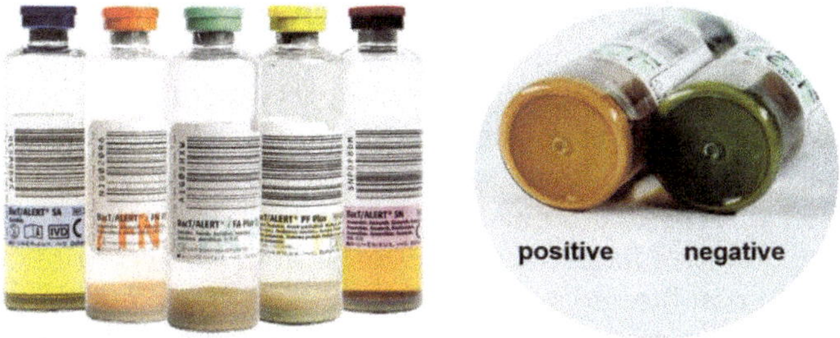

Figure 2.4 Photos of BACT/ALERT® culture bottles from bioMérieux (Marcy-l'Étoile, France). The bottles contain a colorimetric sensor on the bottom to detect the CO_2 level change resulting from bacterial growth. The positive bottle, indicated by a significant colour change on the bottom, tells the presence and sufficient growth of bacteria, *i.e.*, a successful recovery of bacteria from the patient sample.

anticoagulant, and resins or charcoal mixtures to reduce the negative influence of toxic compounds on bacterial growth.[12] The culture bottle offers a comprehensive range of culture media to ensure the recovery of a wide range of bacterial pathogens including Gram-negative, Gram-positive, aerobic, and anaerobic bacteria. In most cases (for instance, more than 98% as determined using the BACT/ALERT® 3D system), bacterial pathogens can be recovered, *i.e.*, the culture bottle turning positive with a significant colour or fluorescence change, after 3–7 days of culture period. A longer time of incubation does not significantly improve the overall sensitivity of culture, but increases the recovery rate of unwanted contaminants. However, a longer incubation time is needed if the infections are predicted to be caused by slowly growing pathogens like *Legionella*, *Brucella*, *Bartonella*, and *Nocardia* spp. For fastidious pathogens like *Mycobacterium* spp., the incubation time might be extended to as long as four weeks.[10]

Once culture bottles become positive, Gram staining is performed, which further confirms the presence of bacterial pathogens and can classify the pathogens as Gram-positive or Gram-negative. The Gram staining procedure was developed in 1884 by Hans Christian Gram. It allows one to distinguish Gram-positive bacteria and Gram-negative bacteria based on differential staining using a crystal violet–iodine complex and a safranin counterstain. Gram-positive bacteria have a thick mesh-like cell wall with a high abundance of peptidoglycan, which can retain the crystal violet–iodine complex after alcohol treatment and be stained purple. On the other hand, Gram-negative bacteria contain a much thinner cell wall, which is composed of only a single layer of peptidoglycan. They will not retain the purple complex after alcohol treatment, and subsequent counterstaining with safranin will give them a new pink colour.[13] After Gram staining, the positive cultures are sent to colony plating in agar plates to prepare pure cultures of the recovered bacterial pathogens and to remove unwanted components contained in the patient samples.[14] This step of subculture normally takes 12–24 h to completion. The purified bacterial strains can then be analysed by MALDI-TOF mass fingerprinting for species or sub-species identification.

The culture and sub-culture of patient samples described earlier often take a few days altogether. In order to achieve early and fast bacterial identification, strategies have been developed to isolate bacterial cells from the patient samples, either directly or after a short time period (for instance, a few hours) of culture. The obtained bacterial isolates or protein extracts from the bacterial isolates were then subjected to MALDI-TOF mass fingerprinting for species identification. Among the various isolation procedures, magnetic separation coupled with affinity recognition of bacteria is one of the most widely used procedures. For instance, iron oxide nanoparticles embedded in graphene nanosheets were decorated with chitosan to capture pathogenic bacteria from aqueous suspension, and as low as 500 CFU per mL *Pseudomonas aeruginosa* and 450 CFU per mL *Staphylococcus aureus* were detected from 1 mL of water by using MALDI-TOF MS.[15] Similarly, commercial anti-*Salmonella* Dynabeads® (Lake Success, NY, USA) can selectively isolate *Salmonella choleraesuis* for MALDI-TOF MS analysis, with the detection limit reaching 10^7 cells per mL in water and 10^9 cells per mL in human urine and chicken blood (1 mL of water, urine, or blood sample).[16] Commercial magnetic beads coated with bacterial species-specific antibodies have also been demonstrated as a type of efficient affinity probe for bacterial identification directly from patient blood samples, with the detection limit reaching 500 cells per mL in human blood serum and 8000 cells per mL in human whole blood (1 mL of sample) for bacterial species like *E. coli*, *Bacillus subtilis*, and *S. aureus*.[17] This immuno-affinity probe also enabled identification of *E. coli* pathogens from patient blood samples after as short as 4 hours of incubation in BACT/ALERT® culture bottles. In clinical practice, magnetic beads coated with a mixture of antibodies against common bacterial pathogen species can be prepared for isolation of unknown pathogens from patient samples. Taking bloodstream infections as an example, bacterial pathogens like *E. coli*, *S. aureus*, *Staphylococcus epidermidis*, *P. aeruginosa*, *Klebsiella* spp., *Enterococcus* spp., and *Enterobacter cloacae* account for more than 70% of the infection episodes. Commercial antibodies against all the aforementioned species are currently available, and a cocktail of these antibodies is efficient for the diagnosis in most cases. The immuno-affinity capture can be further improved by using non-specific affinity probes. Instead of using bacterial species-specific antibodies, ligands bearing affinity to universally all bacterial species or multi-species could be employed to blindly capture all bacterial cells from infected blood. The collected bacteria can then be identified by MALDI-TOF mass fingerprinting. This blind capture strategy is extremely suitable for single-species infections. If infections are caused by multiple species, identification could be achieved by adjusting the mathematical algorithm for the analysis of the obtained mass fingerprints.[18] In such a way, there would be no need to prepare different bacterial antibodies, which will greatly reduce the testing time and cost. Such ligands could be found by systematically analysing the bacterial surface structures and compositions. Attention should be paid to the similarities among different bacterial species, as well as the differences between bacterial cells and blood cells. Ideally, ligands should have affinity to every kind of bacteria, but have no or very minimal interaction with blood components. Recently, such ligands have not yet been reported, but potential candidates could include mannose-binding lectin, DNA-binding ligands, eosinophil cationic protein, Zn(II)-dipicolyl-amine, and mammalian immunoglobulin G.[19–23] All of them have been reported to show affinity to multiple bacterial species.

2.2.2 Preparation of Bacterial Cells for Mass Fingerprinting

Bacterial identification by MALDI-TOF mass fingerprinting is performed with purified bacterial cells derived from a fresh culture, or bacterial cells isolated from fresh patient samples. The fingerprinting can be conducted with either protein extracts from the bacterial cells, or intact whole bacterial cells, or intact whole bacterial cells after organic solvent penetration.

In 1994, Cain *et al.* first reported that MALDI-TOF MS could be used for the differentiation of bacteria based on their water-soluble protein profiles. In this report, water-soluble proteins were extracted from the bacterial cells and thereafter analysed by MALDI-TOF MS, in which a Nafion film was used to enhance the ionisation efficiency of the proteins.[24] In 1996, Liang and colleagues extracted proteins from bacterial cells, and the proteins were separated by capillary liquid chromatography and detected on-line using a UV absorption detector. The eluent was then collected for off-line analysis by MALDI-TOF MS. The authors found bacteria could be discriminated from each other to the species level based on the extracted protein profiles. Only picomolar levels of extracted proteins were required for the measurement.[25] At almost the same time, Krishnamurthy and colleagues reported that proteins extracted from bacterial cells could be directly measured by MALDI-TOF MS without the need for chromatographic separation, and the obtained spectra revealed specific biomarkers for individual bacterial strains. The observed biomarkers enabled them to detect pathogenic bacteria and to distinguish pathogenic bacteria from the corresponding non-pathogenic species.[26] In those experiments, purified bacterial strains were freshly grown to a log phase in culture broth. The bacterial cells were harvested by centrifugation, and washed in Tris or phosphate buffer. Proteins were extracted from bacteria by breaking the bacterial cells using sonication and removing the cell debris using centrifugation. After centrifugation, the extracted proteins were present in the supernatant liquid. The water-soluble proteins were precipitated by chilling the supernatant liquid to ice-cold temperature and adding an appropriate volume of acetone or methanol. The proteins were obtained by centrifugation, and then resuspended in either water or designed aqueous buffer solution for subsequent analysis. Recently, the protein extraction procedures have been simplified and standardised. The most widely used protocol is the ethanol/formic acid/acetonitrile extraction protocol, which is described in the Bruker MALDI Biotyper® 3.0 User Manual (year 2011). Briefly, bacterial cells are harvested from 1 mL of fresh culture (in which bacteria grow at a log phase) by centrifugation at 7500*g* for 3 min and washed twice with deionised water. The cell pellet is suspended in 300 μL of water, followed by adding 900 μL of ethanol and mixing it thoroughly. The mixture of water and ethanol is completely removed by centrifugation at 8500*g* twice (2 min for each time) and air-drying for ~30 min. The obtained cell pellet is resuspended in 50 μL of 70%/30% (volume percentage) formic acid/water. The mixture is vortexed thoroughly and allowed to stand for 5 min, followed by addition of 50 μL of acetonitrile. The final mixture is vortexed thoroughly for another 5 min. Thus, intracellular proteins are extracted into the solvent. After centrifugation at 8500*g* for 2 min, the supernatant containing the extracted proteins is harvested. The product is pipetted onto a MALDI target plate (1–2 μL for each spot), dried naturally at room temperature, and covered with 1–2 μL of MALDI matrix. MALDI-TOF mass fingerprinting is conducted after the mixture is naturally dried.

In 1996, Holland and colleagues found that Gram-negative enterobacteria could be detected by MALDI-TOF MS directly in their intact whole state without the need to perform protein extraction. They collected bacterial cells from colonies on agar plates, mixed the bacteria with a MALDI matrix, air-dried the mixture on a MALDI target plate, and then performed MALDI-TOF MS measurement. The measurement results revealed that each bacterial strain showed a few characteristic high-mass ion peaks which were thought to be derived from bacterial proteins. With the characteristic peaks, unknown bacterial strains were identified correctly by matching their peaks with the peaks generated from previously-known bacterial strains. The detection time for each strain was only a few minutes.[27] Encouraged by this work, Claydon and colleagues demonstrated in the same year that MALDI-TOF MS could generate characteristic spectra, or fingerprints, from both Gram-negative bacteria and Gram-positive bacteria in their intact whole state within minutes. Bacteria could be identified at the genus, species, and even strain levels by comparing the corresponding fingerprints.[28] Although at that time the spectrum quality was much lower than that now, this type of fast identification approach based on whole cell fingerprinting and pattern matching analysis has attracted great attention worldwide. Later studies reveal that the fingerprint peaks mainly come from bacterial cellular proteins (representing 60–70% of the dry weight of bacteria), half of which are highly abundant alkaline ribosomal proteins.[29] After around twenty years of improvement, the bacterial fingerprinting method has been incorporated into commercial systems like VITEK® MS (bioMérieux, France) and MALDI Biotyper® (Bruker, Germany), which have been installed in hospitals for routine diagnosis of bacterial infections. Currently, the most widely used sample preparation protocol for intact cell mass fingerprinting is the matrix covering protocol. Briefly, bacterial cells are harvested from 1 mL of fresh culture (in which bacteria grow at a log phase) by centrifugation at 7500*g* for 3 min and washed twice with deionised water. The cell pellet is suspended in 300 μL of deionised water. The suspension is pipetted onto a MALDI target plate (1–2 μL for each spot), dried naturally at room temperature, and covered with 1–2 μL of MALDI matrix. MALDI-TOF mass fingerprinting is conducted after the mixture is naturally dried. For some Gram-positive bacteria with quite thick cell walls, the intact cell fingerprinting procedure could be enhanced by on-plate cell wall penetration using organic solvents. For instance, the dried bacterial cells on the MALDI target can be covered with 1 μL of 50%/35%/15% acetonitrile/formic acid/deionised water solution to allow penetration and deformation of the bacterial cell wall, to make ionisation of the cellular proteins easier. After being dried naturally at room temperature, the cells are then covered with the MALDI matrix for the subsequent mass fingerprinting.

Typically, the diameter of sample spots on a MALDI target is 3 mm, while the size of bacterial cells is typically 0.5 to 10 μm and the diameter of the laser beam used in the MALDI-TOF MS instrument is around 100 μm. As a result, the measurement of a low number of bacterial cells is often accomplished with many "blind" laser shots, leading to noise accumulation and limited detection sensitivity. In order to improve the fingerprinting sensitivity, the sample spot size can be decreased to increase cell density on the spot and thus increase the chance of efficient laser shots. For example, Girault and his colleagues tried to confine the sample spot size within 800 μm by depositing a volume of bacterial suspension on the target plate using a droplet-by-droplet protocol (*i.e.*, 1.5 μL of bacterial suspension was deposited on the target plate by three repetitions, with 0.5 μL in each droplet), or using a Bruker MSP AnchorChip™ target on which each hydrophilic 800 μm-sized sample spot was surrounded with hydrophobic coating area. With the 800 μm-sized sample spot, a qualified mass fingerprint could be obtained from fewer than 100 bacterial cells, while around 10^5 cells are required when using a 3 mm-sized sample spot.[17]

2.3 Selection of MALDI Matrices for Bacterial Mass Fingerprinting

The choice of matrix is crucial for successful MS measurement. A matrix first functions to separate analyte molecules from one another during the formation of analyte/matrix co-crystals. Generally, a matrix should have low vapor pressure in order not to be volatised under the high-level vacuum conditions in the ion source. A matrix should also have a high ability to take up laser energy. The uptake of energy is based on strong absorption. Thus, the matrix molecule should have a suitable chromophore for the easy generation of excited states upon laser irradiation. When an ultraviolet laser is employed, the matrix often needs to have an aromatic core in the molecular structure together with appropriate functional groups.[30] When an infrared laser is used, the selection of matrix is more flexible, with fewer restrictions applying to the matrix molecular structure, because the infrared laser radiation can be easily absorbed by the stretch vibrations or bending vibrations of O–H, N–H, or C–O.[31] The matrix should have mass low enough for sublimation and good solubility in solvents compatible with the analyte, and, of course, it should not chemically react with the analyte.

Nicotinic acid is the first matrix successfully used for the detection of peptides and proteins. Since then, more matrices have been introduced. Figure 2.5 lists the

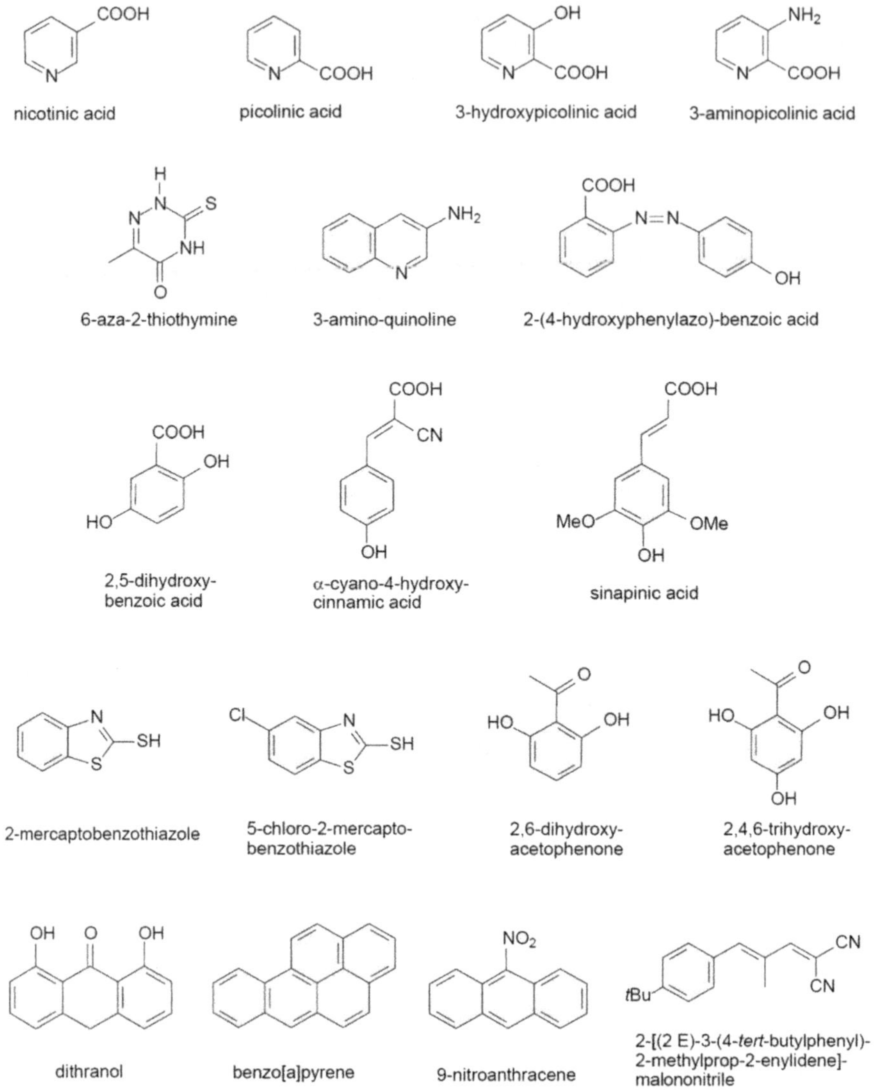

Figure 2.5 Molecular structures of often-used MALDI matrices.

MALDI matrices that are commonly used. Applications of these matrices are illustrated in detail in a textbook on MS written by Jürgen H. Gross.[5] Among them, picolinic acid, 3-hydroxypicolinic acid, 3-aminopicolinic acid, and 6-aza-2-thiothymine are usually employed for the detection of oligonucleotides, which are short DNA or RNA molecules or oligomers. 3-Aminoquinoline is a suitable matrix for oligosaccharides. The detection of synthetic polymers and lipids is achievable with the matrices 2-mercaptobenzothiazole, 5-chloro-2-mercaptobenzothiazole, dithranol, and 9-nitroanthracene. The most widely used matrices for bacterial mass fingerprinting are 2,5-dihydroxybenzoic acid

(DHB), α-cyano-4-hydroxycinnamic acid (CHCA), sinapinic acid (SA), and *trans*-ferulic acid (FA). They are well-suited for the detection of biological macromolecules like oligosaccharides, peptides, and proteins, as well as biological entities like intact bacterial cells, fungi, viruses, and mammalian cells. Generally, DHB and CHCA generate high-quality mass fingerprints in the mass range of 2000–15 000 *m*/*z* from intact bacterial cells, while SA works well over a wider mass range, *e.g.*, 2000–30 000 *m*/*z*. Meanwhile, FA gives more peaks in the mass range of 15 000–60 000 *m*/*z*, but the sensitivity in 2000–15 000 *m*/*z* could be drastically decreased (as shown in Figure 2.6). The selection of a matrix for bacterial mass fingerprinting should be mainly based on the laser system installed in the MALDI-TOF mass spectrometer.

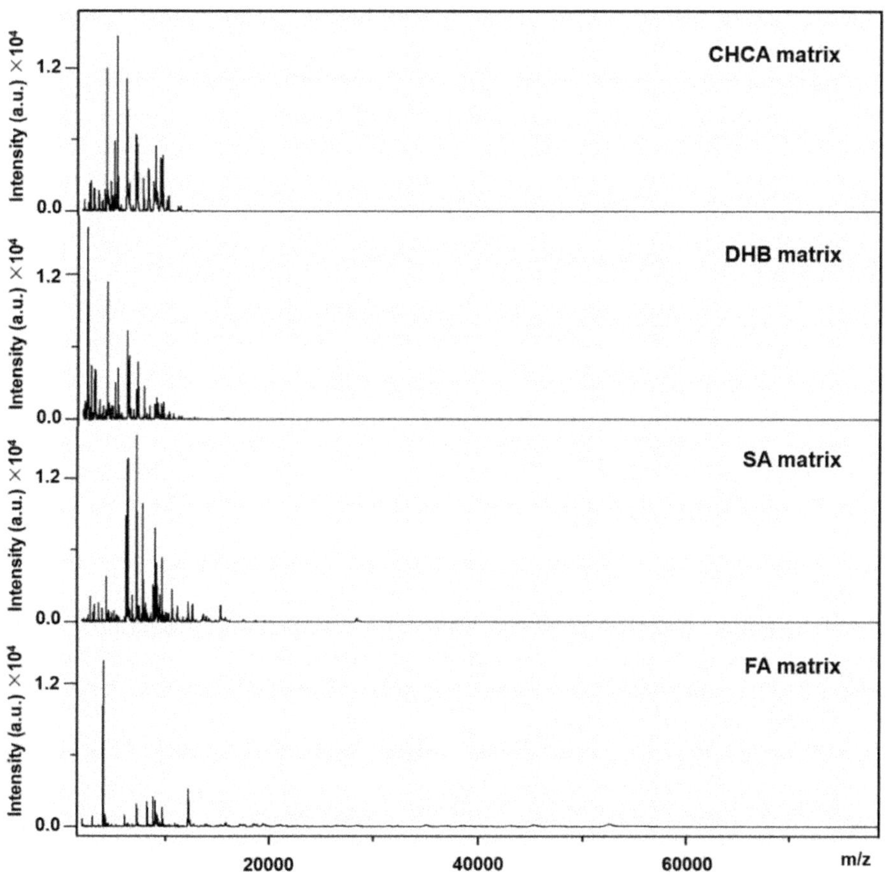

Figure 2.6 Comparison of different matrices for intact bacteria MALDI-TOF mass fingerprinting. Each spectrum was generated from $\sim 5 \times 10^5$ *E. coli* (strain DH5α) cells. The measurements were performed on a Bruker ground steel MSP 96 MALDI target, using a Bruker microflex™ LRF instrument under linear positive mode. α-Cyano-4-hydroxycinnamic acid (CHCA), 2,5-dihydroxybenzoic acid (DHB), sinapinic acid (SA), and *trans*-ferulic acid (FA) are the most widely used matrices for bacterial mass fingerprinting.

Besides the choice of matrix compound, there are also many other experimental parameters that affect the bacterial mass fingerprinting process. For instance, the matrix solution composition, the matrix-to-bacteria molar ratio, and environmental factors like relative humidity, temperature, and air flow rate can substantially influence the bacteria/matrix crystallisation process, the desorption/ionisation process, and ultimately the quality of mass fingerprints.[32] When choosing the matrix solution, several factors should be taken into consideration, including the ability to form a homogenous analyte/matrix mixture, a suitable liquid solvent evaporation speed, and an appropriate solvent pH value in favour of the ionisation process. The matrix-to-bacteria molar ratio greatly influences the initiation of the MALDI ionisation process when the laser fluence value is fixed. A lower abundance of bacteria would decrease the detection efficiency as a large volume of bacteria/matrix mixture is required to be ablated to generate enough protein ions. A too high abundance of bacteria is also unfavourable for the detection, because the dilution of the matrix leads to reduced energy absorption per volume of bacteria/matrix and thus an elevated laser fluence threshold. For intact bacterial mass fingerprinting, typically around 10^5 bacterial cells are spotted on a sample spot, and the optimal matrix solution is usually 10 mg mL^{-1} of CHCA prepared in 50%/47.5%/2.5% (volume percentage) acetonitrile/water/trifluoroacetic acid, 10 mg mL^{-1} of DHB prepared in 50/49.9/0.1% acetonitrile/water/trifluoroacetic acid, 20 mg mL^{-1} of SA prepared in 50/49.9/0.1% acetonitrile/water/trifluoroacetic acid, and 12.5 mg mL^{-1} of FA prepared in 17/33/50% formic acid/acetonitrile/water, with 1–2 μL of volume deposited on each sample spot.

2.4 Instrumental Parameters for Bacterial Mass Fingerprinting

For a selected MALDI-TOF mass spectrometer with a fixed laser system, instrumental parameters like the choice of MALDI target plate, laser intensity, laser frequency, number of laser shots, detector gain, molecular weight suppression threshold, delayed extraction time, positive or negative detection mode, and linear or reflectron detection mode influence the quality of bacterial mass fingerprints.

A MALDI target plate must be electrically conductive to ensure a free flow of electrons. Otherwise, a static charge will build up on the surface of the analyte/matrix condensed phase, which will disrupt the formation of an ion plume and result in a lower spectrum quality. MALDI targets made from stainless steel are often used for bacterial mass fingerprinting. Girault and his colleagues designed a type of photo-reactive titanium dioxide (TiO_2)-modified target plate to improve intact bacterial fingerprinting. This target plate was prepared by depositing an aqueous suspension of TiO_2 nanoparticles on the target spots of a classic stainless steel target plate, or by adding dropwise TiO_2 suspension as an array of spots on stainless steel foil (20 μm thick) which was afterwards affixed onto a bare target plate using adhesive tape. The TiO_2 nanoparticles were subsequently thermally or photonically sintered to obtain a high adherence to the steel substrate and provided a stable support layer (~3 μm thick) for bacteria and matrix. Using this target, substantial improvement of mass fingerprint quality, especially visible in the mass range of 15 000–60 000 *m*/*z*, was observed for bacteria like *E. coli*, *P. aeruginosa*, and *B. subtilis*, with the total mass spectral peak numbers (S/N ≥ 3) increased by 50–70%.[33]

The MALDI ionisation process will not happen if the laser fluence is lower than a threshold value, which is related to the matrix-to-analyte molar ratio, the absolute amount of analyte, and the type of matrix compound.[34] A good quality mass fingerprint with high peak resolution and less (or no) fragmentation is obtained when laser irradiation is performed with the laser fluence slightly above the threshold value. For instance, when the measurement is performed with a Bruker microflex™ LRF instrument under linear positive mode with around 10^5 bacterial cells presented on the target spot together with the application of 10 mg mL^{-1} CHCA matrix prepared in 50%/47.5%/2.5% acetonitrile/water/trifluoroacetic acid, the optimal laser parameters are as follows: laser intensity 70%, laser attenuator offset and range 30% and 40%, respectively, and laser frequency 20.0 Hz.

The dried droplet procedure is often used for the preparation of bacteria and matrix on the target plate, as described earlier. One disadvantage of this droplet deposition procedure is the possible presence of "sweet spots" coming from deposit heterogeneity, which leads to low laser shot-to-shot reproducibility and causes shot-to-shot spectrum variation. Such a kind of negative influence can be overcome by accumulating multiple single-shot spectra (*e.g.*, 2 × 300 laser shots) from an individual sample spot to create the final bacterial mass fingerprints.

The detector gain is adjusted to control the absolute intensity of the detected protein peaks. A too low value of detector gain is unfavourable to the detection of low-abundant bacterial protein ions, while a too high value of detector gain increases the background noise. The value is typically set at <10 for a CHCA matrix and 10–20 for DHB, SA, and FA matrices on the Bruker microflex™ LRF instrument under linear positive mode with around 10^5 bacterial cells spotted on the target spot.

The matrices themselves often generate ions in the mass range of 0–1000 *m*/*z*, coming from the matrix compounds, the organic solutions, and other impurities present in the matrix solution or on the target plate, and a molecular weight suppression of up to 1000 Da is often used to generate bacterial mass fingerprints. The mass window is usually set as 2000–20 000 *m*/*z*, as the window lower than 2000 *m*/*z* suffered from interference from the matrix and impurities, while the window higher than 20 000 *m*/*z* had the distribution of only low-abundance peaks due to the low ionisation efficiency of large proteins from the whole intact cells.

Along with the continuous development of TOF analysers, two techniques have been proposed to enhance the resolution, *i.e.*, delayed pulse extraction and reflectron mode of TOF. Delayed pulse extraction allows ions to fly for a certain period of time with their initial velocity before being accelerated. Ions with the same *m*/*z* can reach the same velocity or quite similar velocities when they leave the acceleration region. In this way, these ions can reach the detector at the same time, resulting in sharp mass peaks and thus improved mass spectrum resolution.[35] The reflectron mode of TOF uses an electrostatic reflector also called a reflectron to improve the resolution.[36] The simplest reflectron consists of a series of equally spaced grid electrodes that are connected through a resistive network of equal-value resistors. The reflectron is placed at the end of the flight tube and pointed towards the ion source. The reflectron functions as a "mirror" by deflecting the ions and sending them back to the flight tube. The detector is placed on the same side with the ion source to capture the reflected ions. For ions with the same *m*/*z*, those with more kinetic energy and thus higher velocity will penetrate the reflectron more deeply than those with lower velocity. Therefore, the faster ions have to pass a longer distance before reaching the detector, while the slower ions fly through a shorter distance. In this way, they can reach the detector at (almost) the same time and result in a sharp mass peak and hence higher resolution. A drawback of the reflectron mode of TOF is that it could lead to the loss of some ions during the flight process and thus a decreased detection sensitivity. For the mass fingerprinting of bacteria, the measurements are usually carried out under linear positive mode with delayed extraction. The positive mode benefits the detection of cellular proteins as they have high proton affinities with the tendency to be ionised through protonation.[37] The linear mode is applied to avoid the missing of low-abundance bacterial protein ions, which could happen in the reflectron detection mode. The delayed extraction involves a time delay, typically set as 400 ns for bacterial mass fingerprinting, between the laser pulse and the ion-accelerating voltage, providing TOF compensation for ion velocity spread so as to improve the mass resolution of the fingerprints.[38]

Abbreviations

Ab	Antibody
CFU	Colony-forming unit
CHCA	α-Cyano-4-hydroxycinnamic acid
DHB	2,5-Dihydroxybenzoic acid
ESI	Electrospray ionisation
FA	*trans*-Ferulic acid
LDI	Laser desorption/ionisation
MALDI	Matrix-assisted laser desorption/ionisation
MB	Magnetic bead
MCP	Microchannel plate
MS	Mass spectrometry
SA	Sinapinic acid
TOF	Time-of-flight
TiO_2	Titanium dioxide

Acknowledgements

We gratefully thank Professor Liang Qiao from the Laboratory of Mass Spectrometry Analytical Chemistry in Fudan University for the suggestions he gave for the manuscript writing as well as the bacterial mass fingerprinting research we conducted from year 2014 to 2021.

References

1. E. Hoffman and V. Stroobant, *Mass Spectrometry Principles and Applications*, John Wiley & Sons, 2007.
2. K. Dreisewerd, *Chem. Rev.*, 2003, **103**, 395.
3. M. Abonnenc, L. A. Qiao, B. H. Liu and H. H. Girault, *Annu. Rev. Anal. Chem.*, 2010, **3**, 231.
4. J. L. Wiza, *Nucl. Instrum. Methods*, 1979, **162**, 587.
5. J. H. Gross, *Mass spectrometry: a textbook*, Springer Science & Business Media, 2006.
6. R. A. Garcia, E. D. Spitzer, J. Beaudry, C. Beck, R. Diblasi, M. Gilleeny-Blabac, C. Haugaard, S. Heuschneider, B. P. Kranz and K. McLean, *Am. J. Infect. Control*, 2015, **43**, 1222.
7. M. L. Wilson, M. Mitchell, A. Morris, P. Murray, L. Reimer, L. Reller, M. Towns, M. Weinstein, S. Wellstood and J. Dunne, *CLSI document M47-A*, Clinical and Laboratory Standards Institute, Wayne, PA, 2007.
8. S. R. Snyder, A. M. Favoretto, R. A. Baetz, J. H. Derzon, B. M. Madison, D. Mass, C. S. Shaw, C. D. Layfield, R. H. Christenson and E. B. Liebow, *Clin. Biochem.*, 2012, **45**, 999.
9. B. Lamy and H. Seifert, *European Manual of Clinical Microbiology (SFM/ESCMID)*, 2012, p. 101.
10. T. Kirn and M. Weinstein, *Clin. Microbiol. Infect.*, 2013, **19**, 513.
11. B. Lamy, S. Dargere, M. C. Arendrup, J. J. Parienti and P. Tattevin, *Front. Microbiol.*, 2016, 7, 697.
12. S. Mirrett, C. Petti, C. Woods, R. Magadia, M. Weinstein and L. Reller, *J. Clin. Microbiol.*, 2004, **42**, 4581.
13. R. Coico, *Curr. Protoc. Microbiol.*, 2006, A.3C.1–A.3C.2.
14. O. Opota, A. Croxatto, G. Prod'hom and G. Greub, *Clin. Microbiol. Infect.*, 2015, **21**, 313.
15. P. Yagupsky and F. Nolte, *Clin. Microbiol. Rev.*, 1990, **3**, 269.
16. A. J. Madonna, F. Basile, E. Furlong and K. J. Voorhees, *Rapid Commun. Mass Spectrom.*, 2001, **15**, 1068.
17. Y. D. Zhu, L. Qiao, M. Prudent, A. Bondarenko, N. Gasilova, S. B. Moller, N. Lion, H. Pick, T. Q. Gong, Z. X. Chen, P. Y. Yang, L. T. Loveyf and H. H. Girault, *Chem. Sci.*, 2016, 7, 2987.
18. Y. Yang, Y. Lin and L. Qiao, *Anal. Chem.*, 2018, **90**, 10400.
19. J. A. Kaizerman, M. L. Gross, Y. G. Ge, S. White, W. H. Hu, J. X. Duan, E. E. Baird, K. W. Johnson, R. D. Tanaka, H. E. Moser and R. W. Burli, *J. Med. Chem.*, 2003, **46**, 3914.

20. M. Torrent, S. Navarro, M. Moussaoui, M. V. Nogués and E. Boix, *Biochemistry*, 2008, **47**, 3544.
21. J. Yi, Q. Qin, Y. Wang, R. Zhang, H. Bi, S. Yu, B. Liu and L. Qiao, *Microchim. Acta*, 2018, **185**, 542.
22. O. Neth, D. L. Jack, A. W. Dodds, H. Holzel, N. J. Klein and M. W. Turner, *Infect. Immun.*, 2000, **68**, 688.
23. J. J. Lee, K. J. Jeong, M. Hashimoto, A. H. Kwon, A. Rwei, S. A. Shankarappa, J. H. Tsui and D. S. Kohane, *Nano Lett.*, 2014, **14**, 1.
24. T. C. Cain, D. M. Lubman and W. J. Weber, *Rapid Commun. Mass Spectrom.*, 1994, **8**, 1026.
25. X. Liang, K. Zheng, M. G. Qian and D. M. Lubman, *Rapid Commun. Mass Spectrom.*, 1996, **10**, 1219.
26. T. Krishnamurthy, U. Rajamani and P. Ross, *Rapid Commun. Mass Spectrom.*, 1996, **10**, 883.
27. R. Holland, J. Wilkes, F. Rafii, J. Sutherland, C. Persons, K. Voorhees and J. LayJr, *Rapid Commun. Mass Spectrom.*, 1996, **10**, 1227.
28. M. A. Claydon, S. N. Davey, V. Edwards-Jones and D. B. Gordon, *Nat. Biotechnol.*, 1996, **14**, 1584.
29. V. Ryzhov and C. Fenselau, *Anal. Chem.*, 2001, **73**, 746.
30. J. Krause, M. Stoeckli and U. P. Schlunegger, *Rapid Commun. Mass Spectrom.*, 1996, **10**, 1927.
31. K. Dreisewerd, S. Berkenkamp, A. Leisner, A. Rohlfing and C. Menzel, *Int. J. Mass Spectrom.*, 2003, **226**, 189.
32. K. Wiangnon and R. Cramer, *Anal. Chem.*, 2015, **87**, 1485.
33. Y. D. Zhu, N. Gasilova, M. Jovic, L. Qiao, B. H. Liu, L. T. Lovey, H. Pick and H. H. Girault, *Chem. Sci.*, 2018, **9**, 2212.
34. G. Westmacott, W. Ens, F. Hillenkamp, K. Dreisewerd and M. Schürenberg, *Int. J. Mass Spectrom.*, 2002, **221**, 67.
35. R. S. Brown and J. J. Lennon, *Anal. Chem.*, 1995, **67**, 1998.
36. V. M. Doroshenko and R. J. Cotter, *J. Am. Soc. Mass Spectrom.*, 1999, **10**, 992.
37. R. Knochenmuss and R. Zenobi, *Chem. Rev.*, 2003, **103**, 441.
38. R. S. Brown and J. J. Lennon, *Anal. Chem.*, 1995, **67**, 1998.

CHAPTER 3

MALDI-TOF Mass Fingerprinting for Rapid Identification of Bacteria: Data Analysis and Algorithm Development

Y. YANG (iD 0000-0002-1330-9985)

ZJU-Hangzhou Global Scientific and Technological Innovation Center, Zhejiang University, Hangzhou 311200, China
Email: y_yi@zju.edu.cn

3.1 Introduction

Over the past decade, matrix-assisted laser desorption/ionization time-of-flight mass spectrometry (MALDI-TOF MS) has gained considerable popularity in the identification of microorganisms and has been approved for the routine identification of pathogenic bacteria in clinical diagnostic laboratories.[1–4] This technique is characterized by many valuable and useful features including simple sample preparation, high automation and throughput, and low cost per sample. In a typical MALDI-TOF MS analysis of microorganisms, biological molecules, typically small proteins originating from cell surfaces, intracellular membranes, and ribosomes, constitute the characteristic mass spectrum, also known as the fingerprint.[5] In a typical MALDI-TOF MS analysis of microorganisms, once spectra are collected, identification is based on spectral pattern matching between the mass

New Developments in Mass Spectrometry No. 13
Detection and Analysis of Microorganisms by Mass Spectrometry
Edited by Liang Qiao and Jia Yi

Published by the Royal Society of Chemistry, www.rsc.org

spectra of unknown samples and a reference library of standard mass spectra of known species. Commercial software tools have provided comprehensive solutions for microorganism identification and reference spectral libraries containing thousands of microorganism species.[6,7] Open-source software tools are also available to facilitate MALDI-TOF MS data analysis.[8–11]

A complete MALDI-TOF MS data analysis workflow can be separated into three distinct steps: (1) preprocessing, (2) spectral matching, and (3) result interpretation. Preprocessing is the first step in MALDI-TOF MS data analysis, where the dense raw spectral profile data are reduced to a list of detected peaks for subsequent analysis. The second step compares the sample spectrum with those in the reference spectral library, quantifies their similarities, and finds the best match. The third step assesses the spectral match with specific criteria for quality control and reports the final identification result.

This chapter focuses on approaches to data analysis for MALDI-TOF MS-based identification of microorganisms. The existing computational methods for the three steps of data analysis and available software solutions are summarized. In addition, bioinformatic methods that do not require a reference spectral library are introduced as alternatives to typical spectral matching approaches. Finally, the current challenges and outlook of MALDI-TOF MS data analysis for microorganism identification are discussed.

3.2 Raw Mass Spectrum Preprocessing

The raw data in a spectrum produced using a MALDI-TOF mass spectrometer usually consist of a vector of 10 000–100 000 entries, and each entry represents the number of ions hitting the detector during a small, fixed interval of time.[12] The observed flight time is transformed to m/z values based on the coefficients pre-determined experimentally by using standard molecules of known masses.[13] As a result, a typical MALDI-TOF mass spectrum produced by running a sample is a series of tens of thousands of intensity measurements, which should undergo the necessary preprocessing steps to remove noise or non-specific baseline signals and localize the peaks accurately (Figure 3.1).

3.2.1 Denoising and Baseline Correction

Raw mass spectral data contain a range of noises, such as electronic and chemical noises. Therefore, denoising is often required to be done before any analysis takes place. As random noise is typically high-frequency components in a spectrum, there are many smoothing approaches that can be used in order to minimize noise in the spectra. Moving average smoothing is one of the simplest methods,[14] where each point in the spectrum is

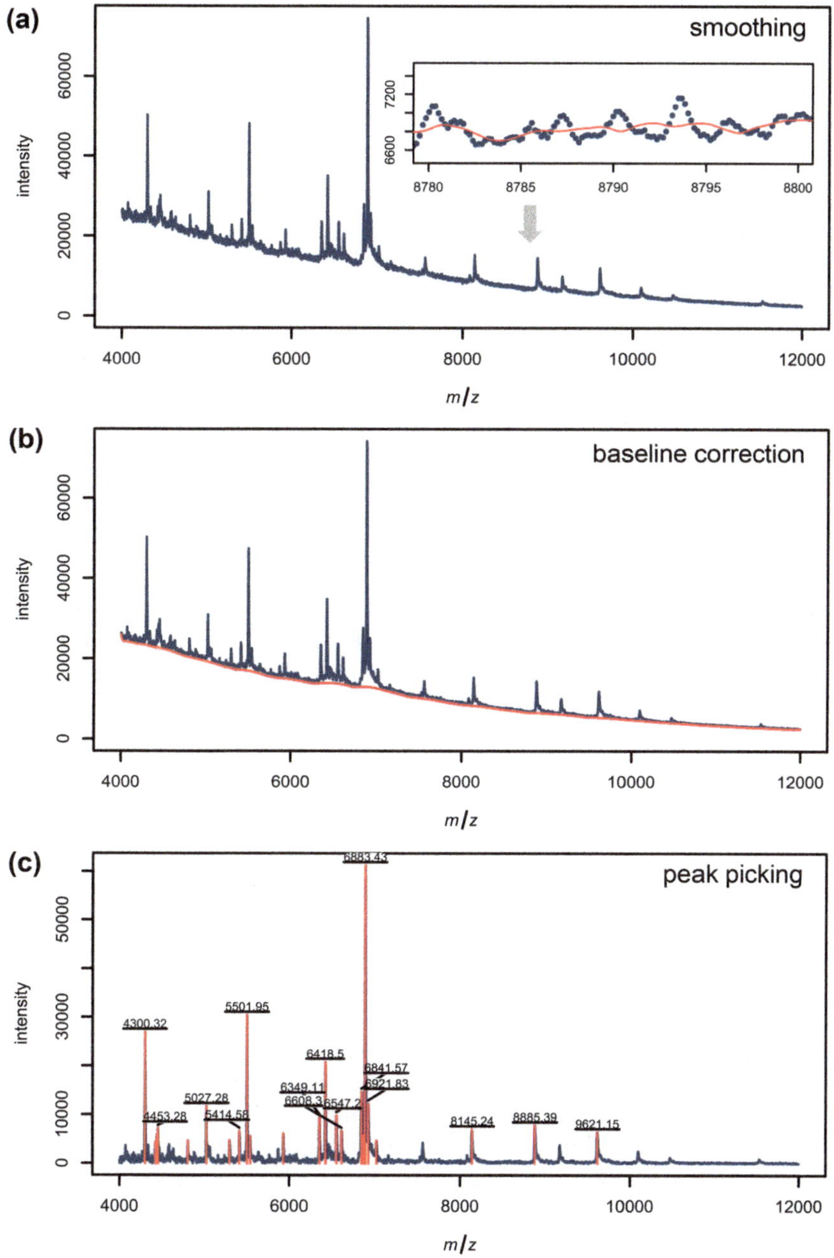

Figure 3.1 Spectrum preprocessing workflow. (a) Smoothing of the raw spectrum. In the zoomed-in view, the blue dots indicate data points in the raw spectrum, and the red curve indicates the smoothed spectrum. (b) Baseline estimation on the spectrum after smoothing. The red curve indicates the baseline. (c) Peaks picked from the spectrum after smoothing and baseline correction. Peaks are marked as red vertical lines. Data are processed and visualized using MALDIquant.[27]

smoothed by taking an average of the point and a number of neighboring points (the so-called window):

$$y_i = \frac{1}{2k+1} \sum_{j=-k}^{k} x_{i+j} \tag{3.1}$$

where x_i and y_i represent the intensities at position i in the input and output spectra, respectively, and $2k+1$ is an odd number that denotes window width. As an alternative, the data within the window can also be fitted with a Gaussian function or a low-order polynomial (Savitzky–Golay filtering).[15] In these methods, the window size must be considerably smaller than the peak width to conserve the peak shape.

Strong baseline distortion of MALDI-TOF mass spectra poses a difficult problem that potentially also introduces artifacts and it is essential to handle it correctly. The intuitive method for baseline estimation uses a moving median, where the baseline is calculated from the spectrum noise as a median of the data points in the moving window. Another commonly-used baseline estimation method is statistics-sensitive non-linear iterative peak-clipping (SNIP).[16] The initial version of SNIP iteratively uses a series of incrementally increased window sizes (*i.e.*, half window size from 1 to the specified maximum size). In each iteration, SNIP takes the element-wise minimum of each value and the average of the values at the left and right edges of a window:

$$y_i = \min\left\{x_i, \frac{x_{i-k} + x_{i+k}}{2}\right\} \tag{3.2}$$

where x_i and y_i represent the intensities at position i in the input and output data in each iteration, respectively, and k denotes half window width. In a later study, a reversed window size range (decreasing from the specified maximum size to 1) has been suggested to get a smoother baseline.[17] In addition to the smoothing-based methods, baseline estimation can be achieved by morphological operations, such as top-hat transformation.[18] Top-hat is more commonly used for separation of small objects from background in image processing. For MS data analysis, it performs a morphological opening on the spectral data, which is a moving minimum (called "erosion") followed by a moving maximum (called "dilation"). After estimating the baseline, the corrected spectrum is obtained by subtracting the baseline from the signal.

3.2.2 Peak Picking

After smoothing and baseline correction, MS peaks are detected by searching for local maxima of the intensity values. To avoid false positive peak detection, several constraints are often adopted for peak picking: (1) a signal-to-noise ratio (SNR) threshold, where noise is estimated as the median of the absolute deviation of data points within a window,[19] and a local maximum must be SNR times higher than noise to be recognized as a peak; (2) an intensity threshold, which is used to filter out small peaks in flat regions that pass the SNR filter; (3) a peak width threshold, where the peaks should be at least several points wide;[20] and (4) a peak slope threshold, where slopes are computed from the left neighboring local minimum up to the local maximum and from the local maximum down to the right neighboring local minimum, and peaks are eliminated if both slopes are less than half the value of the noise.[21]

Although the application of these constraints decreases the false positive rate, it also decreases the sensitivity of peak detection. The width and height of "true" MS peaks can vary a lot and hence fixed peak filters may fail. Continuous wavelet transformation (CWT) provides a method for resolving these problems.[22] The wavelet transform is a method to decompose an input signal into a set of elementary waveforms so that the signal can be analyzed by examining the wavelet coefficients. By applying CWT over the raw spectrum, a two-dimensional coefficient matrix is produced, where one dimension is the scales (wavelet width) and the other is the length of the spectrum (Figure 3.2). The changes of CWT coefficients over scales provide additional information about peak shape. Nearest local maxima across adjacent scales in the CWT coefficient matrix are connected to form "ridge lines", which are correlated with the peaks in the mass spectrum. The CWT coefficients can further be used to estimate peak parameters. In contrast to most other methods that detect peaks after smoothing and baseline correction, CWT does not require the smoothing step and can implicitly remove the baseline if carried out on a raw spectrum.[22] Yang *et al.*[23] benchmarked typical peak detection algorithms on simulated MALDI MS data and real data of known peptide samples, and the results demonstrated that the CWT-based algorithm provides the best average performance in terms of sensitivity.

3.2.3 Software Tools for Mass Spectrum Preprocessing

Algorithms for spectrum preprocessing have been implemented in commercial data analysis software provided by instrument vendors. Also, there are many instrument-independent and open-source tools that support spectrum preprocessing and peak detection (Table 3.1). Mantini *et al.*[24] developed LIMPIC, an automated tool for signal denoising and peak detection in MALDI-TOF-MS data, which was demonstrated to be effective in peak detection from purified protein samples and human plasma samples. Strohalm *et al.*[25,26] developed mMass, a platform-independent software solution that supports various MS data formats and covers a wide range of spectrum preprocessing tasks. In particular, mMass provides a graphical user interface that is friendly to users with no programming background. On the R platform, versatile packages, such as MALDIquant,[27] PROcess,[19] and MassSpecWavelet,[22] have been developed to provide a flexible analysis

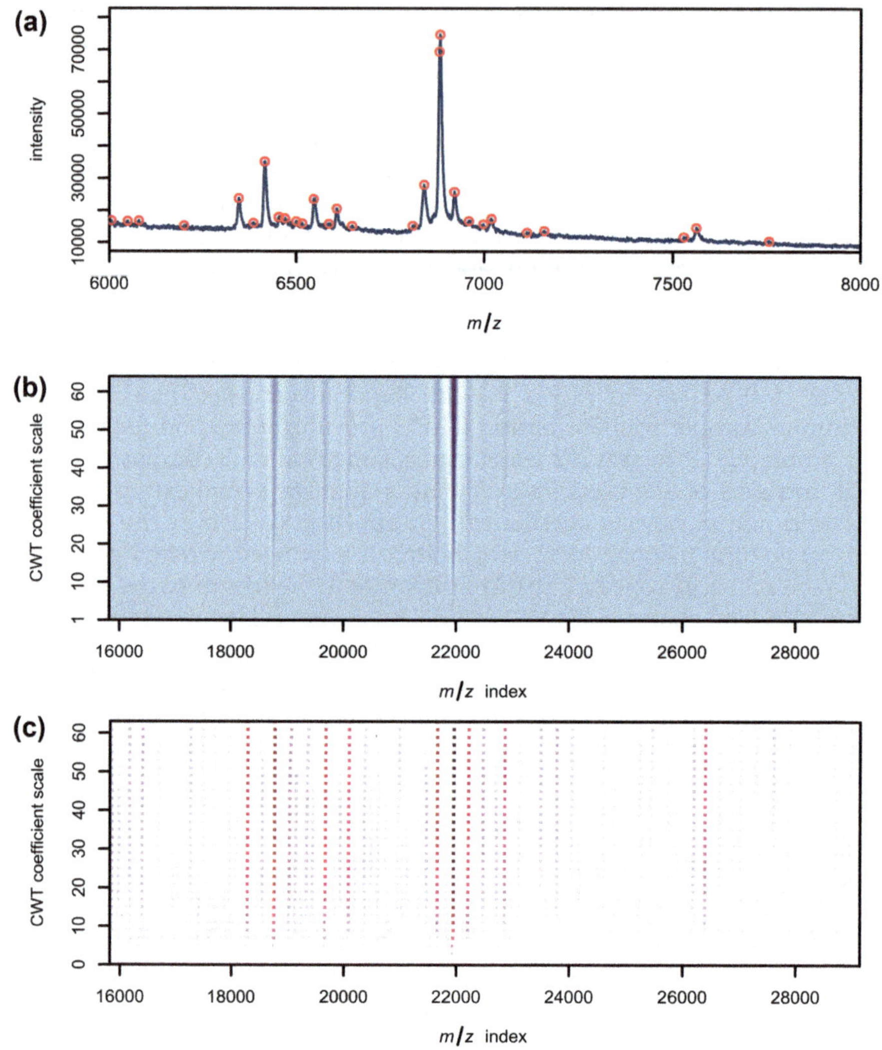

Figure 3.2 Peak detection process based on CWT. (a) The raw spectrum and detected peaks. Peaks are marked as red circles. (b) The CWT coefficient image. The darkness of each color indicates the amplitude. (c) The detected ridge lines based on the CWT coefficient image. The darkness of each color indicates the relative strength of each of the CWT coefficients. Data are processed and visualized using MassSpecWavelet.[22]

pipeline comprising all steps of mass spectrum preprocessing. By integrating MALDIquant, MassSpecWavelet, and other state-of-the-art bioinformatic libraries, Fernández *et al.*[28] developed Mass-Up, an all-in-one software application that covers the whole process of MALDI-TOF MS data analysis, including spectrum preprocessing and subsequent statistical analysis.

Table 3.1 Representative open-source software solutions for MALDI-TOF MS data preprocessing.

Tool	Type	Features
LIMPIC[24]	MATLAB script	Smoothing based on Kaiser filtering; baseline correction based on linear interpolation; peak picking based on local maxima
	https://doi.org/10.1186/1471-2105-8-101	
mMass[25,26]	Standalone GUI application	Smoothing based on moving average, Gaussian filtering, or Savitzky–Golay filtering; baseline correction based on moving median; peak picking based on local maxima
	http://www.mmass.org/	
MALDIquant[27]	R package	Smoothing based on moving average or Savitzky–Golay filtering; baseline correction based on SNIP, top-hat, moving median, *etc.*; peak picking based on local maxima
	https://strimmerlab.github.io/software/maldiquant/	
PROcess[19]	R package	Smoothing based on moving average; baseline correction based on linear interpolation or locally weighted regression; peak picking based on local maxima
	https://www.bioconductor.org/packages/PROcess/	
MassSpecWavelet[22]	R package	Peak detection based on CWT
	https://bioconductor.org/packages/MassSpecWavelet/	
Mass-Up[28]	GUI application	Supporting multiple preprocessing tasks by invoking MALDIquant and MassSpecWavelet
	https://www.sing-group.org/mass-up/	

3.3 Reference Libraries for Spectral Matching

3.3.1 Availability of Reference Spectral Libraries

The identification of microbial mass spectra is achieved by comparing the spectra of the sample with those in reference libraries. Thus, the effectiveness and accuracy of identification of microorganisms directly depend on the availability of a representative reference spectral library.

The comprehensive reference spectral libraries with the most species coverage are proprietary and only commercially available. Biotyper, developed by Bruker Daltonik (Bremen, Germany), is a comprehensive software solution with spectral libraries for identification of microorganisms. In the spectral library of Biotyper, the locations, intensities, and occurrence frequencies of MS peaks of each microorganism strain are stored as a "main spectral profile" (MSP), which comes from multiple MALDI-TOF MS analyses of the strain.[29] In 2013, the Biotyper library covered a broad range of more than 4600 microbial isolates of more than 2000 species from Gram-negative bacteria, Gram-positive bacteria, yeasts, multicellular fungi, and mycobacteria.[8] The library coverage has been extended continuously over the years.

VITEK MS is another commercial system for microorganism identification supplied by bioMérieux (Marcy l'Etoile, France). It consists of a closed library used for *in vitro* diagnosis and a research-use-only library named SARAMIS. The SARAMIS library was originally developed by AnagnosTec (Zossen, Germany), and acquired by bioMérieux in 2010.[8] Different from single-strain MSP in Biotyper, SARAMIS employs the concept of SuperSpectra for identification, which is a composite spectrum containing conserved and specific peaks at different taxonomic levels identified from a population of various well-characterized isolates.[30] In 2012, SARAMIS contained more than 2600 SuperSpectra and 35 000 single fingerprint spectra, representing more than 2000 species and 500 genera.[31]

As a supplement to commercial solutions, researchers have built open-access reference spectral repositories (Table 3.2). SpectraBank is a freely accessible library containing the characteristic mass lists of over 100 bacterial strains.[32] The FoodBIMS library comprises reference mass peak lists of over 50 strains of food-borne bacteria.[33] In 2013, the European Consortium of Microbial Resource Centres (EMbaRC) released an accessible library of MALDI-TOF mass spectra of over 1000 strains. The Hospital-University Institute Méditerranée Infection (France) maintains the URMS library, which contains mass spectra of over 400 strains of clinical microorganisms reported up to now. The Spectra website by the Public Health Agency of Sweden hosts an extended library of reference spectra for bacteria and fungi. In 2016, Robert Koch Institute (RKI, Germany) published a mass spectral library of highly pathogenic bacteria, providing not only the raw mass spectral data but also the peak list data that can be imported into the MicrobeMS software[9] developed at RKI. The latest update of the RKI library in 2018 covers 1064 strains of 171 species.[34]

As commercial spectral libraries heavily focus on medical applications, even organisms commonly found in environmental, agricultural, or food samples are thus often not recognized by standard commercial MALDI-TOF MS systems.[35] Due to this limitation, research groups have built their own "in-house" libraries for particular applications. For example, although the pathogenic *Vibrio* species found in humans are widely represented in commercial spectral libraries, these libraries do not contain sufficient *Vibrio* species found in marine environments.[36] For this reason, Moussa *et al.*[37] created the EnviBase library, which contains 120 strains that describe a high diversity of *Vibrio* species affecting marine organisms. Liu *et al.*[38] developed the PVBase containing 790 clinical and environmental *Vibrio* strains that come from various regions of China, improving the efficiency of identification of closely related *Vibrio* species.

Table 3.2 Open-access spectral libraries of MALDI-TOF mass spectra of microorganisms.

Library	Format	Content
SpectraBank[32]	Mass list text	Mass lists of bacteria with a focus on species of interest in the food sector (119 strains belonging to 59 strains)
	http://www.usc.es/gl/investigacion/grupos/lhica/spectrabank	
FoodBIMS[33]	Peak list text	Peak lists of food contaminant microorganisms that cause human diseases (54 strains belonging to 24 species)
	http://bioinformatica.isa.cnr.it/Descr_Bact_Dbase.htm	
EMbaRC	Bruker flex raw data	Spectra of bacteria (1150 strains belonging to 476 species)
	http://www.embarc.eu/deliverables/EMbaRC_D.JRA2.1.4_D15.33_MALDI-TOF_DB.pdf	
URMS	Bruker flex raw data	Spectra of clinical microorganisms (453 strains belonging to 393 species)
	https://www.mediterranee-infection.com/acces-ressources/base-de-donnees/urms-data-base/	
RKI[34]	Bruker flex raw data; MicrobeMS peak list	Spectra of highly pathogenic bacteria of biosafety level 3 (1064 strains belonging to 171 species)
	https://zenodo.org/record/1880975	
Spectra	Biotyper MSP	Reference spectral library of bacteria and fungi (over 300 strains)
	http://spectra.folkhalsomyndigheten.se/spectra	
EnviBase[37]	Biotyper MSP	Reference spectral library of *Vibrio* species that are potentially responsible for mollusk diseases (120 strains of 25 species)
	https://www.seanoe.org/data/00642/75416/	
PVBase[38]	Biotyper MSP	Reference spectral library of clinical and environmental strains of *Vibrio* species collected from various regions of China (790 strains belonging to 10 species)
	https://doi.org/10.3389/fmicb.2022.872825	

3.3.2 Building Custom Spectral Libraries

MALDI-TOF mass spectra are not universal, but depend on the growth conditions of a microbial culture, cell lysate, and matrix that are applied onto the target plate, as well as the settings of the ionizing laser, flight tube, and mass detector used for recording mass spectra.[35] Thus, custom-made reference spectral libraries enable higher flexibility with respect to different experiment protocols compared with generic libraries.

To compile a reference spectrum, a number of replicate spectra (typically no fewer than 24 valid spectra)[39] of the sample are required. A valid spectrum should satisfy a range of criteria for quality control:[40,41] (1) a good mass resolution of peaks (800–1200); (2) a high SNR of the base peak (more than 2000) after smoothing and baseline correction; (3) a large intensity of the base peak (1000–10 000); and (4) a sufficient number of peaks (100–200). Pairwise correlations of the replicate sample spectra are computed and the spectra dissimilar to others are discarded.[37] For example, the composite correlation index matrix analysis in Biotyper can be used to measure inter-spectral similarities, and the spectra with a correlation index less than 65% can be ruled out.[39] Then the common peaks in the remaining replicate spectra (typically shared in at least 80% spectra)[39] are selected as characteristic mass peaks of the sample, and merged into the final reference spectrum.

Custom-made reference spectra can complement the coverage diversity of commercial libraries. Moreover, it has been reported that the identification scores obtained with these references are usually higher as compared to generic reference libraries, since they can be generated with the same MALD-TOF MS device, settings, and sample preparation as the experimental samples.[35,42–44]

3.4 Spectral Pattern Matching

To achieve microorganism identification, spectra of unknown samples are compared to reference libraries of spectra of known microorganisms by spectral pattern matching. At the heart of spectral pattern matching lies the measurement of spectral similarities, where the likeness of the sample spectrum and each reference spectrum is quantified using a similarity score to find the reference spectrum with top scores. These algorithms are implemented in commercial software solutions like Biotyper. Open-access software tools, including BIOSPEAN,[8] MicrobeMS,[9] and BacteriaMS,[10,11] are also available to facilitate comparison of spectra and quantification of spectral similarities. A variety of similarity metrics can be used to compare two spectra. These similarity scores can be mainly divided into three categories, *i.e.*, those based on shared peak masses, those based on intensity correlations, and those combining both.

3.4.1 Spectral Similarity Scores Based on Shared Peak Masses

The simplest similarity metric for spectrum comparison regards the spectra as binary peak lists that contain only the presence or absence of peaks.[45,46] For each pair of peaks in the two peak lists, if their mass difference is less than a specified value of mass tolerance, then the two peaks are considered to represent the same mass. Then the Jaccard coefficient can be used to measure the similarity between peak lists A and B,[47,48] which is defined as the size of the intersection divided by the size of the union of all the unique peaks:

$$J(\mathrm{A},\mathrm{B}) = \frac{|M_\mathrm{A} \cap M_\mathrm{B}|}{|M_\mathrm{A} \cup M_\mathrm{B}|} \tag{3.3}$$

where M_A and M_B represent the sets of peak masses in peak lists A and B, respectively. Figure 3.3(a) and (b) illustrates an example of the Jaccard coefficient between the two peak lists.

Conway *et al.*[45] proposed another similarity metric, which is defined as the number of peaks in common between the two peak lists divided by the geometric mean of the total number of peaks in each peak list:

$$S(\mathrm{A},\mathrm{B}) = \frac{|M_\mathrm{A} \cap M_\mathrm{B}|}{\sqrt{|M_\mathrm{A}||M_\mathrm{B}|}} \tag{3.4}$$

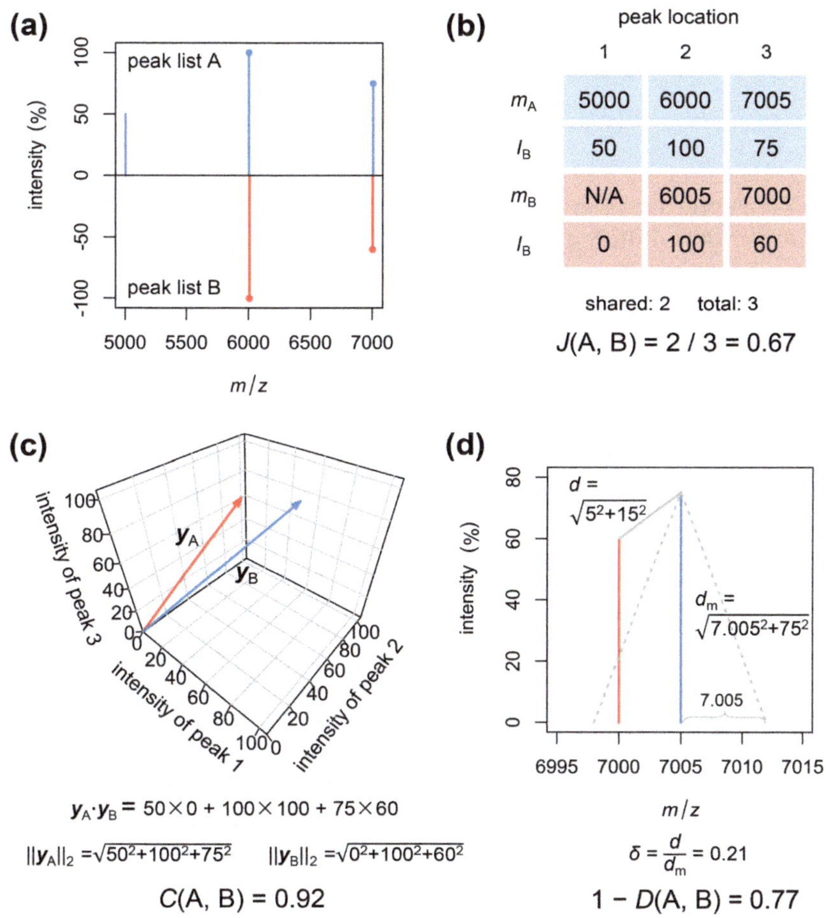

Figure 3.3 Spectral similarity matching. (a) Plots of two peak lists. Only three peaks are considered for simplicity. The dots indicate shared peak pairs between the two peak lists. (b) Peak *m/z* and intensity values, as well as the Jaccard coefficient between the two peak lists. (c) Scheme of vector representation of the two peak lists and their cosine similarity coefficient. (d) Scheme of the relative Euclidean distance between two peaks. The third peak pairs in the two peak lists are shown.

The values of these two metrics are limited to the domain that ranges from 0 (when the masses in the two peak lists are completely dissimilar) to 1 (when the masses in the two peak lists are indistinguishable).

In a binary score, peak matches are successful when the mass difference is within a predefined window or failed otherwise. In contrast, Rikard *et al.*[49] defined a continuous peak match score that allows for smoother inclusion of measurement errors. Given the measured masses m_i and m_j of two peaks from different peak lists as well as the measurement uncertainty σ, the peak match score is defined as the probability that the two peaks originate from the same ion. Assuming that the mass difference is only due to measurement errors and the measurement errors are Gaussian, the peak match score is the probability to get a mass difference larger than the measured difference:

$$P(\delta > |m_i - m_j|) = 1 - \mathrm{erf}\left(\frac{|m_i - m_j|}{2\sigma}\right) \tag{3.5}$$

where erf is the Gaussian error function. The peak match score is 0 for measurements infinitely apart and 1 for measurements being identical. The similarity score between two peak lists is calculated by adding up all scores from individual peak matches. Restrictions are put on the summation that each peak can only be matched to one other peak, and peak order by mass cannot be permuted. Among the possible combinations of peak matches fulfilling these two conditions, the one that maximizes the sum is chosen. Based on the similarity score, the software tool SPECLUST was developed to perform cluster analysis of MALDI-TOF mass spectra,[49] which was used to extract the characteristic mass lists of bacteria in SpectraBank.[32]

3.4.2 Spectral Similarity Scores Based on Intensity Correlations

Whereas several similarity scores have employed a binary peak list, others include peak intensities in calculations. The peak lists are first aligned by matching peak masses as described earlier or grouping peaks into small intervals of mass (called "bins") across the whole mass range. Then the aligned peak lists are represented as vectors of peak intensities. The Pearson correlation coefficient is one of the most commonly used metrics to compare two peak lists.[50–52] It measures the linear correlation between two vectors, and is defined as the covariance of the two vectors divided by the product of their standard deviations:[53]

$$\rho(\mathrm{A}, \mathrm{B}) = \frac{\mathrm{cov}(I_\mathrm{A}, I_\mathrm{B})}{\sigma_\mathrm{A}\sigma_\mathrm{B}} \tag{3.6}$$

where I_A and I_B represent the sequences of the intensities of peak lists A and B, cov represents the covariance, and σ_A and σ_B represent the standard deviations of the intensities. The Pearson correlation coefficient falls in the value range of −1 (indicating perfect negative correlation) to 1 (indicating perfect positive correlation).

The cosine similarity coefficient is analogous to the Pearson correlation coefficient for assessing the similarity between mass spectra.[48] It is also known as the dot product, and is used to obtain the cosine of the contrast angle θ between two sequences of intensities:[53]

$$\cos\theta = C(\mathrm{A},\mathrm{B}) = \frac{\boldsymbol{y}_{\mathrm{A}} \cdot \boldsymbol{y}_{\mathrm{B}}}{\|\boldsymbol{y}_{\mathrm{A}}\|_2 \|\boldsymbol{y}_{\mathrm{B}}\|_2} = \frac{\sum_{i=1}^{p} I_{\mathrm{A}i} I_{\mathrm{B}i}}{\sqrt{\sum_{i=1}^{p} I_{\mathrm{A}i}{}^2}\sqrt{\sum_{i=1}^{p} I_{\mathrm{B}i}{}^2}} \tag{3.7}$$

where $\boldsymbol{y}_{\mathrm{A}} = (I_{\mathrm{A}1},\ldots,I_{\mathrm{A}p})$ and $\boldsymbol{y}_{\mathrm{B}} = (I_{\mathrm{B}1},\ldots,I_{\mathrm{B}p})$ are the intensity vector representations with length p of peak lists A and B, respectively. $\|\boldsymbol{y}_{\mathrm{A}}\|_2$ and $\|\boldsymbol{y}_{\mathrm{B}}\|_2$ represent the Euclidean norms of the intensity vectors. As the intensities are always nonnegative, the cosine similarity coefficient ranges between 0 (indicating a maximal spectral differentiation) and 1 (in the same direction in the space). Figure 3.3(a) and (c) illustrates the vector representation of two peak lists and their contrast angle, where only three peaks are shown in the scheme for simplicity.

A normalized spectral contrast angle can further be derived from the cosine similarity coefficient:[54]

$$\Theta(\mathrm{A},\mathrm{B}) = 1 - \frac{2}{\pi}\cos^{-1}[C(\mathrm{A},\mathrm{B})] \tag{3.8}$$

The normalized spectral contrast angle is more sensitive than the Pearson correlation coefficient when spectra are very similar, *e.g.*, Pearson correlation coefficients of 0.99 and 1.00 are very similar, whereas for the same spectra the normalized spectral contrast angle still captures a difference of 0.05 (0.90 and 0.95, respectively).[55]

3.4.3 Comprehensive Similarity Scores Combining Peak Masses and Intensity Correlations

Yang *et al.*[10] proposed a similarity measure based on relative Euclidean distances of matched peaks. First, peak intensities are normalized with respect to the strongest peak in the corresponding spectrum to 100. Then, peaks in the two peak lists A and B are matched using a specified value of mass tolerance t (relative to the peak mass, *e.g.*, 2000 ppm). For each pair of the peaks shared in the two peak lists, the distance between them is defined as:

$$d_k = \sqrt{(m_{\mathrm{A}k} - m_{\mathrm{B}k})^2 + (I_{\mathrm{A}k} - I_{\mathrm{B}k})^2} \tag{3.9}$$

where m_{Ak} and m_{Bk} represent the m/z values of the k-th pair of peaks, while I_{Ak} and I_{Bk} represent their intensities. Assuming that the peak in peak list A is more intense than the corresponding one in peak list B, a pseudo-peak with zero intensity is placed in peak list B at the border of the mass tolerance window, *i.e.*, m_{Ak} $(1 \pm t)$. The pseudo-peak reflects the maximal possible distance between the peak in peak list A and a matched peak in peak list B. For a pair of peaks in which the one in peak list B is more intense than the other in peak list A, a pseudo-peak is placed in peak list A in the same manner. The relative distance between the pair of peaks is defined as the ratio of their distance to the maximal possible distance:

$$\delta_k = \frac{d_k}{d_{mk}} = \frac{d_k}{\sqrt{(\max\{m_{Ak}, m_{Bk}\} \cdot t)^2 + (\max\{I_{Ak}, I_{Bk}\})^2}} \tag{3.10}$$

For a single peak without a corresponding matched peak, the relative distance is set as 1. Figure 3.3(d) illustrates the concept of the relative Euclidean distance between two peaks. The distance between the two peak lists is calculated by averaging the relative distances of all peak pairs and unmatched single peaks, optionally weighted by their intensities:

$$D(A,B) = \frac{\sum_k [\delta_k(I_{Ak} + I_{Bk})] + \sum_i I_{Ai} + \sum_j I_{Bj}}{\sum_k (I_{Ak} + I_{Bk}) + \sum_i I_{Ai} + \sum_j I_{Bj}} \tag{3.11}$$

where I_{Ai} and I_{Bj} represent the intensities of the unmatched single peaks in peak lists A and B. The final similarity score is given by the distance subtracted from one. In contrast to the widely used spectral similarity measures that consider mainly differences in relative intensities between the common peaks from two spectra, this similarity score further takes the mass difference between each pair of common peaks into consideration. Since linear TOF mass spectra of bacterial whole cells have normally limited resolution (500–1000 ppm), considering the mass differences in the case can help to rule out false positive identification caused by peak mismatch when using a large mass tolerance.

Biotyper uses a comprehensive scoring algorithm that calculates three separate values for the basic characteristics of sample and reference spectra (MALDI Biotyper 3.0 User Manual Revision 2): (1) the proportion of peaks in the reference spectrum that are matched with peaks in the sample spectrum; (2) the proportion of peaks in the sample spectrum that are matched with peaks in the reference spectrum; and (3) a correlation value reflecting the symmetry of relative intensities of the matched peaks in the sample and reference spectra. The three values are then multiplied together and the result is normalized to 1000. The final score is the decimal logarithm of the result, with a maximal possible value of 3.

3.5 Interpretation Criteria and Quality Control of Spectral Matching Results

Since the spectral library may not contain a microbial species of interest, the spectral match with the highest score does not imply exactly the identification of the sample. Even when the library contains such a species, misidentifications can occur. Hence, interpretation criteria are needed after the measurement of similarity levels between the mass spectra from samples and a reference library to determine whether the spectral matching result is an effective identification or a false positive match.

3.5.1 Score Thresholds as Interpretation Criteria

A numerical threshold value of spectral matching scores is usually used as a straightforward criterion for the interpretation of results. For instance, Biotyper suggests two cutoff values for its logarithmic score.[56] Scores larger than or equal to 2.0 are accepted for species identification and scores between 1.7 and 2.0 for identification at the genus level. Scores below 1.7 are considered to be unreliable. Researchers suggested that the cutoff value of Biotyper could be lower or higher depending on spectral quality and analyzed bacterial samples.[57] Several studies have reported that lower cutoffs may be suitable for some groups of microorganisms to increase the identification rate,[58–64] whereas higher thresholds are required to improve the identification accuracy for some species.[65]

3.5.2 Statistical Assessment by Spectrum Resampling

Since fingerprint mass spectra can vary from microorganisms and sample preparation protocols, it is not reasonable to predefine a universal threshold value of similarity for a reliable identification. For this reason, Yang *et al.*[10] developed a method for statistical assessment of the identification results from spectral pattern matching by adopting a resampling technique named "bootstrap", which was introduced by Efron[66] and is used to estimate the confidence limits on a complicated estimator that may be analytically intractable. Given the sample spectrum, a large number of (*e.g.*, 1000) fictional, resampled spectra are constructed by random sampling the peaks with a replacement from the sample spectrum. Each resampled spectrum is then compared with reference spectra in the reference library to find the best match. The proportion of the resampled spectra matched with a specific species or genus is reported as a confidence score of identification, providing statistical assessment of the spectral matching result (Figure 3.4). It has been reported that the confidence score by resampling is an indicator of effective species identification with highly increased sensitivities at a constant error rate compared with primitive similarity scores.[10] Moreover, the resampling-based method has been

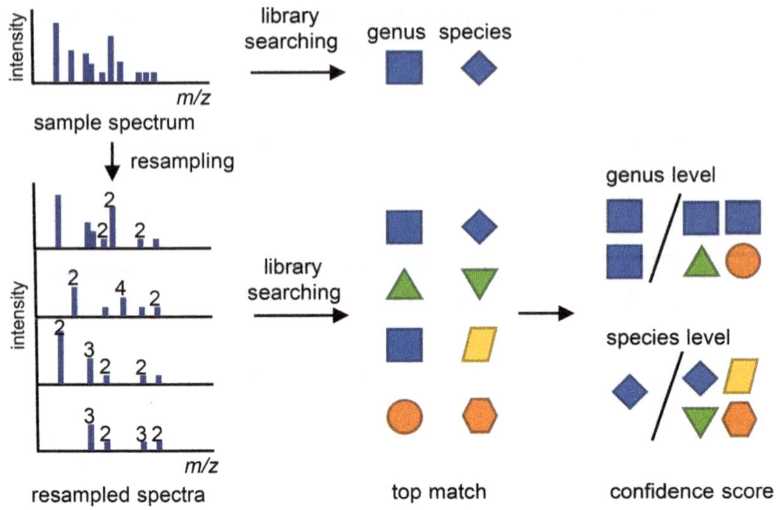

Figure 3.4 Workflow of the spectrum resampling approach for statistical assessment of spectral matching results. Numbers on the resampled spectra indicate the occurrence frequency of each peak. Reproduced from ref. 10 with permission from American Chemical Society, Copyright 2017.

demonstrated to resolve species within the *Bacillus cereus* group, which was previously claimed to be an impossible task for MALDI-TOF MS.[41]

3.5.3 Error Rate Control Using Decoy Spectra

A similarity score alone does not provide statements about the statistical accuracy of the spectral match. In order to set appropriate scoring criteria, statistical techniques are necessary to estimate error rates of identifications. Ryu *et al.*[67] developed a method to measure the uncertainty of bacterial species identifications using decoy mass spectra. The concept of decoy was initially proposed for the identification of tandem mass spectra of peptides in proteomics.[68] In addition to the protein database of interest (*i.e.*, the target database), the sample spectra are also searched against a database of sequences known to be incorrect (*i.e.*, the decoy database). The matching results between the sample spectra and the decoy peptides serve as a surrogate for incorrect matches of the target peptides, revealing the information what scores for wrong answers look like.

In the context of bacterial identification, decoy mass spectra do not belong to any species, and are constructed by extracting a set of (*e.g.*, 50) random peaks from a set of mass spectra of different bacterial species and randomly permuting their intensities.[67] A large number of (*e.g.*, 10 000) decoy spectra are generated and then scored as non-decoy ones. The *p*-value of a spectrum match is the probability to randomly draw a result of this or better quality, under the null hypothesis for which a spectrum has been randomly generated. It is calculated by dividing the number of decoy mass spectra with their top matching score greater than the top matching score of a given non-decoy spectrum by the total number of decoy mass spectra:

$$p = \frac{\left|\{D_i | s_{D_i} > s_0\}\right|}{N_D} \tag{3.12}$$

where D_i represents the i-th decoy mass spectrum, N_D is the total number of decoy mass spectra, s_{D_i} is the top matching score of the i-th decoy mass spectrum, and s_0 is the top matching score of a non-decoy mass spectrum. When multiple spectra are queried for identification, a false discovery rate (FDR) is estimated by multiple testing correction,[68] which represents the number of false positives over total positives that are reported and can be used as a criterion to control the error rates of bacterial identification. For example, an FDR value of 0.01 means that 1% of the reported identification results are incorrect.

It should be noted that the method by Ryu *et al.*[67] is designed for analysis of membrane glycolipid mass spectra rather than the routine MALDI-TOF MS approach for whole-cell bacteria, and bacterial identification is achieved by machine learning models that predict a probability score for each candidate species instead of conventional spectral similarities. Nevertheless, the principle of decoy-based error rate control is still applicable to spectral matching approaches.

3.6 Extending Spectral Matching to Characterization of Bacterial Mixtures

For a typical MALDI-TOF MS identification, bacteria must be isolated in pure culture.[69] The isolation and cultivation procedures are very time consuming, and have been observed to affect spectrum quality and reproducibility in MALDI-TOF MS-based bacterial identification.[70] For this reason, there has been considerable interest in applications of MALDI-TOF MS to poly-bacterial sample characterization without pure culture isolation. Studies have shown that the peaks present in the mass spectra of the bacterial mixtures mainly originate from the spectra of individual bacteria comprising the mixture.[11,71–79] This indicates that a routine reference library of pure bacteria is potentially applicable for characterization of mixtures with appropriate spectral matching methods.

3.6.1 Methods Based on Peak Mass Matching

A probability-based algorithm was originally proposed by Jarman *et al.*[71] and then modified by Wahl *et al.*[72] for bacterial mixture identification. The sample peak list is compared to each characteristic peak list (fingerprint) in the reference library, where the set of fingerprint peaks is extracted from the sample peak list and used to determine the likelihood of the reference microorganism being present in the sample. This approach allows for the capability of identifying individual components present in a mixed bacterial sample.

Consider the following hypotheses: (1) the null hypothesis H_0, the reference microorganism is not present in the sample and (2) the alternate hypothesis H_A, the reference microorganism is present in the sample. The hypothesis test is a likelihood ratio test for H_0 *versus* H_A. Under the alternate hypothesis H_A, the probability that the sample peak list contains a peak at fingerprint peak location j is given by its frequency of appearance p_j estimated from the reference fingerprint. Under the null hypothesis H_0, it is estimated to be the probability of q_j of spurious peaks appearing due to chemical or background noise. Assuming that the appearance of peaks at different locations is independent, the likelihood ratio for H_0 *versus* H_A is given by the probability of observing the outcome under H_A divided by the probability of observing the outcome under H_0:

$$\lambda = \frac{\prod_{j \in M} p_j \prod_{j \in M^c} (1 - p_j)}{\prod_{j \in M} q_j \prod_{j \in M^c} (1 - q_j)} \tag{3.13}$$

The log-likelihood ratio is:

$$\ln \lambda = \sum_{j \in U} \ln \frac{1 - p_j}{1 - q_j} + \sum_{j \in M} \ln \frac{p_j(1 - p_j)}{q_j(1 - p_j)} \tag{3.14}$$

where U represents the universal set of all fingerprint peaks, M represents the set of fingerprint peaks observed in the sample peak lists, and M^c represents the complementary set, *i.e.*, fingerprint peaks not observed in the sample peak lists. When the number of fingerprint peaks is large, the statistical distribution of the log-likelihood ratio $\ln \lambda$ approximates a normal distribution, which can be used to determine the critical threshold K_c by setting the desired significance level α, *i.e.*, the probability of falsely rejecting the null hypothesis (type I error):

$$\alpha = P(\ln \lambda > K_c | H_0) \tag{3.15}$$

If the log-likelihood ratio $\ln \lambda$ is larger than the critical threshold K_c, the null hypothesis H_0 is rejected and the reference microorganism is determined to be present in the sample.

This method was validated using 50 different microbial mixtures that contained 2–4 species. The sample spectra were searched against a reference library containing fingerprints for over 30 bacterial strains of 14 species. The bacterial components were correctly identified at the genus level in all samples and to the species level in the majority of the samples.[72]

3.6.2 Methods Based on Peak Intensity Regression

In the methods based on peak mass matching, component bacteria in mixtures are identified mainly dependent on species-specific peaks using an approach similar to a biomarker-based approach.[77] When a large reference library is considered, few species-specific peaks can be found for each species. Therefore, species-specific peak matching-based methods can only be employed when the number of candidate species is limited, which requires good pre-knowledge of analytes. In addition to the peak locations, the relative intensities of the peaks can be taken into account in the development of an identification algorithm. Zhang *et al.*[77] applied both a routine biomarker-based method and a similarity coefficient-based approach to a model mixture sample containing up to six bacterial species that are mixed with equal volumes. In the similarity coefficient-based approach, a synthetic mixture spectrum was constructed by averaging the respective signal intensities in the spectra of the pure cultures, and the similarity of the synthetic mixture spectrum was compared to the sample mixture spectra using the Pearson correlation coefficient. The results showed that the biomarker-based strategy might result in misidentification for species with limited numbers of biomarker peaks, while the similarity coefficient-based strategy could provide a correct identification.[77]

Indeed, the synthetic mixture spectrum in the method by Zhang *et al.*[77] is a special case of a linear combination where the components have equal coefficients. In a more general manner, given the vectors representing the peak intensity in the mass spectra of the mixture and individual bacteria, the mixture spectrum can be considered as a linear combination of the spectra of individual bacteria in the reference library:

$$\boldsymbol{y} = \sum_{i=1}^{n} \beta_i \boldsymbol{x}_i + \boldsymbol{\varepsilon} = \boldsymbol{X\beta} + \boldsymbol{\varepsilon} \tag{3.16}$$

where $\boldsymbol{y}$ is an intensity vector representing the mixture spectrum, $\boldsymbol{x}_i$ is an intensity vector representing species i, the coefficient β_i accounts for the contribution of $\boldsymbol{x}_i$ in explaining $\boldsymbol{y}$, $\boldsymbol{\varepsilon}$ is a vector of normally distributed residuals, $\boldsymbol{X} = (\boldsymbol{x}_1, \ldots, \boldsymbol{x}_n)$ is a matrix containing the intensity vectors $\boldsymbol{x}_i$, and $\boldsymbol{\beta} = (\beta_1, \ldots, \beta_n)^{\mathrm{T}}$ is a vector containing the coefficients β_i. The objective is to find the coefficients $\boldsymbol{\beta}$ that minimize the residual sum of squares:

$$\widehat{\boldsymbol{\beta}} = \operatorname{argmin} \|\boldsymbol{y} - \boldsymbol{X\beta}\|_2{}^2 \tag{3.17}$$

An additional constraint is required on the solution that the coefficients are not allowed to be negative. This is a non-negative least squares problem, which can be solved using many algorithms since Lawson and Hanson published the first one in 1974.[80]

In typical clinical specimens, the number of detected species is expected to be relatively small. Therefore, the decomposition of the mixture spectrum with the minimal residual sum of squares may not be the best solution, when it consists of many components with small coefficients. Instead, the sample is preferably interpreted by only a few species. For this purpose, Mahé *et al.*[78] proposed a penalized non-negative linear regression framework for mixture spectrum decomposition. A penalty is appended to the objective function:

$$\widehat{\boldsymbol{\beta}} = \text{argmin} \|\boldsymbol{y} - \boldsymbol{X\beta}\|_2{}^2 + \lambda \|\boldsymbol{\beta}\|_1 \quad (3.18)$$

where $\|\boldsymbol{\beta}\|_1$ is the L1 norm of $\boldsymbol{\beta}$, *i.e.*, the sum of the absolute values of the coefficients. The parameter λ affects the number of non-null coefficients in the solution, where small λ values lead to many positive coefficients and large λ values lead to few positive coefficients. Varying the parameter λ provides a collection of decomposition models achieving various trade-offs between the reconstruction error and sparsity of the solution. This is a LASSO problem with a positivity constraint,[81] which can be solved using many efficient algorithms, such as LARS-EN,[82] with access to the whole collection of solutions that can be reached when the parameter λ is varied (referred to as the regularization path). The optimal decomposition model on the regularization path is selected using the Bayesian information criterion (BIC). The BIC is a well-known general approach to model selection that favors more parsimonious models over more complex models.[83,84] For regression models, the formula of the BIC is defined as:

$$\text{BIC} = -2 \ln L + k \ln p \quad (3.19)$$

The first part of the BIC formula is derived from the log-likelihood of a standard linear regression model. The log-likelihood $\ln L$ is used as the objective function in maximum likelihood estimation, an alternative approach to the least squares. In brief, the solution of maximizing the log-likelihood is proven identical to minimizing the residual sum of squares. In the second part of the BIC, p is the number of data points (the length of intensity vectors, expected to be a constant value given the sample and reference spectra), and k is the number of parameters, which is directly related to the number of non-null coefficients in $\boldsymbol{\beta}$. Among various alternative models on the regularization path, the model that is preferred is the one with the minimum BIC value, which achieves the appropriate trade-off between the spectrum reconstruction error and sparsity of the solution. For a detailed explanation of the least squares, maximum likelihood estimation, and BIC, the readers are referred to literature on statistics.[85]

This method was evaluated using a mass spectral dataset of 180 bi-microbial samples mixed *in vitro* with different concentration ratios. Provided a reference library containing mass spectra for over 200 bacterial strains of 20 species, the method correctly identified more than 60% of the mixture samples. The mixture samples were always at least partially identified, and only about 5% were misidentified.[78]

3.6.3 Statistical Assessment of Spectral Matching Results

In the same vein as identification of pure bacteria, a statistical assessment is necessary to set appropriate result interpretation criteria for bacterial mixture identification. Yang *et al.*[11] developed a framework to characterize poly-bacterial samples using a model based on a non-negative linear combination of reference spectra and a statistical assessment by spectrum resampling (Figure 3.5). First, the sample spectrum is searched preliminarily against the reference library, and a number of candidate reference spectra are selected according to their similarity to the sample spectrum based on Jaccard coefficients. Given the maximum possible number of reference bacterial strains present in the sample, combinations of candidate reference spectra are enumerated. For each candidate combination, a mixture spectrum is synthesized *in silico* from the reference spectra by linear combination, where the coefficients are determined by non-negative least square algorithms. Then, all the synthesized spectra are compared with the sample spectrum by calculating the cosine similarity scores between them to find the combination with the highest similarity scores.

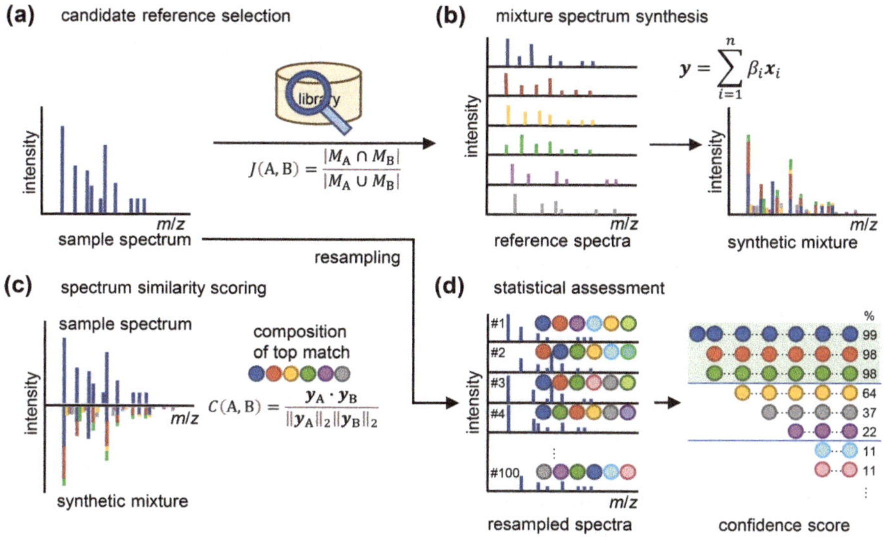

Figure 3.5 Workflow of the approach for bacterial mixture identification with statistical assessment. (a) Preliminary searching to select candidate reference spectra of individual bacterial species by the Jaccard coefficients. (b) *In silico* synthesis of a mixture spectrum using a non-negative linear combination of the candidate reference spectra. (c) Searching against all the synthetic mixture spectra by spectral matching using the cosine similarity coefficients. Colored spots represent the components in the top match. (d) Statistical assessment of the matched components by spectrum resampling. Reproduced from ref. 11 with permission from American Chemical Society, Copyright 2018.

For statistical assessment of the identification result, a resampling technique named the "jackknife" method is adopted. The idea of jackknife resampling was first introduced by Quenouille[86,87] and further developed by Tukey,[88] who also gave it its name. It is an alternative resampling scheme used for bias correction and variance estimation that has some similarities to the bootstrap method. Shao *et al.*[89] found that a generalized version of the jackknife method can rectify the deficiency of the original jackknife method. In the context of mass spectral matching, the sample spectrum is resampled by randomly omitting a fixed percentage of the peaks. Similar to the resampling method described in Section 3.5.2, a large number of resampled spectra are constructed, and each of them is searched against the reference library to find the best linear combination. For a given identified bacterial component from the sample spectrum, its detectable rate from the resampled spectra is reported as a confidence score of identification, providing statistical assessment of the spectral matching result.

This method was tested using a mass spectral dataset of nearly 100 mixture samples containing 2–6 bacterial species and a reference library containing mass spectra of pure cultures of over 1000 strains of 480 species. More than 95% species in binary mixtures and 85% species in ternary mixtures with balanced concentration of each species were correctly identified. Dominant species were always correctly identified for mixtures with asymmetric concentrations. Incorrect identification could be ruled out with statistical assessment by spectrum resampling, improving the reliability of identification results.[11]

3.7 Bioinformatics Solutions Independent of Reference Spectral Libraries

The availability of a comprehensive reference spectral library is the prerequisite for bacterial identification by MALDI-TOF mass spectral matching. The reference spectra are acquired from known monoculture bacterial samples, which limits the method only to cultivable bacteria. Meanwhile, scientists throughout the world are sequencing new genomes of bacteria at a prodigious rate, and many of these genomes and translated protein sequences are uploaded to public repositories, such as the National Center for Biotechnology Information (NCBI)[90] and Ensembl.[91] Since the early 2000s, it has been recognized that bacterial identification is possible by mapping directly to the masses deduced from genomic sequences.[92–94] More recently, new algorithms have been developed to annotate MALDI-TOF spectral peaks with genomic or proteomic sequences, as an alternative approach for bacterial identification to the spectral matching method using reference spectral libraries.

Cheng *et al.*[95] developed a strategy for bacterial typing by MALDI-TOF using protein sequences from the UniProt database.[96] Ten genes that encode proteins most often observed by MALDI-TOF MS were discovered by machine learning. The selected genes correspond to ribosomal subunit proteins and DNA-binding proteins, and are used to annotate peaks in MALDI-TOF spectra for bacterial identification. This approach was evaluated using a dataset of mass spectra of over 400 strains belonging to 81 species and 14 genera. By searching against a protein sequence database containing over 1200 species and 42 genera of bacteria from UniProt, more than 80% of the sample spectra were identified correctly at the genus level.[95] However, this approach cannot provide sufficient resolving power in bacterial identification at the species level within a genus.

Kostas and Parker[97] developed a genome matching approach in which MALDI spectra are mapped directly to the publicly available translated genomes housed at NCBI. The spectrum-to-genome matches are scored based on a variety of parameters, including the percentage of intensity matched, the intensity-weighted mass accuracy of the match, and the percentage of singly or doubly charged proteins that are matched within the mass range of interest. The database contains 65 conserved protein families, which have greater species discriminating power than the 10 proteins used in the study by Cheng *et al.*[95]

Among the highly abundant bacterial proteins, many ribosomal protein subunits have low molecular weights and are highly positively charged,[98] which makes them readily detectable in the mass range of MALDI-TOF spectra. The sequences of many ribosomal protein subunits are often invariant among closely related bacterial species, while the remaining ribosomal subunits often contain substitutions that distinguish each species from one another.[99] For these reasons, the characteristic pattern in MALDI-TOF spectra for bacterial identification is largely based on ribosomal protein polymorphism.[100] Tomachewski *et al.*[101] present Ribopeaks, a wide-ranging bacterium classifier tool created based on ribosomal protein data from GenBank.[102] It compares the sample MALDI-TOF peak mass with models stored in a database of more than nearly 30 000 bacterial taxonomic records by machine learning and then taxonomically classifies the bacteria.

Although these bioinformatic methods are still not as accurate as the spectral matching approaches based on reference spectral libraries, they expand bacterial identification by MALDI-TOF MS to a much wider range of species. Moreover, they have one key advantage that they are independent of the exact protocol used to gather the spectra and in particular independent of the choice of MALDI matrix.[97] Further efforts are needed to improve the peak annotation rate in the mass spectra and thus the identification accuracy. The remaining challenges include:[95] (1) the availability of more complete genome or proteome sequence databases for bacterial species; (2) a comprehensive consideration of post-translational modifications and sequence variations; and (3) accommodation with the mass change due to chemical modifications or fragmentations that intact bacterial proteins undergo during the sample preparation or MALDI procedure.

3.8 Summary and Outlook

This chapter covered several aspects of data analysis for MALDI-TOF MS-based identification of microorganisms, including algorithms for spectrum preprocessing, spectral pattern matching, and result interpretation. Currently, both commercial and open-access software tools are available for these tasks. With regard to open-access software, there are many tools for individual tasks (*e.g.*, those for spectrum preprocessing listed in Table 3.1) but only a limited number of solutions cover the whole data analysis workflow, as summarized in Table 3.3. From a user perspective, most of the tools are packages or scripts that can be accessed only *via* the command line interface; however, there are also tools that provide an online web service or a graphical user interface (GUI), which is more friendly to biologists, biochemists, and clinical scientists with no programming background.

Table 3.3 Open-access software tools of MALDI-TOF MS identification of microorganisms.

Tool	Type	Feature
BIOSPEAN[8]	Web service	Identification by spectral similarity matching (based on shared peak mass); construction of custom libraries
	http://software.cr-hana.upol.cz/biospean/	
MicrobeMS[9]	Standalone GUI application; MATLAB toolbox	Identification by spectral similarity matching (based on intensity correlations); cooperation with the RKI library; construction of custom libraries
	https://www.microbe-ms.com/	
BacteriaMS[10,11]	Web service; R script	Identification by spectral similarity matching (combining peak masses and intensity correlations); identification of bacterial mixtures (based on linear regression); quality control of spectral matching results by statistical assessment (based on spectrum resampling)
	https://bacteriams.net/ https://github.com/lmsac/BacteriaMS	
BacteriaMSLF[95]	Python script	Genus-level identification by matching MS peak mass with protein sequences of 10 genes selected from UniProt
	https://github.com/dipcarbon/BacteriaMSLF	
Ribopeaks[101]	Web service	Taxonomical classification by matching MS peak mass with a database of ribosomal proteins covering 30 000 bacterial taxonomic records from GenBank
	http://www.ribopeaks.com	

Although library-free bioinformatic methods are promising in the future, spectral library coverage determines the limit of the identification capacity by MALDI-TOF spectral pattern matching in the current stage. The majority of spectral libraries, reported to date, are commercial proprietary, and only a few are openly accessible for academic use. Public data repositories are necessary to allow data to be shared by the community. On the other hand, spectral libraries are compatible with only specific instruments and cannot be used across platforms due to the inherent variability introduced by using different experimental protocols, instruments, and algorithms for identification in data analysis software. Efforts are needed to develop an open and vendor-neutral data format, like mzML to store MS information,[103] for analysis, comparison, and integration of spectral libraries originating from different MS experiments.

With large amounts of MALDI-TOF mass spectral data generated routinely in microbiology labs, novel approaches based on machine learning have emerged to investigate MS big data. Machine learning, particularly deep learning, can uncover the information that is embedded in MALDI-TOF mass spectra but not exploited by conventional analysis algorithms. These techniques have been demonstrated to provide improved resolution power for identification at the species level and even the subspecies level, and facilitate prediction of phenotypes beyond species identification, such as antimicrobial susceptibility testing.[104–107] We see a lot of possibilities for MALDI-TOF MS microorganism analysis with advanced algorithms in the future.

Abbreviations

BIC	Bayesian information criterion
CWT	Continuous wavelet transformation
EMbaRC	European Consortium of Microbial Resource Centres
FDR	False discovery rate
GUI	Graphical user interface
MALDI	Matrix-assisted laser desorption/ionization
MS	Mass spectrometry
MSP	Main spectral profile
NCBI	National Center for Biotechnology Information
RKI	Robert Koch Institute
SNIP	Statistics-sensitive non-linear iterative peak-clipping
SNR	Signal-to-noise ratio
TOF	Time-of-flight

References

1. T. R. Sandrin, J. E. Goldstein and S. Schumaker, *Mass Spectrom. Rev.*, 2013, **32**, 188–217.
2. A. Croxatto, G. Prod'hom and G. Greub, *FEMS Microbiol. Rev.*, 2012, **36**, 380–407.

3. P. Dixon, P. Davies, W. Hollingworth, M. Stoddart and A. MacGowan, *Eur. J. Clin. Microbiol. Infect. Dis.*, 2015, **34**, 863–876.
4. A. van Belkum, M. Welker, M. Erhard and S. Chatellier, *J. Clin. Microbiol.*, 2012, **50**, 1513–1517.
5. R. Giebel, C. Worden, S. M. Rust, G. T. Kleinheinz, M. Robbins and T. R. Sandrin, in *Advances in Applied Microbiology*, ed. A. I. Laskin, S. Sariaslani and G. M. Gadd, Academic Press, 2010, vol. 71, pp. 149–184.
6. C. Lohmann, M. Sabou, W. Moussaoui, G. Prévost, J.-M. Delarbre, E. Candolfi, A. Gravet and V. Letscher-Bru, *J. Clin. Microbiol.*, 2013, **51**, 1231–1236.
7. A. Sampedro, J. Ceballos Mendiola and L. Aliaga Martínez, in *The Use of Mass Spectrometry Technology (MALDI-TOF) in Clinical Microbiology*, ed. F. Cobo, Academic Press, 2018, pp. 47–57.
8. M. Raus and M. Šebela, *J. Proteom. Bioinform*, 2013, **6**, 283–287.
9. P. Lasch, T. Wahab, S. Weil, B. Pályi, H. Tomaso, S. Zange, B. Kiland Granerud, M. Drevinek, B. Kokotovic, M. Wittwer, V. Pflüger, A. Di Caro, M. Stämmler, R. Grunow and D. Jacob, *J. Clin. Microbiol.*, 2015, **53**, 2632–2640.
10. Y. Yang, Y. Lin, Z. Chen, T. Gong, P. Yang, H. Girault, B. Liu and L. Qiao, *Anal. Chem.*, 2017, **89**, 12556–12561.
11. Y. Yang, Y. Lin and L. Qiao, *Anal. Chem.*, 2018, **90**, 10400–10408.
12. K. R. Coombes, K. A. Baggerly and J. S. Morris, in *Fundamentals of Data Mining in Genomics and Proteomics*, ed. W. Dubitzky, M. Granzow and D. Berrar, Springer US, Boston, MA, 2007, pp. 79–102.
13. J. Gobom, M. Mueller, V. Egelhofer, D. Theiss, H. Lehrach and E. Nordhoff, *Anal. Chem.*, 2002, **74**, 3915–3923.
14. S. W. Smith, in *Digital Signal Processing*, ed. S. W. Smith, Newnes, Boston, 2003, pp. 277–284.
15. A. Savitzky and M. J. E. Golay, *Anal. Chem.*, 1964, **36**, 1627–1639.
16. C. G. Ryan, E. Clayton, W. L. Griffin, S. H. Sie and D. R. Cousens, *Nucl. Instrum. Methods Phys. Res. B*, 1988, **34**, 396–402.
17. M. Morháč, *Nucl. Instrum. Methods Phys. Res. A*, 2009, **600**, 478–487.
18. M. van Herk, *Pattern Recognit. Lett.*, 1992, **13**, 517–521.
19. X. Li, R. Gentleman, X. Lu, Q. Shi, J. D. Iglehart, L. Harris and A. Miron, in *Bioinformatics and Computational Biology Solutions Using R and Bioconductor*, ed. R. Gentleman, V. J. Carey, W. Huber, R. A. Irizarry and S. Dudoit, Springer New York, New York, NY, 2005, pp. 91–109.
20. W. Yu, B. Wu, N. Lin, K. Stone, K. Williams and H. Zhao, *Comput. Biol. Chem.*, 2006, **30**, 27–38.
21. K. R. Coombes, H. A. Fritsche Jr., C. Clarke, J.-N. Chen, K. A. Baggerly, J. S. Morris, L.-C. Xiao, M.-C. Hung and H. M. Kuerer, *Clin. Chem.*, 2003, **49**, 1615–1623.
22. P. Du, W. A. Kibbe and S. M. Lin, *Bioinformatics*, 2006, **22**, 2059–2065.
23. C. Yang, Z. He and W. Yu, *BMC Bioinform.*, 2009, **10**, 4.

24. D. Mantini, F. Petrucci, D. Pieragostino, P. Del Boccio, M. Di Nicola, C. Di Ilio, G. Federici, P. Sacchetta, S. Comani and A. Urbani, *BMC Bioinform.*, 2007, **8**, 101.
25. M. Strohalm, M. Hassman, B. Košata and M. Kodíček, *Rapid Commun. Mass Spectrom.*, 2008, **22**, 905–908.
26. M. Strohalm, D. Kavan, P. Novák, M. Volný and V. Havlíček, *Anal. Chem.*, 2010, **82**, 4648–4651.
27. S. Gibb and K. Strimmer, *Bioinformatics*, 2012, **28**, 2270–2271.
28. H. López-Fernández, H. M. Santos, J. L. Capelo, F. Fdez-Riverola, D. Glez-Peña and M. Reboiro-Jato, *BMC Bioinform.*, 2015, **16**, 318.
29. S. Sauer, A. Freiwald, T. Maier, M. Kube, R. Reinhardt, M. Kostrzewa and K. Geider, *PLoS One*, 2008, **3**, e2843.
30. Å. Lindgren, N. Karami, R. Karlsson, C. Åhrén, M. Welker, E. R. B. Moore and L. S. Stadler, *Eur. J. Clin. Microbiol. Infect. Dis.*, 2018, **37**, 57–68.
31. H. L. Köhling, A. Bittner, K.-D. Müller, J. Buer, M. Becker, H. Rübben, A. W. Rettenmeier and F. Mosel, *J. Med. Microbiol.*, 2012, **61**, 339–344.
32. K. Böhme, I. C. Fernández-No, J. Barros-Velázquez, J. M. Gallardo, B. Cañas and P. Calo-Mata, *Electrophoresis*, 2012, **33**, 2138–2142.
33. F. Mazzeo Maria, A. Sorrentino, M. Gaita, G. Cacace, M. Di Stasio, A. Facchiano, G. Comi, A. Malorni and A. Siciliano Rosa, *Appl. Environ. Microbiol.*, 2006, **72**, 1180–1189.
34. P. Lasch, M. Stämmler and A. Schneider, *Zenodo*, 2018, DOI: 10.5281/zenodo.1880975.
35. D. Drissner and F. M. Freimoser, *Chem. Biol. Technol. Agric.*, 2017, **4**, 13.
36. R. Erler, A. Wichels, E.-A. Heinemeyer, G. Hauk, M. Hippelein, N. T. Reyes and G. Gerdts, *Syst. Appl. Microbiol.*, 2015, **38**, 16–25.
37. M. Moussa, E. Cauvin, A. Le Piouffle, O. Lucas, A. Bidault, C. Paillard, F. Benoit, B. Thuillier, M. Treilles, M. A. Travers and C. Garcia, *Appl. Microbiol. Biotechnol.*, 2021, **105**, 2527–2539.
38. T. Liu, L. Kang, J. Xu, J. Wang, S. Gao, Y. Li, J. Li, Y. Yuan, B. Yuan, J. Wang, B. Zhao and W. Xin, *Front. Microbiol.*, 2022, **13**, 872825.
39. Y. Luo, Y. Xu, H. Wang, H. Yu, S. Cui, X. Lu, Y. Ma, J. Hu, J. Su, C. Chen, J. Zhao, B. Gu, X. Ma, Y. Lin, M. Wang, H. Wang, Y. Zhang, W. Zhou, R. Chen, S. Li and W. Yang, *Chin. J. Lab. Med.*, 2019, **42**, 414–419.
40. X. Li, Y. Li and X. Lu, *Chin. J. Clin. Lab. Sci.*, 2018, **36**, 297–303.
41. J. A. Branda, *M58 Methods for the Identification of Cultured Microorganisms Using Matrix-Assisted Laser Desorption/Ionization Time-of-Flight Mass Spectrometry*, Clinical and Laboratory Standards Institute, Wayne, PA, United States, 2017.
42. P. Rahi, O. Prakash and Y. S. Shouche, *Front. Microbiol.*, 2016, **7**, 1359.
43. P. T. Becker, A. de Bel, D. Martiny, S. Ranque, R. Piarroux, C. Cassagne, M. Detandt and M. Hendrickx, *Med. Mycol.*, 2014, **52**, 826–834.
44. M. E. Turvey, F. Weiland, J. Meneses, N. Sterenberg and P. Hoffmann, *Appl. Microbiol. Biotechnol.*, 2016, **100**, 2761–2773.

45. G. C. Conway, S. Smole, D. A. Sarracino, R. D. Arbeit and P. E. Leopold, *J. Mol. Microbiol. Biotechnol.*, 2001, **3**, 103–112.
46. T. J. Siegrist, P. D. Anderson, W. H. Huen, G. T. Kleinheinz, C. M. McDermott and T. R. Sandrin, *J. Microbiol. Methods*, 2007, **68**, 554–562.
47. K. V. Starostin, E. A. Demidov, N. I. Ershov, A. V. Bryanskaya, V. M. Efimov, V. N. Shlyakhtun and S. E. Peltek, *Front. Microbiol.*, 2020, **11**, 609033.
48. M. G. LaMontagne, P. L. Tran, A. Benavidez and L. D. Morano, *PeerJ*, 2021, **9**, e11359.
49. R. Alm, P. Johansson, K. Hjernø, C. Emanuelsson, M. Ringnér and J. Häkkinen, *J. Proteome Res.*, 2006, **5**, 785–792.
50. M. Lundquist, M. B. Caspersen, P. Wikström and M. Forsman, *FEMS Microbiol. Lett.*, 2005, **243**, 303–310.
51. M. Vargha, Z. Takáts, A. Konopka and C. H. Nakatsu, *J. Microbiol. Methods*, 2006, **66**, 399–409.
52. R. A. Giebel, W. Fredenberg and T. R. Sandrin, *Water Res.*, 2008, **42**, 931–940.
53. S. Kim and X. Zhang, *Comput. Math. Methods Med.*, 2013, **2013**, 509761.
54. U. H. Toprak, L. C. Gillet, A. Maiolica, P. Navarro, A. Leitner and R. Aebersold, *Mol. Cell. Proteom.*, 2014, **13**, 2056–2071.
55. S. Gessulat, T. Schmidt, D. P. Zolg, P. Samaras, K. Schnatbaum, J. Zerweck, T. Knaute, J. Rechenberger, B. Delanghe, A. Huhmer, U. Reimer, H.-C. Ehrlich, S. Aiche, B. Kuster and M. Wilhelm, *Nat. Methods*, 2019, **16**, 509–518.
56. B. Schulthess, K. Brodner, V. Bloemberg Guido, R. Zbinden, C. Böttger Erik and M. Hombach, *J. Clin. Microbiol.*, 2013, **51**, 1834–1840.
57. A. Mather Cheryl, F. Rivera Sheila and M. Butler-Wu Susan, *J. Clin. Microbiol.*, 2014, **52**, 130–138.
58. A. Alatoom Adnan, J. Cazanave Charles, A. Cunningham Scott, M. Ihde Sherry and R. Patel, *J. Clin. Microbiol.*, 2012, **50**, 160–163.
59. N. Dhiman, L. Hall, L. Wohlfiel Sherri, P. Buckwalter Seanne and L. Wengenack Nancy, *J. Clin. Microbiol.*, 2011, **49**, 1614–1616.
60. B. Schulthess, V. Bloemberg Guido, R. Zbinden, C. Böttger Erik and M. Hombach, *J. Clin. Microbiol.*, 2014, **52**, 1089–1097.
61. M. Ercibengoa Arana, M. Alonso, P. Idigoras, D. Vicente and J. M. Marimón, *AMB Express*, 2018, **8**, 121.
62. B. Rodríguez-Sánchez, M. J. Ruiz-Serrano, A. Ruiz, M. Timke, M. Kostrzewa and E. Bouza, *J. Clin. Microbiol.*, 2016, **54**, 1144–1147.
63. G. Saleeb Paul, K. Drake Steven, R. Murray Patrick and M. Zelazny Adrian, *J. Clin. Microbiol.*, 2011, **49**, 1790–1794.
64. M. C. Mediavilla-Gradolph, I. De Toro-Peinado, M. P. Bermúdez-Ruiz, M. D. L. Á. García-Martínez, M. Ortega-Torres, N. Montiel Quezel-Guerraz and B. Palop-Borrás, *BioMed Res. Int.*, 2015, **2015**, 854078.
65. D. Rodriguez-Temporal, B. Rodríguez-Sánchez and F. Alcaide, *J. Clin. Microbiol.*, 2020, **58**, 01103–01120.

66. B. Efron, *Ann. Stat.*, 1979, 7, 1–26.
67. S. Y. Ryu, G. A. Wendt, C. E. Chandler, R. K. Ernst and D. R. Goodlett, *Anal. Chem.*, 2019, **91**, 11482–11487.
68. N. Gupta, N. Bandeira, U. Keich and P. A. Pevzner, *J. Am. Soc. Mass Spectrom.*, 2011, **22**, 1111–1120.
69. L. Zhang, C. M. Borror and T. R. Sandrin, *PLoS One*, 2014, **9**, e92720.
70. J. E. Goldstein, L. Zhang, C. M. Borror, J. V. Rago and T. R. Sandrin, *Lett. Appl. Microbiol.*, 2013, **57**, 144–150.
71. K. H. Jarman, S. T. Cebula, A. J. Saenz, C. E. Petersen, N. B. Valentine, M. T. Kingsley and K. L. Wahl, *Anal. Chem.*, 2000, **72**, 1217–1223.
72. K. L. Wahl, S. C. Wunschel, K. H. Jarman, N. B. Valentine, C. E. Petersen, M. T. Kingsley, K. A. Zartolas and A. J. Saenz, *Anal. Chem.*, 2002, **74**, 6191–6199.
73. E. Mandrell Robert, A. Harden Leslie, A. Bates, G. Miller William, F. Haddon William and K. Fagerquist Clifton, *Appl. Environ. Microbiol.*, 2005, **71**, 6292–6307.
74. S.-Y. Hsieh, C.-L. Tseng, Y.-S. Lee, A.-J. Kuo, C.-F. Sun, Y.-H. Lin and J.-K. Chen, *Mol. Cell. Proteom.*, 2008, 7, 448–456.
75. R. D. Holland, J. G. Wilkes, W. M. Cooper, P. Alusta, A. Williams, B. Pearce, M. Beaudoin and D. Buzatu, *Rapid Commun. Mass Spectrom.*, 2014, **28**, 2617–2626.
76. D. Barbano, R. Diaz, L. Zhang, T. Sandrin, H. Gerken and T. Dempster, *PLoS One*, 2015, **10**, e0135337.
77. L. Zhang, S. Smart and T. R. Sandrin, *Sci. Rep.*, 2015, **5**, 15834.
78. P. Mahé, M. Arsac, S. Chatellier, V. Monnin, N. Perrot, S. Mailler, V. Girard, M. Ramjeet, J. Surre, B. Lacroix, A. van Belkum and J.-B. Veyrieras, *Bioinformatics*, 2014, **30**, 1280–1286.
79. W. Cheng, H. Shi, M. Teng, M. Yu, B. Feng, C. Ding, S. Yu and F. Yang, *Analyst*, 2022, **147**, 443–449.
80. C. L. Lawson and R. J. Hanson, *Solving Least Squares Problems*, Society for Industrial and Applied Mathematics, Philadelphia, PA, USA, 1995.
81. E. Bradley, H. Trevor, J. Iain and T. Robert, *Ann. Stat.*, 2004, **32**, 407–499.
82. H. Zou and T. Hastie, *J. R. Stat. Soc. Series B Stat. Methodol.*, 2005, **67**, 301–320.
83. S. Gideon, *Ann. Stat.*, 1978, **6**, 461–464.
84. A. E. Raftery, *Soc. Methodol.*, 1995, **25**, 111–163.
85. P. K. Bhattacharya and P. Burman, in *Theory and Methods of Statistics*, ed. P. K. Bhattacharya and P. Burman, Academic Press, 2016, pp. 309–382.
86. M. H. Quenouille, *Ann. Math. Stat.*, 1949, **20**, 355–375.
87. M. H. Quenouille, *Biometrika*, 1956, **43**, 353–360.
88. J. W. Tukey, *Ann. Math. Stat.*, 1958, **29**, 614–623.
89. S. Jun and C. F. J. Wu, *Ann. Stat.*, 1989, **17**, 1176–1197.
90. E. W. Sayers, R. Agarwala, E. E. Bolton, J. R. Brister, K. Canese, K. Clark, R. Connor, N. Fiorini, K. Funk, T. Hefferon, J. B. Holmes, S. Kim,

A. Kimchi, P. A. Kitts, S. Lathrop, Z. Lu, T. L. Madden, A. Marchler-Bauer, L. Phan, V. A. Schneider, C. L. Schoch, K. D. Pruitt and J. Ostell, *Nucleic Acids Res.*, 2019, **47**, D23–D28.

91. F. Cunningham, J. E. Allen, J. Allen, J. Alvarez-Jarreta, M. R. Amode, I. M. Armean, O. Austine-Orimoloye, A. G. Azov, I. Barnes, R. Bennett, A. Berry, J. Bhai, A. Bignell, K. Billis, S. Boddu, L. Brooks, M. Charkhchi, C. Cummins, L. Da Rin Fioretto, C. Davidson, K. Dodiya, S. Donaldson, B. El Houdaigui, T. El Naboulsi, R. Fatima, C. G. Giron, T. Genez, J. G. Martinez, C. Guijarro-Clarke, A. Gymer, M. Hardy, Z. Hollis, T. Hourlier, T. Hunt, T. Juettemann, V. Kaikala, M. Kay, I. Lavidas, T. Le, D. Lemos, J. C. Marugán, S. Mohanan, A. Mushtaq, M. Naven, D. N. Ogeh, A. Parker, A. Parton, M. Perry, I. Piližota, I. Prosovetskaia, M. P. Sakthivel, A. I. A. Salam, B. M. Schmitt, H. Schuilenburg, D. Sheppard, J. G. Pérez-Silva, W. Stark, E. Steed, K. Sutinen, R. Sukumaran, D. Sumathipala, M.-M. Suner, M. Szpak, A. Thormann, F. F. Tricomi, D. Urbina-Gómez, A. Veidenberg, T. A. Walsh, B. Walts, N. Willhoft, A. Winterbottom, E. Wass, M. Chakiachvili, B. Flint, A. Frankish, S. Giorgetti, L. Haggerty, S. E. Hunt, G. R. Iisley, J. E. Loveland, F. J. Martin, B. Moore, J. M. Mudge, M. Muffato, E. Perry, M. Ruffier, J. Tate, D. Thybert, S. J. Trevanion, S. Dyer, P. W. Harrison, K. L. Howe, A. D. Yates, D. R. Zerbino and P. Flicek, *Nucleic Acids Res.*, 2022, **50**, D988–D995.
92. F. J. Pineda, J. S. Lin, C. Fenselau and P. A. Demirev, *Anal. Chem.*, 2000, **72**, 3739–3744.
93. P. A. Demirev, J. S. Lin, F. J. Pineda and C. Fenselau, *Anal. Chem.*, 2001, **73**, 4566–4573.
94. F. J. Pineda, M. D. Antoine, P. A. Demirev, A. B. Feldman, J. Jackman, M. Longenecker and J. S. Lin, *Anal. Chem.*, 2003, **75**, 3817–3822.
95. D. Cheng, L. Qiao and P. Horvatovich, *J. Proteome Res.*, 2018, **17**, 2124–2130.
96. The UniProt Consortium, *Nucleic Acids Res.*, 2019, **47**, D506–D515.
97. J. Kostas and K. C. Parker, *J. Am. Soc. Mass Spectrom.*, 2020, **31**, 73–84.
98. R. J. Arnold and J. P. Reilly, *Anal. Biochem.*, 1999, **269**, 105–112.
99. S. Suarez, A. Ferroni, A. Lotz, K. A. Jolley, P. Guérin, J. Leto, B. Dauphin, A. Jamet, M. C. J. Maiden, X. Nassif and J. Armengaud, *J. Microbiol. Methods*, 2013, **94**, 390–396.
100. N. Singhal, M. Kumar, P. K. Kanaujia and J. S. Virdi, *Front. Microbiol.*, 2015, **6**, 791.
101. D. Tomachewski, C. W. Galvão, A. de Campos Júnior, A. M. Guimarães, J. C. Ferreira da Rocha and R. M. Etto, *Bioinformatics*, 2018, **34**, 3058–3060.
102. D. A. Benson, M. Cavanaugh, K. Clark, I. Karsch-Mizrachi, D. J. Lipman, J. Ostell and E. W. Sayers, *Nucleic Acids Res.*, 2013, **41**, D36–D42.
103. L. Martens, M. Chambers, M. Sturm, D. Kessner, F. Levander, J. Shofstahl, W. H. Tang, A. Römpp, S. Neumann, A. D. Pizarro,

L. Montecchi-Palazzi, N. Tasman, M. Coleman, F. Reisinger, P. Souda, H. Hermjakob, P.-A. Binz and E. W. Deutsch, *Mol. Cell. Proteom.*, 2011, **10**, R110.000133.

104. T. Mortier, A. D. Wieme, P. Vandamme and W. Waegeman, *Comput. Struct. Biotechnol. J.*, 2021, **19**, 6157–6168.
105. C. Weis, A. Cuénod, B. Rieck, O. Dubuis, S. Graf, C. Lang, M. Oberle, M. Brackmann, K. K. Søgaard, M. Osthoff, K. Borgwardt and A. Egli, *Nat. Med.*, 2022, **28**, 164–174.
106. C. V. Weis, C. R. Jutzeler and K. Borgwardt, *Clin. Microbiol. Infect.*, 2020, **26**, 1310–1317.
107. H.-Y. Wang, T.-T. Hsieh, C.-R. Chung, H.-C. Chang, J.-T. Horng, J.-J. Lu and J.-H. Huang, *Front. Microbiol.*, 2022, **13**, 821233.

CHAPTER 4

Ambient Ionisation Mass Spectrometry for Microbial Analysis

N. STRITTMATTER

Department of Biosciences, School of Natural Sciences, Technical University of Munich, 85748 Garching, Germany
Email: nicole.strittmatter@tum.de

4.1 Introduction and General Applications

Ambient mass spectrometry or Ambient Ionisation Mass Spectrometry (AIMS) refers to mass spectrometry (MS) techniques which operate under atmospheric pressure and allow the analysis of sample material in its native state, meaning without any sample pretreatment (ambient). The first such technology was Desorption Electrospray Ionisation (DESI, Figure 4.1a),[1] followed soon after by Direct Analysis in Real-Time (DART, Figure 4.1b).[2] Since then, an explosion of AIMS techniques has occurred, with nearly 50 different methods reported to date.[3] Broadly, they can be divided into liquid extraction-based, plasma-based and laser-ablation based techniques. Other desorption modes exist and can be thermal, vibrational and acoustic. The reader is referred to a recent review for a wider overview of AIMS technologies.[3]

Applications to profile bacterial biomass have been introduced swiftly and continuously ever since the first introduction of AIMS techniques.[3–8] In the case of most non-ambient technologies bacterial biomass has to be at least harvested (pyrolysis-based workflows) or harvested and extracted (ESI-based workflows) or plated and mixed with matrix (MALDI-based workflows).

New Developments in Mass Spectrometry No. 13
Detection and Analysis of Microorganisms by Mass Spectrometry
Edited by Liang Qiao and Jia Yi

Published by the Royal Society of Chemistry, www.rsc.org

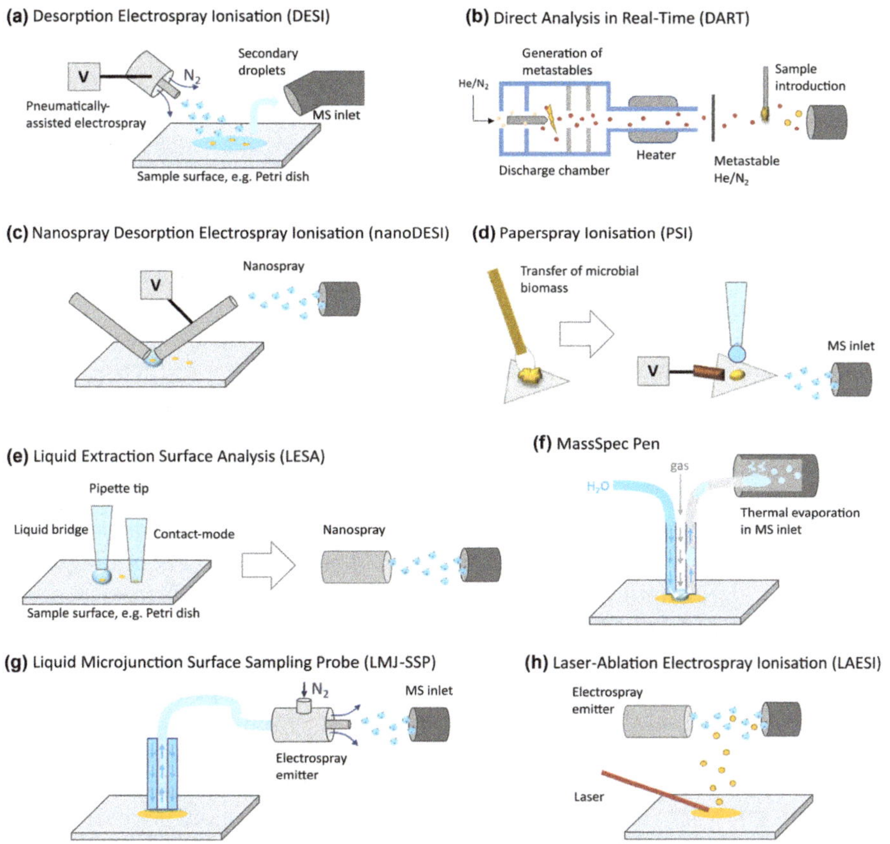

Figure 4.1 Experimental AIMS setups for microbial analysis: (a) DESI, (b) DART, (c) nanoDESI, (d) PSI, (e) LESA, (f) MassSpec Pen, (g) LMJ-SSP and (h) LAESI.

Ambient methods, on the other hand, allow analysis of bacteria and other microorganisms directly from agar plates. Although almost all ambient MS methods have been deployed for bacterial analysis to some extent or other, the most commonly deployed techniques are those that allow analysis of a wide variety of comparatively large (up to *m*/*z* 2000) molecules as this enables the generation of spectra with more specificity to differentiate bacterial species. In practice, these requirements are met especially but not exclusively by solvent extraction- or electrospray-based technologies.

DESI remains the flagship method for solvent-based methods and AIMS methods in general. Here, a pneumatically-assisted electrospray is directed at the sample surface at an angle towards an extended MS inlet (often referred to as a sniffer). A small liquid pool forms on the sample surface which dissolves analytes from the sample. Upon impact of further primary droplets, secondary droplets get ejected and then sampled by the sniffer through the inherent instrument vacuum system.[1] The suggestion to use ambient MS methods for bacterial identification directly from agar plates or clinical specimens was made in 2005 and first MS spectra of bacterial biomass of *Escherichia coli* and *Pseudomonas aeruginosa* deposited on a PTFE carrier were obtained.[8] In 2007, Meetani *et al.* showed that mass spectral fingerprints of seven bacterial species differed dramatically, although sample preparation was still tedious, involving several steps from harvesting to washing, resuspension and spotting onto the target plate.[6] In 2009, Song *et al.* showed that DESI-MS can be performed directly on an agar plate to detect excreted surfactin in *Bacillus subtilis* as well as minor signals originating from membrane phospholipids.[9]

However, due to the pneumatic assistance of the electrospray in DESI, indentation of the agar surface occurs, which results in a reduction of sensitivity due to the loss of secondary droplets containing analyte ions.[10] To maintain a flat sampling surface, imprinting and the use of membranes to transfer bacterial biomass to a more solid sampling surface are commonly deployed strategies.[10–13] Another workflow that was found to minimise the problem was to grow cultures on thin agar and then vacuum-desiccate the samples prior to analysis.[14] An alternative solvent-based technology that avoids agar deformations due to the absence of pneumatic assistance is nanospray DESI (nanoDESI, Figure 4.1c).[15] Methodologically more similar to liquid microjunction techniques than DESI, nanoDESI uses a continuous liquid microjunction between the sample and the self-aspirating nanospray ionisation source and avoids the problem of pneumatically-caused depression of the sampled surface. Watrous *et al.* first deployed nanoDESI to the analysis of microorganisms in 2012 to monitor metabolite production from live colonies of *B. subtilis*, *Streptomyces coelicolor*, *Mycobacterium smegmatis* and *P. aeruginosa*.[16] They recorded MS/MS data which were then used to construct a spectral network in which the individual compounds are nodes in a network of chemical (fragmentation) similarity. The authors demonstrated the diversity of chemical structures produced by the tested microorganisms, while these networks can also be used to identify unknown *m*/*z* values based on similar fragmentation behaviour. Lanekoff *et al.* used nanoDESI in 2013 for spatially-resolved analysis of glycolipids and metabolites in living *Synechococcus* sp. PCC 7002 and the surrounding agar.[17] Also in 2013, nanoDESI was used for *in vivo* metabolic profiling directly from an agar dish to analyse single colonies of *Shewanella oneidensis*, *B. subtilis* and *S. coelicolor*, as well as to image biofilms using a mixed biofilm of *S. oneidensis* and *B. subtilis* as an example.[18]

In Liquid Extraction Surface Analysis (LESA, Figure 4.1e), a pipette tip is positioned over the sample of interest and a liquid microjunction is established with the sample. Analysis is subsequently performed *via* chip-based nanospray ionisation, most commonly using the Advion TriVersa NanoMate (Ithaca, NY, USA) system. Randall *et al.* deployed LESA to extract proteins from *E. coli* K12 colonies directly on an agar plate.[19] They further introduced "contact mode" LESA, which led to higher protein yields while simultaneously improving spatial resolution. Instead of having a liquid bridge between the pipette tip and sample surface, here the rim of the pipette tip touches the sample directly. The authors performed top-down proteomics to identify bacterial proteins, including DNA binding proteins and shock response proteins. They later expanded this work to include both Gram-positive and Gram-negative bacteria, detecting a total of 39 proteins, including 16 previously unreported microbial proteins.[20] The authors analysed three species of *Streptococcus* – *S. pneumoniae* D39, *S. oralis* ATCC 35037 and *S. gordonii* ATCC 35105 – and found five, six and 15 protein masses, respectively. Of these, only one was shared among all three isolates, enabling straightforward differentiation between the three strains. However, the reproducibility of these protein masses among several strains of the same species was not tested. In 2019, LESA was further coupled with high-field asymmetric waveform ion mobility separation (FAIMS), leading to an increase in the number of detected proteins by removing interferences in the low *m*/*z* range, enabling the analysis of previously inaccessible proteins.[21] LESA was further deployed to study cell wall lipids in *Mycobacterium smegmatis*[22] using a chloroform:methanol mixture with added sodium or potassium acetate salts as a solvent system. The authors detected a range of acyl-phosphatidylinositol mannoside (Acyl-PIM) lipids, alterations of which might be implicated in drug-resistance mechanisms.[23]

Liang *et al.* developed and tested different designs for Standing Wave Surface Acoustic Wave Nebulisation (SW-SAWN) MS for the analysis of Bacterial Lipid A, based on which five different bacterial species could be differentiated.[24] In SAWN, surface acoustic waves are generated through RF signals applied to piezoelectric substrates patterned with metalised interdigitated transducers. When a droplet of a liquid sample is placed in their travel path, the energy is transferred to the droplet, leading to rippling and nebulisation, resulting in gas phase analytes with low internal energies.[25] This makes SAWN a promising technology for analysis of liquid cultures. To analyse Lipid A, the authors performed a rapid ammonium isobutyrate lipid micro-extraction process.[24] Only Gram-negative bacteria contain Lipid A; however, it is known to vary between bacterial species or even within the same species; for instance, *P. aeruginosa* has been shown to use Lipid A modifications to escape innate immunity in cystic fibrosis infections.[26] This has important implications for the use of this technology for bacterial phenotyping.

Nemes *et al.* first used laser-ablation electrospray ionisation (LAESI-MS, Figure 4.1h) for the spatially-resolved analysis of biofilms.[27] LAESI utilises the beam of a mid-infrared laser to excite water molecules present in the biological sample, which leads to the production of an aerosol plume above the sample which is then ionised through interaction with multiply charged electrosprayed droplets. Li *et al.* used LAESI-IMS-MS and LAESI-MSI to characterise the distributions of bacterial metabolites and lipids in *E. coli* and *B. subtilis* interacting with antibiotics in Kirby–Bauer type experiments.[28] From over 400 deisotoped spectral features, the authors were able to identify 66 compounds based on exact masses, fragmentation patterns and collisional cross sections. Crude 3D depth profiling was achieved by probing each sampling position with 3–4 consecutive laser pulses, each ablating approx. 25 μm depth of material. Figure 4.2 shows the three-dimensional distributions of the streptomycin fragment and bacterial metabolites glutathione, PE(16:0/16:1) and PE(16:0/17:1) in *B. subtilis*. While glutathione and the antibiotic streptomycin are detected at similar levels at increasing depths, the two phospholipid species are in significantly lower abundance at the bottom layers.

Another solvent-based technology that was deployed for the analysis of microorganisms is paperspray ionisation (PSI, Figure 4.1d), in which bacterial colonies are deposited onto filter paper pre-cut to a sharp point.[29] The filter paper is then wetted with solvent and held at potentials in the range of several kV. This induces the release of charged droplets through field emission which are then sampled using a mass spectrometer. The mass spectra obtained by PSI are similar to those of other solvent-based AIMS

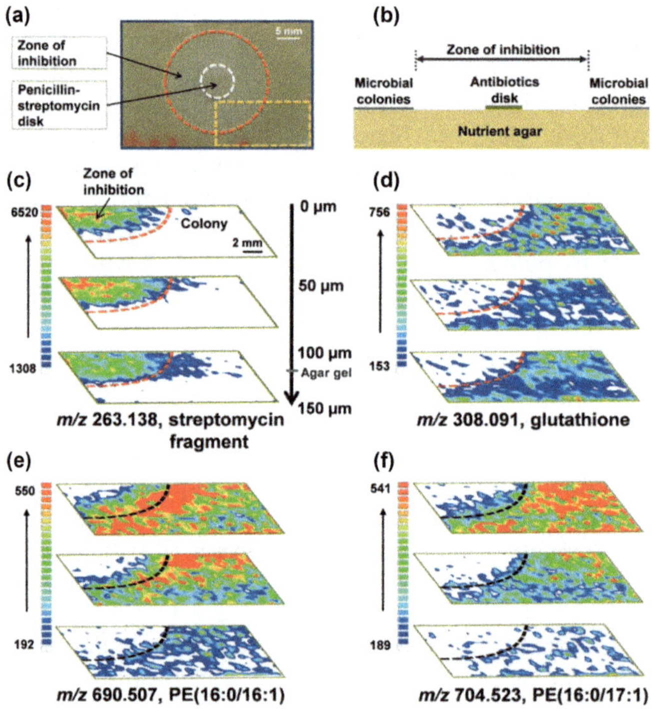

Figure 4.2 LAESI-MSI of *B. subtilis* colonies in a Kirby–Bauer type experiment. (a) Agar plate with a zone of inhibition (red), a penicillin–streptomycin disc (white), and an imaged region of interest (orange). (b) Side-view schematic of the agar plate after the antibiotic disc was removed. (c–f) 3D distributions of (c) the streptomycin fragment, (d) glutathione, (e) PE(16:0/16:1) and (f) PE(16:0/17:1) in 3 different depths obtained using LAESI-MSI. Reproduced from ref. 28 with permission from John Wiley & Sons, Copyright 2016.

technologies and feature predominantly glycerophospholipids and fatty acids. Another variation of the electrospray method is touch spray ionisation (TSI), in which the sampling device itself is subsequently used as an electrospray source. In TSI, sampling probes generally are metallic objects with a sharp tip to which solvent and a high voltage are applied to initiate field emission of analyte-containing charged droplets. Kerian *et al.* used such a probe to sample colonies of *E. coli* and generate lipid profiles from intact bacteria.[30] A medical swab type sampling probe was also presented and shown to be capable of generating spectra from single sampled *Streptococcus* colonies.[31]

The most widely used non-solvent based AIMS method is DART.[2] Analysis using plasma-based techniques usually relies on analysis of smaller lipid-originating constituents, most frequently fatty acids and analogues. Fatty acid profiles are highly reproducible and species-specific and were deployed in previous commercial systems that relied on profiling of fatty acid methyl esters (FAMEs) using pyrolysis–GC–MS, such as the Microbial ID, Inc. (MIDI, Newark, DE, USA) system.[32] In DART, reagent ions such as protonated water clusters are generated by subjecting a carrier gas, typically helium or nitrogen, to corona discharge from a needle electrode. These excited gas molecules are then directed onto the sample surface, where they desorb and ionise molecules which are then sampled using a mass spectrometer.[2] Heating of the reagent gas can further improve desorption characteristics.

Pierce *et al.* showed that DART could be used to profile bacterial FAMEs, resembling data acquired using the MIDI instrument.[5] Tetramethylammonium hydroxide (TMAH) was used to generate FAMEs *in situ* from intact bacteria by setting the DART ionisation gas heater to 400 °C. While the GC step was skipped, resulting in non-resolved isobaric species such as branched and straight chain fatty acids, these profiles still contained sufficient information to distinguish different bacterial species. More recently, fatty acids and other small metabolites were detected directly as negative ions without requiring thermal hydrolysis and methylation.[33–35] Other plasma-based techniques such as Atmospheric Solids Analysis Probe (ASAP) and low temperature plasma (LTP) methods were also used to analyse microorganisms. While the ASAP method was used to rapidly determine which of the multiple steps in the fungal ergosterol pathway were interrupted by an inhibitor,[7] the LTP method was used to generate fatty acid ethyl ester (FAEE) profiles of bacteria suspended in 70% ethanol.[36]

4.2 AIMS Methods as Bacterial Identification Tools

Shortcomings of MALDI-based microbial analysis, such as the requirement of sample preparation, its limited ability to distinguish certain bacterial genera/species and lack of direct antimicrobial resistance profiling, have resulted in continued interest in the development of alternative microbial identification methods. AIMS methods are of particular interest here due to the possibility to analyse bacteria directly, without or with only limited sample preparation. The general workflow for AIMS methods is similar to that used in MALDI-based studies of microorganisms. First, these are plated out in solid cultures before they are analysed for database generation. Subsequently, multivariate statistics can be used to classify unknown bacteria. As such, these methods come with similar limitations, such as the need for the isolate to be in the database for successful identification as well as a restriction in applicability due to the bacteria having to be cultivatable.

Due to the usually limited mass range of AIMS methods, analytes used for microbial identifications are generally small molecule metabolites ranging from primary and secondary metabolites to complex structural lipids up to and including Lipid A. The first AIMS application that deployed a concept similar to ribotyping by MALDI-MS was demonstrated in 2011 by Zhang *et al.* using DESI-MS.[36] The authors showed that bacteria can be distinguished based on their lipid spectral profiles in both positive and negative ion modes using multivariate statistical methods. Four different bacterial species (including four different *Salmonella* strains) could be clearly separated using principal component analysis (PCA).

Bacterial fatty acid profiling by esterification and pyrolysis followed by GC–MS analysis is a well-established technique for bacterial identification. Pierce *et al.* showed that DART could be used to profile bacterial FAMEs.[5] Cody *et al.* used DART in 2015 to successfully differentiate 10 different bacterial species based on their fatty acid profiles, detecting them directly as negative ions without requiring thermal hydrolysis and methylation with 100% correct classification using PCA.[35] However, only one strain per species was used, so the intra-species variance is not well represented.

Several strong strain typing applications have been published using DART. Cody used DART to differentiate 17 different strains of dried *Saccharomyces* yeasts (12 *S. cerevisiae* and 5 *S. pastorianus* strains).[34] The yeast cells were suspended in 50% methanol and sampled by dipping the sealed end of a melting point capillary into the solution before subjecting them to DART MS in positive and negative ion modes. Most metabolic features detected were below m/z 500. The negative ion mode was found to contain more distinctive metabolites and was able to differentiate both species and all strains with 97.7% accuracy. It could further differentiate the analysed strains into top- *vs.* bottom-fermenting yeasts, beer and baking yeasts and red-wine and champagne yeasts. More recently, Tata *et al.* used DART for the identification and serotyping of the waterborne pathogen *Legionella pneumophila* after a simple extraction procedure involving methanol, followed by sonication and centrifugation.[33] The authors assessed profiling by DART-MS and subsequent partial least squares discriminant analysis to differentiate *Legionella* spp. from other unspecified bacteria identified from water samples and found an accuracy of 98.1%. A second classifier differentiating *L. pneumophila* sg. 1 (responsible for 85% of infections worldwide) from the other serotypes (sg. 2–15) and *L.* non-*pneumophila* spp. performed with an overall accuracy of 95.93%, while classifying all sg. 1 strains correctly.

The most differentiating compounds were tentatively assigned as hydroxylated fatty acids, typical LPS fragments which are known to be abundant in *Legionella* spp. and important in serogroup variations.[37] The analysis times are in the order of approximately 1 min per sample, making it a rapid and cost-effective approach.

Zhang *et al.* deployed the LTP method to differentiate 16 bacterial species (*B. subtilis*, *S. aureus*, *E. coli* and 13 strains of *Salmonella enterica*) based on their FAEE profiles.[36] In this workflow, bacteria were harvested, mixed with 70% ethanol, deposited onto glass slides and dried before being subjected to LTP analysis. Using PCA, Gram-positive and Gram-negative species were clearly separated, as were *E. coli* from the *Salmonella* strains. In a separate model of the *S. enterica* spp. only, all 11 of the 13 strains produced distinct clusters based on their FAEE profiles.

We introduced Rapid Evaporative Ionisation Mass Spectrometry (REIMS) in 2013 for the sample-preparation-free identification of bacteria and yeasts.[38,39] REIMS was originally developed as a tool for intra-surgical identification of tissues, allowing the differentiation of healthy and pathological tissues, by subjecting cellular biomass to rapid heating with radiofrequency (RF) alternating electric current.[40,41] This leads to thermal disintegration of the sample and production of an aerosol of lipid-covered droplets containing both intracellular and extracellular metabolites that are then subjected to characterisation using MS.[38,40,41] As no solvent or matrix is used in the desorption and ionisation process, mass spectral profiles are very clean and reproducible, containing only peaks that originate from the sample itself.

For microbial analysis, we used two hand-held electrodes in the form of a forceps with an integrated aerosol line (so-called irrigated bipolar forceps) to sample bacterial biomass directly from solid culturing plates.[38] Spectral profiles contain information-rich lipid fingerprints ranging from fatty acids over sphingolipids and glycerophospholipids to cardiolipins and large glycosylated species such as Lipid A. The method is further suitable for other amphiphilic molecules such as biosurfactants (including rhamnolipids and lipopeptides), quorum-sensing molecules and biopolymers such as polyhydroxybutyrates (PHBs).[38,42] Analysis of liquid cultures is equally possible but requires centrifugation prior to REIMS analysis to generate a cell pellet. This workflow has also been demonstrated for analysis of mammalian cell lines.[43]

15 isolates of each of 28 different clinically-relevant bacterial species isolated from blood cultures were subjected to REIMS analysis, which led to 95.9%, 97.8% and 100% correct classification on species, genus and Gram levels, respectively. Mass spectra were reproducible over three different experimental platforms, while the ability to perform in-depth phenotyping was shown on seven different strains of *E. coli* which were grown on different solid culturing media for up to three days and still classified 87.3% correctly at the strain level. Since then, several steps have been taken to move this setup towards automation, ranging from the development of an *ex vivo*

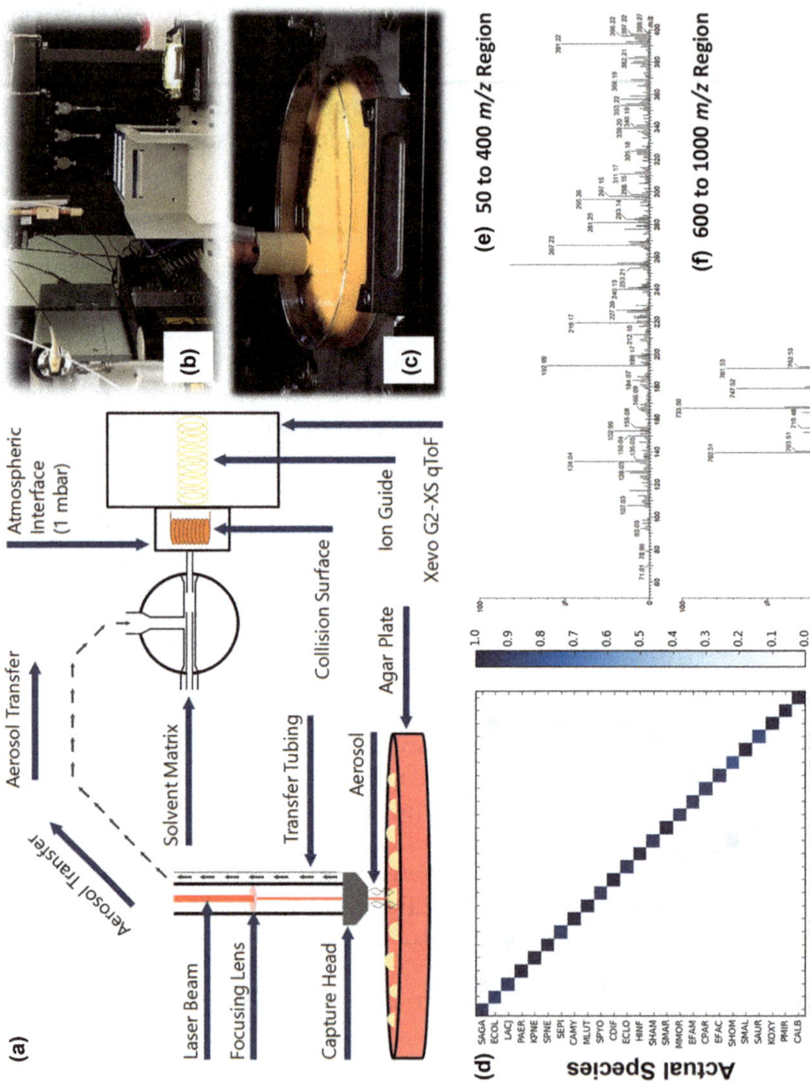
(a)
Laser Beam
Focusing Lens
Capture Head
Aerosol Transfer
Aerosol Transfer
Solvent Matrix
Transfer Tubing
Aerosol
Agar Plate
Collision Surface
Atmospheric Interface (1 mbar)
Ion Guide
Xevo G2-XS qToF
(b)
(c)
(d)
Actual Species
Predicted Species
(e) 50 to 400 m/z Region
(f) 600 to 1000 m/z Region

imaging stage to fully automated sample handling and analysis based on a Tecan Evo 75 robotic platform, as shown in Figure 4.3.[44–47] Automated REIMS sampling probes have further moved from diathermy tools towards laser-based setups, which in a set of ten bacterial species led to a classification accuracy of 99.4% in comparison to 98.8% for the bipolar diathermy REIMS probe.[48] This high-throughput tool is a promising alternative to protein-based identification using MALDI in clinical microbiology practice for the identification of pure cultures and without loss in classification accuracy, classifying 25 bacterial species correctly at 97.2% accuracy.

Using PSI, Hamid *et al.* showed the differentiation of six Gram-positive and ten Gram-negative bacterial species in PCA based on their lipid profiles.[29] To increase specificity, the authors fused data from positive and negative ion modes, which significantly improved separation observed in PCA. A similar approach was also proven successful in differentiating eight species of the yeast *Candida*.[49] Using PCA followed by linear discriminant analysis for classification, an accuracy of 90% was obtained on the species level. Zero-volt PSI was used to analyse eight common bacterial species (four Gram-positive, four Gram-negative). Mass spectra were found to be almost identical to those of conventional PSI, although with three orders of magnitude lower absolute intensity values, which were, however, partially offset by the significant reduction in spectral noise.[50] PSI implemented on a portable, miniature mass spectrometer, Mini 12, yielded data with a similar spectral appearance to a linear ion trap (Thermo LTQ) instrument,[51] allowing the separation of seven different bacterial species in both cases, although with smaller intra-group variance in the LTQ data. This was also observed in the case of REIMS; even with a low bin size of 5 Da, 28 bacterial species can be differentiated without large losses in identification accuracy (94.5% at 5 Da *vs.* 96.1% at 0.1 Da bin size) while pointing out that analysis on high resolution mass analysers facilitates the identification of unknown spectral features and leads to more compact data groups in PCA.[38]

Recently, Olajide *et al.* further developed the PSI approach and successfully showed species-level discrimination of five *Bacillus* species with an accuracy of 99% by using instrumentation capable of high mass resolution and ion mobility separation[52] after only 4 h of incubation time by filtering 8 mL of liquid bacterial culture onto glass filter paper with pores sufficiently

Figure 4.3 (a) Automated high-throughput REIMS platform using a CO_2 laser sampling probe. The resulting aerosol is mixed with 2-propanol containing internal standard before passing through an atmospheric pressure interface containing a heated collision surface. (b) Modified Tecan Freedom Evo 75 platform performing analysis of a bacterial culture (c). (d) Confusion matrix of 25 microbial species (15 isolates from 24 bacteria and 1 yeast each). The overall species-level accuracy is 97.2%. (e and f) Representative spectra of *E. coli* for the lower weight metabolite (50 to 400 *m*/*z*, (e)) region and the complex lipid (600 to 1000 *m*/*z*, (f)) region. Reproduced from ref. 44 with permission from American Chemical Society, Copyright 2021.

small to collect bacteria (0.3 μm). They detected species-level differences between isobaric metabolites such as differences in fatty acid chains in glycerophospholipids. Replicate cultures of a single strain of each *Bacillus* species were used under homogeneous culturing conditions, leaving the variance of these isomers in a more diverse dataset uncharacterised.

Liu *et al.* used easy ambient sonic-spray ionisation mass spectrometry (EASI-MS) to investigate the changes of acidic lipids in unicellular (*Synechocystis* 6803, *Synechococcus* 7002) and filamentous (*Anabaena* 7120) cyanobacteria during different growth phases.[53] Due to the absence of a voltage between the sample surface and EASI source tip, it is one of the simplest and softest AIMS technologies. Using PCA, three different cyanobacteria could readily be distinguished based on their lipid profiles. However, the authors also showed that the spectral profiles varied with culture age and when changing from exponential into stationary growth. This highlights a possible shortcoming of metabolic identification technologies as culturing conditions might interfere with spectral patterns obtained, possibly requiring defined culturing conditions. On the other hand, this sensitivity can also serve as an advantage for microbial phenotyping of, for instance, more virulent or antimicrobially-resistant species.

Recently, Povilaitis *et al.* applied the MassSpec Pen (Figure 4.1f), another ionisation source originating from the intra-surgical tissue identification field to bacterial identification.[54] The MassSpec Pen deploys a liquid microjunction to sample bacterial biomass, with the extract subsequently directly introduced into the heated MS inlet where the solvent evaporates and ions are generated.[55] The MassSpec Pen generally uses water at a predefined liquid volume (4–10 μL), which is delivered to the sample surface and subsequently back-aspirated for analysis. To avoid the interference of signals from the culturing medium, the bacterial biomass was transferred from the agar plate onto a glass slide and dried prior to analysis. Using this workflow, 43 strains belonging to eight different bacterial species were analysed. The authors reported 96% accuracy on the Gram-stain level over all eight bacterial species using a LASSO classification model, and built further well-performing LASSO models to distinguish closely related bacterial strains such as *Streptococcus* Grp A *versus* Grp B and *Staphylococcus aureus versus S. epidermis*. However, a species-level classification over all eight species analysed in a single model was not presented.

Also in 2022, Su *et al.* presented an approach based on thermal desorption-electrospray ionisation/mass spectrometry (TD-ESI/MS).[56] Here, a metallic inoculating probe is used to collect the sample, such as a single bacterial colony. This is then inserted into a preheated oven at 280 °C, where lipids are desorbed and delivered *via* nitrogen carrier gas into an electrospray plume for ionisation. The metallic sampling probe can be reused without memory effects after burning off residual biomass. A ten-fold loss in ionisation efficiency was found for Gram-positive bacteria in comparison to Gram-negative bacteria, which was associated with the peptidoglycan layer surrounding the cell membranes of Gram-positive bacteria. To alleviate the loss in sensitivity, acid hydrolysis (5% TFA) followed by solvent extraction similarly to MALDI-MS protocols for the analysis of Gram-positive bacteria was found to be appropriate. Mass spectra were obtained from *m*/*z* 400 to 1000 and were dominated by ions from glycerophospholipids and especially lyso-lipids, indicating the thermal breakdown of heavier lipid species. Spectral profiles in positive ion mode of ten bacterial species were then analysed using PCA and hierarchical cluster analysis (HCA) and were found to yield similar separation as MALDI-based ribosomal protein profiling, with more closely related species grouping more closely together.

4.3 Ambient MS for Natural Product Research/ Biotechnology

4.3.1 General/Natural Products

Microbial natural products have become promising sources of antibiotics and anticancer agents in human medicine. Increasing our knowledge of microbial metabolite production might further lead to improved understanding of intermicrobial communication, while also advancing areas such as biofuel production and clinical research into host–microbe interactions. Considering natural product research, AIMS methods can be used to study large libraries of microbial strains to identify those producing specific compounds or compound classes of interest or to screen large numbers of culturing conditions for those with highest yield. This was exemplified by Bacon *et al.* who used LAESI as a rapid method for screening *Bacillus mojavensis* strains for surfactin content as well as characterised the effects of medium composition and assay time on biosurfactant production.[57]

Although AIMS techniques are often limited in their mass range to about 2000 Da and rely predominantly on the lipid compartment for speciation, they still offer plenty of possibilities to study microbial natural compounds. In particular, imaging setups are valuable tools in natural product research by allowing the correlation of the location of measured metabolites with the observed microbial morphologies and functional roles of the observed compounds. For instance, Ellis *et al.* used Membrane-Scaffolded DESI-MSI to probe intercellular interactions in a microbial predator–prey scenario. The prey *E. coli* was cocultured with the predator *Myxobacterium* and the resulting MSI data were analysed using unsupervised segmentation, leading to the discovery of distinct metabolic segments corresponding to intermicrobial predation and communication. Secondary metabolism was found to be activated at the location of predation, with 52 unknown metabolites located to this predation zone.[11] The high number of unknown metabolites necessitates workflows that can be used to identify unknowns, for instance based on spectral similarities in fragmentation patterns. Such a workflow to organise MS/MS spectra was suggested by Watrous *et al.*, who deployed a spectral networking method of deconvoluted fragmentation features where structurally related molecules (represented as nodes) are mapped based on spectral similarity.[16] The antifungal effect of the *Pseudomonas* sp. SH-C52 could be attributed to thanamycin, a predicted lipopeptide encoded by a non-ribosomal peptide synthase gene cluster. The approach made it possible to partially characterise thanamycin, which showed that it is a monochlorinated lipopeptide of the syringomycin family of antifungal agents. This method was further suggested as a possible extension to the metabolomics profiling of microbial colonies with a liquid microjunction surface sampling probe (LMJ-SSP, Figure 4.1g), which was used to perform real-time metabolomics on a diverse set of 30 microorganisms including yeast, filamentous fungi and marine-derived bacteria.[58] Here, analyte extraction is performed *via* a pneumatic nebuliser rather than a self-aspirating nanospray, which enables the use of higher solvent flow rates as compared to nanoDESI. A wide range of metabolites was observed: from small redox compounds to quorum-sensing molecules, glycolipidic antifungals and non-peptidic antibiotics.

Sica *et al.* used DESI-MSI to study the interactions of two fungi of the phylum *Ascomycota* producing the antifungal phomopsinone A and mycotoxin T-2 toxin, respectively. The authors compared the spatial and temporal distributions of those two secondary metabolites between both isolated cultures and cocultures. While phomopsinone A was readily detected at the point of inoculation and outwards (but not at the growth front), T-2 toxin could only be observed once the fungus was cross-sectioned, proving the compound to be localised to the interior of the fungus.[59] A similar approach using a tape-imprint method[12] was used to visualise the metabolic exchange during the antagonistic interaction between the phytopathogen *Moniliophthora roreri* and the biocontrol agent *Trichoderma harzianum*.[60] The authors compared the secondary metabolites produced when cultured in isolation *versus* a coculture to identify four pythopathogen-dependent metabolites including T39 butenolide, harzianolide, sorbicillinol and an unknown substance.

Rath *et al.* performed metabolite identification with a data-independent nanoDESI workflow and the previously established spectral networking algorithm to explore the metabolic output of the gut microbiota in host–microbe interaction studies on colon tissue sections and on pure bacterial cultures in combination with MALDI-MSI. This work allowed the identification of altered glycan distribution patterns based on microbiota status and microbial transformations of host-derived bile acids.[61]

4.3.2 Biotechnology

Microorganisms are of high interest for the production of metabolites of industrial and pharmaceutical use. One of the most widely used approaches in the development of novel biocatalysts is the expression of gene variants expressed by non-pathogenic, easily-grown bacterial colonies on solid agar plates.[62] With advances in synthetic biology, huge strain libraries can be produced easily, generating library sizes of up to 10^{10} genetic variants.

Unless the reaction of interest can be measured colorimetrically or fluorescently through intrinsic characteristics of the target molecule, readout is likely to be performed using LC–MS/MS-based analysis which has extensive requirements for sample preparation and limited sample throughput in the range of hundreds of samples per day only. In order to address these shortcomings of limited chemical scope and throughput, several workflows using AIMS have been proposed.

Yan *et al.* developed a screening approach for biotransformation in live bacterial colonies that deploys DESI-MSI and ion mobility separation to distinguish colonies expressing products from those that do not, termed Direct BioTransformation-Mass Spectrometry (DiBT-MS).[63] In the proposed workflow (shown in Figure 4.4), bacteria are plated onto a nylon membrane laid on top of an agar plate and incubated. Subsequently, the membrane is transferred to another plate containing the inducer for expression and incubated on filter paper soaked with substrate solution. In subsequent DESI-MSI analysis, colony positions can be visualised using an *m*/*z* marker of the host bacterium, such as FA(18:1) in this study. Biotransformation products can be found at the predicted *m*/*z* value, which will be blank for non-producing variants. The method was demonstrated for ammonia lyases and P450 monooxygenases and enables multiplexing of enzyme variants and substrate libraries simultaneously. While the calculation of substrate to product conversion rates was feasible using DiBT-MS, this approach currently lacks absolute quantification capabilities. However, colonies of interest can be recovered post-sampling and further analysed to perform absolute quantification and/or identify the mutations of interest, although cross-contamination caused by the pneumatically-assisted electrospray and washout are likely.

Based on their prior analysis of microbes deploying a microporous membrane scaffold method,[11] Ellis *et al.* developed a DESI-MSI-based workflow to characterise metabolic phenotypes associated with two different *E. coli* strains engineered to overproduce free fatty acids with defined chain lengths by expressing different thioesterase (TesA) enzymes.[64] Using untargeted metabolomics including unsupervised segmentation allowed identification of strains with the highest production of target molecules, while simultaneously enabling monitoring of metabolites outside the engineered product pathway. This can be useful to analyse off-target products or to characterise the overall metabolic fitness of host cells. For instance, alterations were found in amino acid levels and membrane lipid saturation, with the latter possibly implicated in a reduction of cell viability. The authors also used a common lipid species, PG(16:0/17:1), to visualise all colonies while separately monitoring for dodecanoic and octanoic acid production in specific colonies.

Most biotechnology-based applications, to date, have deployed rather time-consuming imaging workflows, in which several datapoints are recorded per colony and most of the time is actually spent analysing blank spaces between the individual colonies. Gowers *et al.* demonstrated that an

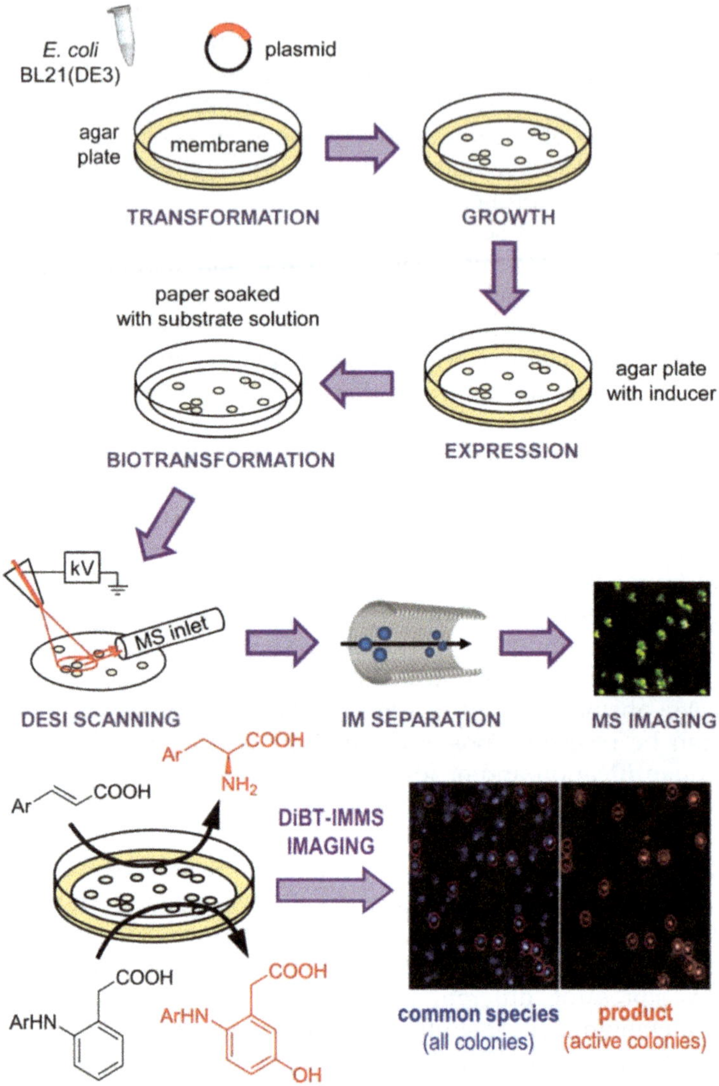

Figure 4.4 Workflow of DiBT-IMMS imaging of bacterial colonies expressing biocatalysts on agar plates under ambient conditions. Reproduced from ref. 63 with permission from American Chemical Society, Copyright 2017.

automated laser-assisted REIMS (LA-REIMS) technique was suited for biotechnological applications by screening engineered yeast strain libraries directly from agar to assess the production levels of violacein and betulinic acid at a rate of six colonies per minute (450 colonies in 4 h), producing a single datapoint per sample.[65] Results of this analysis were compared to colorimetric readout and a correlation was found between the high producers in both assays. The authors developed a sampling robot that enables automatic colony detection, followed by collection of a portion of the colony into growth medium stocks in a 96-well plate format before deploying a high-powered laser to evaporate an area corresponding to 0.8 mm^2 of each colony for MS analysis. The 96-well plate was then incubated overnight, before adding glycerol and freezing, resulting in each LA-REIMS result being linked to a well ID with stock to recover the corresponding strain at a later stage. The purity of the strains in the recultured samples, however, was not assessed.

Finally, Sarsby *et al.* deployed monopolar REIMS to monitor recombinant protein expression of QconCATs in *E. coli* from liquid culture by centrifugation and analysis of the resulting cell pellet.[66] QconCATs show a variable degree of expression while lacking intrinsic biological activity to directly impact host metabolism. Using multivariate statistics, the authors found REIMS to be predictive of expressing *vs.* non-expressing strains, even indicative of high *vs.* low protein expression due to metabolic changes in the mass spectra associated with diversion of metabolites into recombinant protein production.

4.4 Analysis of Biofilms

Biofilms are complex microbial samples in which bacterial cells are embedded in an extracellular matrix with gradients in pH, nutrient composition and oxygen supply.[67] In most scenarios, biofilms will contain multitudes of bacterial species, adding further complexity and local gradients in bacterial composition.[67] MSI is an interesting technology to study bacterial biofilms, as it allows the multiplexed and untargeted acquisition of spatial heterogeneity in microbial metabolites in 2D and 3D. MALDI-MSI is the dominant technology also in the field of MSI; however, it comes with several disadvantages for microbial biofilm analysis such as the requirement of a conductive target surface for analysis, the limited thickness of the sample and matrix application which results in the biofilm being unrecoverable. Due to its high spatial resolution and 3D imaging capabilities, secondary ion mass spectrometry (SIMS) imaging is also a popular technology for biofilm imaging; however, SIMS is performed under high vacuum, restricting its use to wet samples such as biofilms which contain >90% water.[68] This can be partially overcome by using cryogenic sample handling; however, the field of view remains small (3 mm^2) and molecular information is limited to small molecules.[69] Ambient technologies, on the other hand, enable *in vivo* biofilm analysis from biofilms of any thickness and size and on any target surface.

The first such application deployed nanoDESI to analyse a mixed species biofilm in which *B. subtilis* was grown in the centre and *S. oneidensis* was grown on the periphery of the biofilm to test how the metabolite repertoire changes in coculture. Watrous *et al.* observed several metabolites produced especially at the interface between the two species and also changes in expression of metabolites, such as reductions in putrebactin and plipastatin.[10] Nguyen *et al.* introduced constant-distance mode nanoDESI in 2017 by integrating a shear-force probe with the nanoDESI probe.[70] Using this setup, simultaneous acquisition of chemical and topographical images was demonstrated for *B. subtilis* biofilms grown on agar plates. For this sample, height variations of up to 0.8 mm were measured which without distance control would lead to breaking of the liquid microjunction. Ravindran *et al.* used DESI-MSI to identify microbial signatures of bacterial fouling for early detection of biofouling on RO water purification membranes by inoculating static and continuous flow units with *B. subtilis*, *P. aeruginosa* and mixtures thereof.[71] They showed that biosurfactants produced by the bacteria can be indicator molecules that biofilm buildup is initiated, for instance rhamnolipids in the case of *P. aeruginosa*. They further used complexes of these biosurfactants to indicate heavy metal contamination in the water.

Due to the thick cell walls of bacteria, especially of Gram-positive bacteria, DESI and nanoDESI of biofilms cannot readily and reliably provide readouts on intracellular metabolites, making them particularly suitable to visualise the distribution of excreted molecules like quorum-sensing molecules, biosurfactants or scavenging molecules. They are also surface analysis tools, making sectioning of the biofilms necessary in order to analyse lower layers of the biofilm. However, they are comparatively non-invasive, theoretically enabling longitudinal biofilm sampling.

More promising technologies for intracellular biofilm analysis and depth profiling are laser-based technologies such as LAESI, which was shown to be capable of crude 3D profiling of bacterial cultures in Kirby–Bauer type experiments, in which the plate is commonly covered in a continuous film of bacterial biomass, similar to a biofilm.[28] Ablation craters with a depth of 25 μm were produced, enabling visualisation of the bacterial film in three different layers. LAESI was used for 2D profiling of mixed biofilms of *S. aureus* and *P. aeruginosa* in order to assess colocalisation patterns between the two bacteria and their reaction to treatment with an antimicrobial peptide.[72] The authors used species-specific fatty acid distributions to show that the two bacterial species were largely colocalised, while the colocalisation was found to decrease upon antimicrobial treatment. REIMS was also shown to be suitable to efficiently disrupt bacterial cell walls for analysis, yielding higher spectral intensity for most Gram-positive bacteria over Gram-negative species.[38] A REIMS-based imaging stage was developed and shown to be suitable for imaging of biofilms directly from an agar plate, yielding 2D distribution of bacterial metabolites in a coculture of *B. subtilis*, *P. aeruginosa* and *S. aureus* using a monopolar needle electrode.[47] Later developed laser-based REIMS sources are likely to yield better spatial resolution and might have potential for 3D analysis, similarly to LAESI.

A nice example of multimodal imaging for the study of biofilms was presented by Brauer *et al.*, who used light profilometry together with ambient laser ablation and solvent capture by aspiration (LASCA) MSI to investigate the microbially-influenced corrosion of 1018 carbon steel surfaces by pure cultures of *Marinobacter* spp. or a wild population of bacteria present in coastal seawater.[73] Profilometry data of biofilm-free surfaces demonstrated heterogeneous distributions of corrosion damage. Correlating the degree of corrosion with the metabolomics data obtained using LASCA data revealed differences in chemical composition within the corresponding biofilm regions. The authors further used HPLC-based metabolomics for metabolite identification. They suggested that such a workflow can be used to increase our understanding about the fundamentals of microbially-influenced corrosion and might be used for early detection of corrosion processes.

4.5 Clinical Applications

Detection of bacteria in a sample is most commonly attempted by using a biomarker approach in which the presence or absence of individual molecular species or a combination of these is used to determine the presence or absence of a bacterium as a causative agent. This approach requires building of a bacterial database on pure cultures from which species-specific markers can be derived which are sufficiently specific that they can also be detected in bacterial mixtures and complex matrices. This approach is promising as it has been established in many different studies that there is a huge variety in microbial molecular features and proof-of-principle studies of small numbers of individual markers that can be used to distinguish between species or detect bacteria directly in clinical samples that have been published. Bacterial lipids also differ greatly from mammalian lipids, with many lipid classes being known to exhibit specificity on the bacterial level as well as being absent from mammalian hosts, such as different versions of Lipid A in Gram-negative bacteria, mycolic acids among *Mycobacteria* spp, lipopolypeptides such as surfactin for *Bacillus* spp. and rhamnolipids for *P. aeruginosa*. Such an approach also has the advantage that markers can be transferred between methodologies.

Povilaitis *et al.* used the MassSpec Pen to identify the causative agent in clinical specimens of synovial fluid, sacral tissue and a tissue smear, which were all identified as MRSA (methicillin-resistant *S. aureus*) positive. The authors derived species-specific molecular features using LASSO classifiers on the Gram level (from a set of eight bacterial species), *Staphylococcus vs. Streptococcus* level and *S. aureus vs. S. epidermis* level.[54] While all pure cultures isolated from these samples were identified correctly as *S. aureus*, on the clinical sample only the Gram classification worked reliably in the three examples presented. No distinction between MRSA and MSSA (methicillin-sensitive) was attempted.

Deploying swab touch spray ionisation (TSI), a medical grade swab with a metallic handle was developed from which an electrospray can directly be emitted. Spectra were shown to be able to differentiate *Streptococcus pyogenes* from *S. agalactiae* based on their spectral fingerprint from a model of these two species only. Jarmusch *et al.* further suggested that the approach could detect *S. pyogenes* on throat swabs for clinical diagnostics of Strep throat.[31] For that, swabs were soaked in a human oral fluid sample prior to picking up a single culture of the respective *Streptococcus* sp. and TSI analysis. The *Streptococcus* agent was identified correctly among a model of Grp B *vs.* Grp A *Streptococcus*; however, it was not assessed how the local bacterial concentrations compare to those in a real clinical specimen.

While all of these methods were based on the detection of small molecule markers of bacterial origin, Jana *et al.* presented a method to detect ESKAPE pathogens from 3D wound cultures based on the analysis of bacterial proteins using LESA-MS.[74] Wound infections can lead to the requirement to amputate affected limbs, highlighting the importance of detecting infections early and accurately. ESKAPE pathogens (*Enterococcus faecium*, *S. aureus*, *Klebsiella pneumoniae*, *Acinetobacter baumannii*, *P. aeruginosa* and *Enterobacter* spp.) are a leading group of nosocomial infectious agents and particularly problematic due to the increased prevalence of antibiotic resistance.[75] Using LESA-MS, bacterial proteins could be detected as soon as visible signs of infection were apparent on skin culture (approx. 48 h after inoculation with a single bacterial species). Two proteins of bacterial origin each could be identified from infection models of *S. aureus* and *K. pneumoniae*, while three proteins were identified from *P. aeruginosa*. Strain differentiation of *S. aureus* was demonstrated by differentiation of allelic variants of the protein δ-hemolysin. In addition, seven proteins of human origin were identified, enabling simultaneous monitoring of host response to infection.

While several applications of AIMS methods to detect bacteria directly in clinical specimens have been proposed, it seems to be seen how widely applicable these are in realistic clinical settings where the bacterial cell numbers can vary widely and the range of possible causative agents is significantly larger than any of the datasets that have been used to derive bacterial species-specific markers to date, which in most cases have been less than ten bacterial species. Bacterial markers derived from insufficiently sized datasets might not be as specific or as conserved as initially suggested, making further analyses strictly necessary before meaningful clinical applications can be developed.

4.6 Conclusions

AIMS methods are widely used within the analytical research community to characterise microbial specimens for their metabolite and protein content, to perform bacterial speciation and possible *in situ* detection of infections directly from clinical samples. Most techniques such as REIMS and solvent-based techniques rely on the analysis of small molecular weight, mostly lipid-based spectral features; however, some liquid microjunction technologies lend themselves more readily for protein analysis. Plasma-based technologies are even further restricted in their chemical scope to molecules below 500 Da. While many techniques show great potential, the developments to this point rarely exceed the proof-of-concept stage and currently no method is commercially developed. Some of the most advanced methods are currently within the realms of biotechnology for screening recombinant libraries, enabling colony analysis directly from solid agar plates. Some of the biggest opportunities are within biofilm analysis, where ambient methods are more promising over techniques such as MALDI due to their ability to analyse samples of any solid (or even liquid) target material without the need to place the biofilm into vacuum or destroy it *via* matrix application. More work needs to be performed to test the specificity of bacterial markers on species, genus and Gram levels and to identify all of the metabolic microbial features which can already be detected, yet are largely unknown, especially since analysis largely concentrates on repeated analysis of better-known model organisms such as *S. aureus*, *P. aeruginosa* and *B. subtilis*.

References

1. Z. Takats, J. M. Wiseman, B. Gologan and R. G. Cooks, *Science*, 2004, **306**, 471–473.
2. R. B. Cody, J. A. Laramee and H. D. Durst, *Anal. Chem.*, 2005, 77, 2297–2302.
3. C. L. Feider, A. Krieger, R. J. DeHoog and L. S. Eberlin, *Anal. Chem.*, 2019, **91**, 4266–4290.
4. K. Chingin, J. Liang and H. Chen, *RSC Adv.*, 2014, **4**, 5768–5781.
5. C. Y. Pierce, J. R. Barr, R. B. Cody, R. F. Massung, A. R. Woolfitt, H. Moura, H. A. Thompson and F. M. Fernandez, *Chem. Commun.*, 2007, 807–809.
6. M. A. Meetani, Y. S. Shin, S. Zhang, R. Mayer and F. Basile, *J. Mass Spectrom.*, 2007, **42**, 1186–1193.
7. C. McEwen and S. Gutteridge, *J. Am. Soc. Mass Spectrom.*, 2007, **18**, 1274–1278.
8. Z. Takats, J. M. Wiseman and R. G. Cooks, *J. Mass Spectrom.*, 2005, **40**, 1261–1275.
9. Y. Song, N. Talaty, K. Datsenko, B. L. Wanner and R. G. Cooks, *Analyst*, 2009, **134**, 838–841.
10. J. Watrous, N. Hendricks, M. Meehan and P. C. Dorrestein, *Anal. Chem.*, 2010, **82**, 1598–1600.
11. B. M. Ellis, C. N. Fischer, L. B. Martin, B. O. Bachmann and J. A. McLean, *Anal. Chem.*, 2019, **91**, 13703–13711.
12. A. Tata, C. J. Perez, M. O. Ore, D. Lostun, A. Passas, S. Morin and D. R. Ifa, *RSC Adv.*, 2015, **5**, 75458–75464.
13. E. C. Cabral, M. F. Mirabelli, C. J. Perez and D. R. Ifa, *J. Am. Soc. Mass Spectrom.*, 2013, **24**, 956–965.
14. C. F. Angolini, P. H. Vendramini, F. D. Araujo, W. L. Araujo, R. Augusti, M. N. Eberlin and L. G. de Oliveira, *Anal. Chem.*, 2015, **87**, 6925–6930.
15. P. J. Roach, J. Laskin and A. Laskin, *Analyst*, 2010, **135**, 2233–2236.

16. J. Watrous, P. Roach, T. Alexandrov, B. S. Heath, J. Y. Yang, R. D. Kersten, M. van der Voort, K. Pogliano, H. Gross, J. M. Raaijmakers, B. S. Moore, J. Laskin, N. Bandeira and P. C. Dorrestein, *Proc. Natl. Acad. Sci. U. S. A.*, 2012, **109**, E1743–E1752.
17. I. Lanekoff, O. Geydebrekht, G. E. Pinchuk, A. E. Konopka and J. Laskin, *Analyst*, 2013, **138**, 1971–1978.
18. J. Watrous, P. Roach, B. Heath, T. Alexandrov, J. Laskin and P. C. Dorrestein, *Anal. Chem.*, 2013, **85**, 10385–10391.
19. E. C. Randall, J. Bunch and H. J. Cooper, *Anal. Chem.*, 2014, **86**, 10504–10510.
20. K. I. Kocurek, L. Stones, J. Bunch, R. C. May and H. J. Cooper, *J. Am. Soc. Mass Spectrom.*, 2017, **28**, 2066–2077.
21. K. I. Kocurek, R. C. May and H. J. Cooper, *Anal. Chem.*, 2019, **91**, 4755–4761.
22. A. N. Conner, J. R. Jarvis, L. J. Alderwick and R. L. Griffiths, *Rapid Commun. Mass Spectrom.*, 2019.
23. V. Jarlier and H. Nikaido, *FEMS Microbiol. Lett.*, 1994, **123**, 11–18.
24. T. Liang, T. Schneider, S. H. Yoon, B. L. Oyler, L. M. Leung, W. E. Fondrie, G. Yen, Y. Huang, R. K. Ernst, E. Nilsson and D. R. Goodlett, *Int. J. Mass Spectrom.*, 2018, **427**, 65–72.
25. Y. Huang, S. H. Yoon, S. R. Heron, C. D. Masselon, J. S. Edgar, F. Turecek and D. R. Goodlett, *J. Am. Soc. Mass Spectrom.*, 2012, **23**, 1062–1070.
26. C. Cigana, L. Curcuru, M. R. Leone, T. Ierano, N. I. Lore, I. Bianconi, A. Silipo, F. Cozzolino, R. Lanzetta, A. Molinaro, M. L. Bernardini and A. Bragonzi, *PLoS One*, 2009, **4**, e8439.
27. P. Nemes and A. Vertes, *J. Visualized Exp.*, 2010, **43**, e2097.
28. H. Li, P. Balan and A. Vertes, *Angew. Chem., Int. Ed.*, 2016, **55**, 15035–15039.
29. A. M. Hamid, A. K. Jarmusch, V. Pirro, D. H. Pincus, B. G. Clay, G. Gervasi and R. G. Cooks, *Anal. Chem.*, 2014, **86**, 7500–7507.
30. K. S. Kerian, A. K. Jarmusch and R. G. Cooks, *Analyst*, 2014, **139**, 2714–2720.
31. A. K. Jarmusch, V. Pirro, K. S. Kerian and R. G. Cooks, *Analyst*, 2014, **139**, 4785–4789.
32. S. K. Haack, H. Garchow, D. A. Odelson, L. J. Forney and M. J. Klug, *Appl. Environ. Microbiol.*, 1994, **60**, 2483–2493.
33. A. Tata, F. Marzoli, A. Massaro, E. Passabi, M. Bragolusi, A. Negro, I. Cristaudo, R. Piro and S. Belluco, *J. Appl. Microbiol.*, 2022, **132**, 1479–1488.
34. R. B. Cody, *Rapid Commun. Mass Spectrom.*, 2020, **34**, e8835.
35. R. B. Cody, C. R. McAlpin, C. R. Cox, K. R. Jensen and K. J. Voorhees, *Rapid Commun. Mass Spectrom.*, 2015, **29**, 2007–2012.
36. J. I. Zhang, N. Talaty, A. B. Costa, Y. Xia, W. A. Tao, R. Bell, J. H. Callahan and R. G. Cooks, *Int. J. Mass Spectrom.*, 2011, **301**, 37–44.
37. A. Sonesson, H. Moll, E. Jantzen and U. Zahringer, *FEMS Microbiol. Lett.*, 1993, **106**, 315–320.

38. N. Strittmatter, M. Rebec, E. A. Jones, O. Golf, A. Abdolrasouli, J. Balog, V. Behrends, K. A. Veselkov and Z. Takats, *Anal. Chem.*, 2014, **86**, 6555–6562.
39. N. Strittmatter, E. A. Jones, K. A. Veselkov, M. Rebec, J. G. Bundy and Z. Takats, *Chem. Commun.*, 2013, **49**, 6188–6190.
40. J. Balog, T. Szaniszlo, K. C. Schaefer, J. Denes, A. Lopata, L. Godorhazy, D. Szalay, L. Balogh, L. Sasi-Szabo, M. Toth and Z. Takats, *Anal. Chem.*, 2010, **82**, 7343–7350.
41. K. C. Schafer, J. Denes, K. Albrecht, T. Szaniszlo, J. Balog, R. Skoumal, M. Katona, M. Toth, L. Balogh and Z. Takats, *Angew. Chem. Int. Ed.*, 2009, **48**, 8240–8242.
42. E. E. Bardin, S. J. S. Cameron, A. Perdones-Montero, K. Hardiman, F. Bolt, E. W. F. W. Alton, A. Bush, J. C. Davies and Z. Takáts, *Sci. Rep.*, 2018, **8**, 10952.
43. N. Strittmatter, A. Lovrics, J. Sessler, J. S. McKenzie, Z. Bodai, M. L. Doria, N. Kucsma, G. Szakacs and Z. Takats, *Anal. Chem.*, 2016, **88**, 7507–7514.
44. S. J. S. Cameron, A. Perdones-Montero, L. Van Meulebroek, A. Burke, K. Alexander-Hardiman, D. Simon, R. Schaffer, J. Balog, T. Karancsi, T. Rickards, M. Rebec, S. Stead, L. Vanhaecke and Z. Takáts, *J. Am. Soc. Mass Spectrom.*, 2021, **32**, 1393–1401.
45. Z. Bodai, S. J. S. Cameron, F. Bolt, D. Simon, R. Schaffer, T. Karancsi, J. Balog, T. Rickards, A. Burke, K. Hardiman, J. Abda, M. Rebec and Z. Takats, *J. Am. Soc. Mass Spectrom.*, 2018, **29**(1), 26–33.
46. F. Bolt, S. J. Cameron, T. Karancsi, D. Simon, R. Schaffer, T. Rickards, K. Hardiman, A. Burke, Z. Bodai, A. Perdones-Montero, M. Rebec, J. Balog and Z. Takats, *Anal. Chem.*, 2016, **88**, 9419–9426.
47. O. Golf, N. Strittmatter, T. Karancsi, S. D. Pringle, A. V. Speller, A. Mroz, J. M. Kinross, N. Abbassi-Ghadi, E. A. Jones and Z. Takats, *Anal. Chem.*, 2015, **87**, 2527–2534.
48. S. J. S. Cameron, Z. Bodai, B. Temelkuran, A. Perdones-Montero, F. Bolt, A. Burke, K. Alexander-Hardiman, M. Salzet, I. Fournier, M. Rebec and Z. Takáts, *Sci. Rep.*, 2019, **9**(1), 3006.
49. A. M. Hamid, P. Wei, A. K. Jarmusch, V. Pirro and R. G. Cooks, *Int. J. Mass Spectrom.*, 2015, **378**, 288–293.
50. P. Wei, S. Bag, C. J. Pulliam, D. T. Snyder, R. M. Pielak and R. G. Cooks, *Anal. Methods*, 2016, **8**, 1770–1773.
51. C. J. Pulliam, P. Wei, D. T. Snyder, X. Wang, Z. Ouyang, R. M. Pielak and R. Graham Cooks, *Analyst*, 2016, **141**, 1633–1636.
52. O. E. Olajide, Y. Yi, J. Zheng and A. M. Hamid, *Int. J. Mass Spectrom.*, 2022, **478**, 116871.
53. Y. Liu, J. Zhang, H. Nie, C. Dong, Z. Li, Z. Zheng, Y. Bai, H. Liu and J. Zhao, *Anal. Chem.*, 2014, **86**, 7096–7102.
54. S. C. Povilaitis, A. Chakraborty, L. M. Kirkpatrick, R. D. Downey, S. B. Hauger and L. S. Eberlin, *Clin. Chem.*, 2022, **68**(11), 1459–1470.
55. J. Zhang, J. Rector, J. Q. Lin, J. H. Young, M. Sans, N. Katta, N. Giese, W. Yu, C. Nagi, J. Suliburk, J. Liu, A. Bensussan, R. J. DeHoog,

K. Y. Garza, B. Ludolph, A. G. Sorace, A. Syed, A. Zahedivash, T. E. Milner and L. S. Eberlin, *Sci. Transl. Med.*, 2017, **9**(406), eaan3968.

56. S. Hung, J. Zong-Han, C. Shu-Fen, S. Jentaie, W. Deng-Chyang, T. Sung-Pin, J. Shu-Huei, C. Chung-Yu and L. Po-Liang, *Molecules*, 2022, **27**(9), 2772.
57. C. W. Bacon, D. M. Hinton and T. R. Mitchell, *J. Appl. Microbiol.*, 2018, **125**, 867–875.
58. C. C. Hsu, M. S. ElNaggar, Y. Peng, J. Fang, L. M. Sanchez, S. J. Mascuch, K. A. Moller, E. K. Alazzeh, J. Pikula, R. A. Quinn, Y. Zeng, B. E. Wolfe, R. J. Dutton, L. Gerwick, L. Zhang, X. Liu, M. Mansson and P. C. Dorrestein, *Anal. Chem.*, 2013, **85**, 7014–7018.
59. V. P. Sica, H. A. Raja, T. El-Elimat and N. H. Oberlies, *RSC Adv.*, 2014, **4**, 63221–63227.
60. A. Tata, C. Perez, M. L. Campos, M. A. Bayfield, M. N. Eberlin and D. R. Ifa, *Anal. Chem.*, 2015, **87**, 12298–12305.
61. C. M. Rath, T. Alexandrov, S. K. Higginbottom, J. Song, M. E. Milla, M. A. Fischbach, J. L. Sonnenburg and P. C. Dorrestein, *Anal. Chem.*, 2012, **84**, 9259–9267.
62. J. Becker and C. Wittmann, *Angew. Chem., Int. Ed.*, 2015, **54**, 3328–3350.
63. C. Yan, F. Parmeggiani, E. A. Jones, E. Claude, S. A. Hussain, N. J. Turner, S. L. Flitsch and P. E. Barran, *J. Am. Chem. Soc.*, 2017, **139**, 1408–1411.
64. B. M. Ellis, P. K. Babele, J. C. May, C. H. Johnson, B. F. Pfleger, J. D. Young and J. A. McLean, *Proc. Natl. Acad. Sci. U. S. A.*, 2021, **118**(49), e2109633118.
65. G. F. Gowers, S. J. S. Cameron, A. Perdones-Montero, D. Bell, S. M. Chee, M. Kern, D. Tew, T. Ellis and Z. Takats, *ACS Synth. Biol.*, 2019, **8**, 2566–2575.
66. S. Joscelyn, M. Lynn, M. H. Victoria and J. B. Robert, *Rapid Commun. Mass Spectrom.*, 2021, **35**(S2), e8670.
67. J. Jo, A. Price-Whelan and L. E. P. Dietrich, *Nat. Rev. Microbiol.*, 2022, **20**, 593–607.
68. J. Schmitt and H.-C. Flemming, *Water Sci. Technol.*, 1999, **39**, 77–82.
69. J. Zhang, J. Brown, D. J. Scurr, A. Bullen, K. MacLellan-Gibson, P. Williams, M. R. Alexander, K. R. Hardie, I. S. Gilmore and P. D. Rakowska, *Anal. Chem.*, 2020, **92**, 9008–9015.
70. S. N. Nguyen, A. V. Liyu, R. K. Chu, C. R. Anderton and J. Laskin, *Anal. Chem.*, 2017, **89**, 1131–1137.
71. S. Jakka Ravindran, R. Kumar, A. Srimany, L. Philip and T. Pradeep, *Anal. Chem.*, 2018, **90**, 988–997.
72. S. N. Dean, C. Walsh, H. Goodman and M. L. van Hoek, *Biofouling*, 2015, **31**, 151–161.
73. J. I. Brauer, Z. Makama, V. Bonifay, E. Aydin, E. D. Kaufman, I. B. Beech and J. Sunner, *Biointerphases*, 2015, **10**, 019003.
74. H. Jana, C. M. Robin, B. S. Iain and J. C. Helen, *J. Am. Soc. Mass Spectrom.*, 2021, **32**(6), 1345–1351.
75. S. Santajit and N. Indrawattana, *BioMed Res. Int.*, 2016, **2016**, 2475067.

CHAPTER 5

Nucleic Acid Amplification Coupled with Mass Spectrometry Detection for Microbial Identification

AIZHONG HU,[a] JUN HU,[b] HONG XIE[c] AND QINGWEI MA*[c]

[a] VacTx, 13 Melody Lane, Harriman, NY10926, USA; [b] Ningbo University, China; [c] Bioyong Co. Ltd, China
*Email: maqw@clintof.com

5.1 Introduction

Infectious diseases caused by microbes, such as bacteria, viruses, and fungi, remain a major cause of morbidity and mortality worldwide. The COVID-19 pandemic reminds us of the microbial infection challenges we are facing. Fortunately, advancements in molecular biology allow us to understand the microbial genome sequence, gene structure, and gene function. Studies on microbial genomes using high-throughput sequencing technologies, such as next generation sequencing (NGS), have generated a significant amount of information on microbial genomes. Extensive investigations, such as single nucleotide polymorphism (SNP) genotyping and multi-locus sequence typing (MLST), have identified specific genetic features of individual microbes. The genetic profiles of the various microbes provide us with the foundation for identifying target pathogens.

Antimicrobial agents are the key to controlling microbial infection. However, because of the massive and often untimely use of antimicrobial

New Developments in Mass Spectrometry No. 13
Detection and Analysis of Microorganisms by Mass Spectrometry
Edited by Liang Qiao and Jia Yi

Published by the Royal Society of Chemistry, www.rsc.org

agents, pathogenic microbes have developed resistance mechanisms against most classes of antimicrobial agents currently available. The possible transmission of genes encoding mechanisms of resistance between different microbial species has led to the emergence of drug-resistant microbes particularly in hospitals, which leads to increasing difficulties in therapeutic management.

Mass spectrometry (MS) has brought us into a new era of clinical microbiology. The promising new broad-based MS technology can rapidly identify many microorganisms, including microbes that are difficult to culture and newly emerging strains. MS techniques can also be used as epidemiological tools for following disease transmission. The benefit of rapid reporting of isolates and/or identification of resistance of microorganisms can potentially impact patient outcome, improve the use of antibiotics, and reduce the emergence of drug-resistant bacteria and mortality rates.

In clinical microbiology laboratories, traditional culture-based methods take days to generate results. While improvements in rapid identification of infectious agents have been made, there are needs to provide test results as rapidly as possible while maintaining accuracy and quality. With advancements in microbiological nucleic acid amplification (NAA) and analytical MS technologies for large biomolecules, it is proven feasible to couple the two technologies and take advantage of the high amplification power of the former and the quick and high mass-resolving power of the latter, and thus generate high-quality test results rapidly and improve patient care.[1]

5.2 NAA

Since the first NAA technique, PCR, was developed by Mullis,[2–4] NAA technology has advanced significantly. Besides PCR, a variety of alternative NAA methods, such as ligation mediated amplification,[5] loop mediated isothermal amplification (LAMP),[6] nucleic acid sequence-based amplification (NASBA),[7] rolling circle amplification (RCA),[8] and transcription-based amplification (TBA),[9] have been established. The NAA technologies have revolutionized the diagnostic and research fields, allowing researchers and clinical technicians to carry out a wide range of clinical and laboratory applications, such as cloning, genotyping, sequencing, mutagenesis, forensics, and detection of pathogens to diagnose infectious diseases.

The detection of amplified nucleic acid products is an essential step. The amplified products can be evaluated by electrophoresis, enzyme-linked immunosorbent assay (ELISA) like a microplate procedure, or real-time monitoring for quantitative or qualitative analysis. To quantitatively determine the amplified nucleic acid products, a fluorescent dsDNA dye, a fluorescent reporter probe, or a molecular beacon can be included in the NAA reaction. Fluorescence is detected and measured using a real-time PCR machine. The geometric increase of the product, corresponding to an exponential increase of the fluorescence intensity, is used to determine the range of quantification/threshold cycle (Ct) in each reaction. Although quantitative PCR has advantages over conventional PCR, it has several limitations, such as its costly instrument, expensive labelled probe, and limited level of multiplexing. MS has the ability to overcome the aforementioned limitations.

5.3 MS

Since its initial organic chemistry applications, MS technology has evolved immensely and become one of the most effective tools for the analysis of large biomolecules, including nucleic acids,[10,11] due to the development of "soft" ionization methods.[12–14] There are two main MS technologies for microbial identification: PCR matrix-assisted laser desorption/ionization time-of-flight MS (PCR MALDI-TOF MS) and PCR electrospray ionization MS (PCR ESI MS). The microbial identification can be based on the detection of the ion profile of biomacromolecules including proteins (peptides) and nucleic acids. In this chapter, we focus on the nucleic acid method for microbial identification. The principle of MS detection is illustrated in Figure 5.1.

5.4 NAA Coupled with MS Detection for Microbial Identification

5.4.1 Workflow

The workflow of NAA MS includes three key procedures: sample preparation, NAA (PCR), and MS detection. They are illustrated in Figure 5.2.

5.4.1.1 Sample Preparation

Theoretically, any sample can be used for NAA MS. The commonly used samples for microbial identification include whole blood specimens, sputum, saliva, faeces, and urine. Sample collection, transportation, and storage are critical preanalytical steps, which can affect the validity and usefulness of data for downstream analyses. In order to obtain the most clinically relevant information, it is imperative to ensure that there are minimal sample loss, high consistency in collection, and preservation of sample integrity. It is also especially important to prevent potential NA contamination while handling samples.

In principle, low levels of microbes can be detected quantitatively in complex samples, owing to the efficient and specific amplification of targeted regions of microbial genomes by NAA. However, false-positive amplicons can interfere with the analysis, as even a single copy of a microbial genome can potentially generate notable signals. Moreover, the detection of microbial DNA cannot differentiate between dead and alive bacterial cells. Purification of NA from microbes is required for efficient amplification. Although it is feasible with crude mixtures of biological samples, in practice the efficiency and the reproducibility of specific amplification

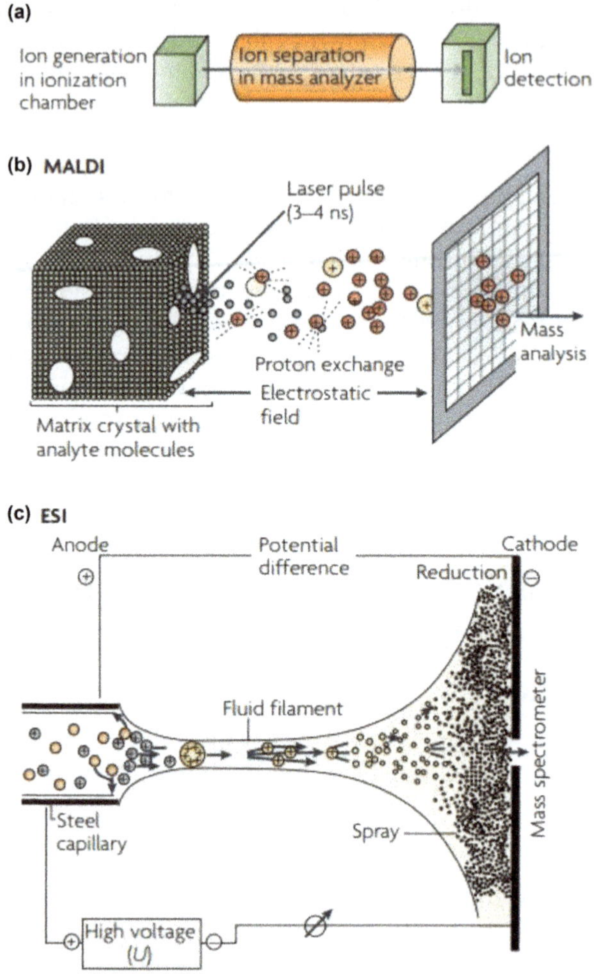

Figure 5.1 Principle of MS for NA detection. (a) Three parts of MS: (1) an ion source, to ionize and transfer analyte ions into the gas phase; (2) a mass analyser, to separate ions by their mass-to-charge ratio (m/z); and (3) a detection device, to monitor the quantities of ions. (b) MALDI ion source: Samples are prepared by embedding analyte molecules in a laser absorbing organic matrix, which desorb laser energy. A nano second UV laser pulse irradiates the sample, resulting in ablation of a small volume of the matrix and desorption of the embedded analyte molecules. Analyte molecules are then ionized, which leads to the formation of predominantly singly charged analyte ions. (c) ESI ion source: samples are dissolved in a polar solvent, which is guided through a fine capillary. On the capillary, 2–4 kV voltage is applied, which results in the emission of an aerosol of charged droplets of the sample. Using heated gas, the spray is directed through a series of sections that have successively decreased pressures, resulting in the formation of gas-phase analyte ions. For both procedures, TOF mass analysers are generally employed. Reproduced from ref. 15 with permission from Springer Nature, Copyright 2007.

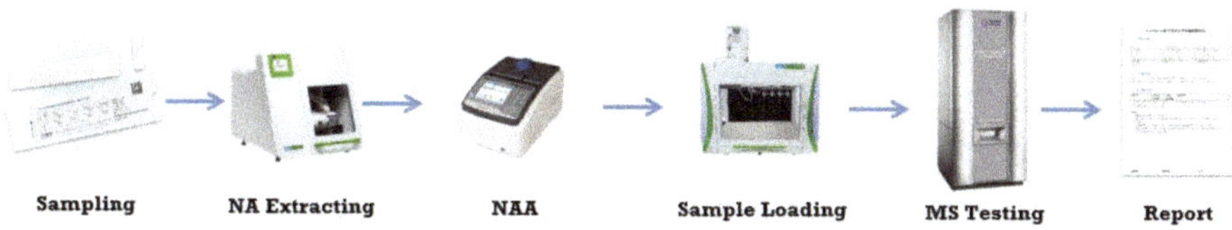

Figure 5.2 Workflow of NAA MS.

from crude mixtures are low. Therefore, NA purification from crude samples is usually included in sample preparation. Purification procedures for nucleic acids are well established. High-throughput NA purification procedures are readily accessible and are commercially available for MS detection and analysis.

Another important consideration for sample preparation is MS detection. The sample should be optimized for MS. Each MS method has certain factors that must be considered for it to be successful, such as the volume, concentration, sample phase, and composition of the analyte solution. One of the most important considerations in sample preparation is to determine what condition the sample must be in for MS detection and analysis to be successful. In some cases, the analyte itself must be purified before it enters the ion source. In other situations, the MALDI matrix, or the solution surrounding the analyte, is the most important factor to consider and adjust.

5.4.1.2 NAA

As mentioned earlier, many NAA technologies can be employed to amplify NA for subsequent MS detection and analysis. One of the most popular approaches is PCR. For many microbes, genetic information is available, and broad-range primers can be designed, which are species-specific or strain-specific and have other characteristic features to target identifiable genes for key pathogenic features, such as antibacterial, antiviral, and antifungal resistance.

In general, MS detection of RNA is more efficient than that of DNA. RNA is more stable than DNA during ionization, owing to the additional 2′-hydroxyl group that stabilizes the glycosidic bond, resulting in substantially reduced depurination and fragmentation of the polymer. Two stages of primer-specific amplification, PCR and RNA transcription followed by base-specific cleavage called MassCLEAVE, were developed and applied for detection of RNA derived from targeted microbial genomes by Honisch and her colleagues[15] (Figure 5.3). The assay showed high analytical sensitivity and specificity. The concordance observed between this method and Sanger sequencing-based MLST was 98.9%.

Another technique for NAA is PCR coupled with primer extension (Figure 5.4). This format can be used for the analysis of deletion, insertion, and point mutations; short tandem repeats; and SNPs.

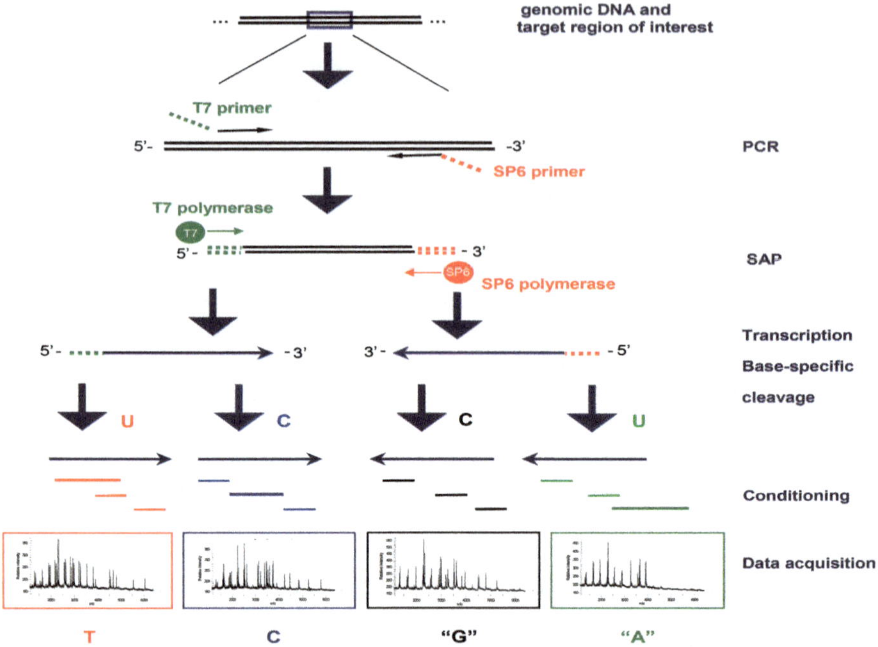

Figure 5.3 MassCLEAVE technology: PCR and transcription biochemistry. The microbial genome of interest is generated by PCR with primers tagged with a 5′-tail containing a promoter site for T7 or SP6 polymerase. The fragment is amplified by PCR. Transcription of both strands of the PCR amplicon is performed, so that both sequences of the genome are used for further processing. After transcription, the RNA strands are cleaved specifically using RNases; 2′-deoxycytidine-5′-triphosphate (dCTP) or 2′-deoxyuridine-5′-triphosphate (dUTP) or 2′-deoxythymidine-5′-triphosphate (dTTP) that resists hydrolysis and RNase A are used to generate efficiently C- or U- or T-specific cleavage products. MS analysis is carried out. Single-stranded RNA cleavage products (4 to 30 nucleotides long) are simultaneously detected in the range of 1000 to 9000 *m*/*z*. Reference sequences that can be uploaded using conventional formats are digested *in silico*, and the resulting masses of the products are computed. The experimentally generated mass patterns are matched with the *in silico* patterns to identify the microbes under investigation.

5.4.1.3 MS Detection and Analysis

After the purification and amplification steps, the resulting NA samples are analysed by MS. The entire process can be fully automated with regard to sample tracking, instrument control and robotics, data processing and analysis, and comparison with the curated database.

A key element of the MS system is the curated database of genomic information that associates base composition with the identity of microbes. Despite the enormous class diversity of organisms, the database, containing carefully selected and curated genetic sequences, allows the interrogation to produce relevant and accurate results. The database is populated based on

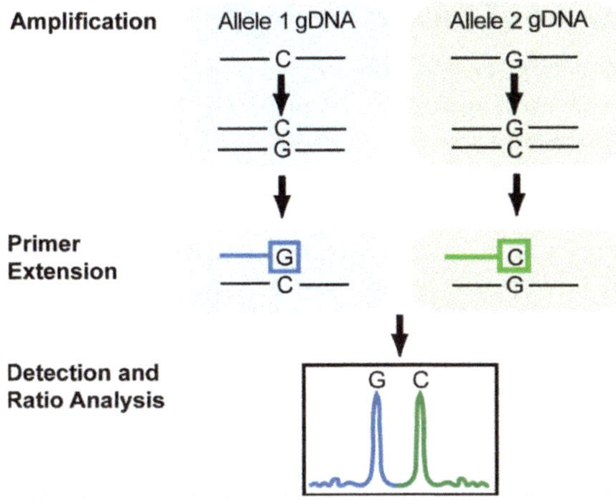

Figure 5.4 Primer single-base extension assay. The target fragment is amplified by PCR. The amplified PCR fragment serves as a template for primer extension reaction, which involves annealing of an extension primer adjacent to the mutated site. The reaction contains Thermo Sequenase and a deoxyribonucleotide dNTP/dideoxynucleotide (ddNTP) mixture. Depending on the chosen nucleotide mixture, two products of single-base extension with distinct masses are generated and detected by MS.

the needs of infectious disease surveillance, diagnostic testing, forensic sciences, and biological research. The database should be regularly updated with information from newly identified data. Novel base counts pass through the curation process to determine whether they are representative of a new microbe or strain. The design of primers and the determination of base compositions for individual microorganisms to be entered into the database are performed by scientists using sequence alignments and extensive primer scoring over many sequences for target microorganisms. Once primers are validated in the laboratory, they are included in the database. With advancements in bioinformatics, continuous improvement of the database is feasible and it is possible to instantly update the microbe gene information, which ensures the identification of different microbes and their relevant features in a timely fashion.

At this moment, there are two major commercially available MS platforms for NA detection and analysis: Clin-TOF (Bioyong) and MassARRAY (Agena Bioscience). Both have been used in various genetics and epigenetics research, and in clinical diagnosis of genetic disorders, including the quantitative analysis of gene expression, analysis of gene copy number variation (CNV), SNP genotyping, MLST, and DNA methylation identification. For microbe detection, Bioyong provides complete tools for both peptide-based methods and nucleic acid-based assays in the identification of a variety of pathogens, including tuberculosis drug-resistant strains and 8 different respiratory pathogens (www.bioyong.com).

5.4.2 Advantages of the Technology

Coupling of NAA with MS takes advantage of the high amplification power of the former and the high mass-resolving power of the latter. NAA MS is the most promising technology for microbial identification. Table 5.1 summarizes the advantages and disadvantages among the commonly used genetic test technologies.

NAA MS requires a minimal amount of sample to start with. It is not necessary to have positive cultures of a microbe. The assay consumes a small quantity of sample, is simple, can be easily used, and has low running costs and low maintenance needs. The test is automated, shows high throughput, and is highly versatile. It is feasible to carry out simultaneous detection of multiple mutations with rapid results. NAA MS exhibits robust analytical sensitivity, specificity, accuracy, reproducibility, and quantification. The assay has the potential to change the practice of clinical microbiology. Based on the aforementioned advantages, this technology has already replaced many of the biochemical tests used for bacterial identification in routine laboratories,[16] such as catalase, oxidase, identification card or API gallery, and latex agglutination tests. This method has proven to be reliable and safe for the identification of clinically relevant bacteria,[17] such as Enterobacteriaceae, Staphylococci, and Streptococci.

Table 5.1 Comparison between NAA MS and other genetic tests.

Platform	Advantage(s)	Disadvantage(s)
Microarray	High throughput	Expensive, easy contamination, missing SNP
1st generation sequencing	Relative quantitation	Low sensitivity, low throughput, detects only one sequence, expensive
Next generation sequencing	Detects multiple loci, tests multiple samples, high throughput	Time consuming, requires highly trained personnel, complicated bioinformatic analyses
Real-time PCR	Rapid, simple, low cost, quantitation	Needs a fluorescent label, channel restriction, only 2–3 loci each time
Digital PCR	Rapid, high sensitivity, initial sample quantitation	Only 2 channels, only 1–2 loci each time
NAA MS	High sensitivity (0.1%), relative quantitation, tests multiple loci (5 to 200 loci), simple operation, rapid, easy bioinformatic analyses, and instant reporting	Requires NAA, determines only the known locus

5.4.3 Limitations of the Technology

Despite the advantages of MS as a rapid technique for identifying microorganisms, the current state of the technology will not completely replace culture-based testing in the near future. Clearly, MS reduces a portion of the laboratory's workload; however, a strong need remains for culture-based testing to assess antimicrobial susceptibility. Moreover, culture-based testing may still be advantageous for identifying mutated strains that may not be detected by this molecular technique. Currently, the cost of implementing this emerging technology and the familiarity and track record of culture-based testing create barriers for laboratories to clear. As the technique advances and an affordable benchtop instrument is developed, the system will have the cost–benefit advantage and have the potential of wider applications.

5.4.4 Applications

NAA MS has become an invaluable tool for identifying different types of microbes, including bacteria, viruses, and fungi. Genetic information about most of the important or relevant microbes is well established. It is feasible to employ the technology to carry out microbial detection, identification, genotyping, and quantitation, and antimicrobial resistance (AMR) determination.

5.4.4.1 Bacterial Identification

NAA MS was employed to identify bacteria at an early stage of the technology development. PCR MALDI-TOF was accurate in detecting and monitoring Legionella species, the causative agents of Legionnaires' and other diseases.[18] Rapid and accurate detection of ligation products generated in ligase chain reactions (LCR) by MALDI-TOF was demonstrated.[19] Since then, the technology has been widely implemented in bacterial identification.

Using *Neisseria meningitidis* as a model pathogen, Honisch and colleagues developed a base-specific cleavage MALDI-TOF MS test called MassCLEAVE for microbial typing.[14] PCR-amplified sequences are subjected to *in vitro* transcription and base-specific RNA cleavage. MS signal patterns of the resulting cleavage products are compared with reference sequences in the system database, generating MLST profiles. One hundred *N. meningitidis* samples were typed by comparing MS fingerprints of the standard MLST loci to reference sequences available in the public MLST database. Compared with traditional microbial typing, MS provides automated high-throughput analysis with 98% correct identification.

Streptococcus pneumoniae is an important human pathogen that causes respiratory tract infections, bacteraemia, and meningitis. *S. pneumoniae* is a diverse species, with over 90 immunologically distinct serotypes. The changing prevalence in serotype carriage following vaccination and the emergence of antibiotic resistance highlight the need for increased molecular epidemiology and surveillance of *S. pneumoniae*. MLST characterizes bacterial isolates on the basis of the sequences of designated housekeeping genes, and in the case of *S. pneumoniae*, the sequences of seven housekeeping genes are used. The 7 housekeeping genes are aroE, gdh, gki, recP, spi, xpt, and ddl, and they are amplified by PCR. The amplified PCR products are subjected to *in vitro* transcription and base-specific RNA cleavage. Analysis of cleavage fragments by MALDI-TOF generates a mass signal pattern that serves as a fingerprint of the sample isolate. Spectral data are then compared with data in a reference library created by *in silico* digestion of known allele types to identify the unknown sample and any possible single-base deviation. This PCR MALDI-TOF procedure was used to analyse and compare the MLST profile for *S. pneumoniae*.[20] A 99% concordance rate between MS and traditional sequence typing was achieved.

Early and accurate aetiological determination provides effective guidance about precision patient care. Ventilator associated pneumonia (VAP) is a very common nosocomial infection in patients who require mechanical ventilation in intensive care units. PCR ESI was employed for diagnosis and surveillance of bacteria associated with VAP.[21] The assay detected 51 bacterial species. The positive rate of the PCR ESI was 88%, while that of routine clinical culture was 38%, demonstrating enhanced assay sensitivity. Further, the PCR ESI MS identified pathogens prior to the occurrence of clinical symptoms. The assay provided correct and early prediction of potential bacterial infections, guiding timely and appropriate treatment.

Sepsis is a life-threatening condition that can lead to tissue damage, organ failure, and death. Bacterial, viral, and fungal infections can trigger sepsis. Identification of pathogens that are associated with sepsis is critical for patient care. PCR ESI MS has the potential of detecting hundreds of different microorganisms in whole blood.[22] Based on a study carried out in 16 teaching hospitals in the US and Europe, it has been shown that the assay demonstrates high specificities and sensitivities and has the potential for further development to be used for detection of bloodstream microorganisms in sepsis.[22]

Infective endocarditis is a serious infection that is associated with patients with underlying chronic rheumatic heart disease, prosthetic heart valves, injection drug use, and intravascular catheters. Accurate and prompt detection of infecting microbes is essential to ensure appropriate management. Diagnosis requires evidence of bloodstream infection. Determination of infecting microbes guides antibiotic treatment. A challenge in the diagnosis of the disease is the occurrence of negative blood and valve cultures, due to the use of antibiotics prior to cultures being collected or the presence of fastidious organisms. In such circumstances, pathogen detection by molecular methods might be helpful. A PCR ESI MS assay was developed for direct detection of pathogens and determination of AMR in heart valves in patients with infective endocarditis.[23] The assay decreases the turnaround time to 12 to 16 h compared to 24 to 48 h for standard tissue processing and culture (provided that it is performed on fresh tissue). Beyond organism detection and classification, select AMR is also defined. Another benefit of this method is the ability to detect more than one organism simultaneously in a sample. The assay detects not only common infectious endocarditis pathogens (*Staphylococcus aureus*, *Staphylococcus epidermidis*, and *Enterococcus faecalis*), but also not-so-common pathogens (*Tropheryma whipplei*, *Granulicatella adiacens*, *Gemella morbillorum*, *Cryptosporidium hominis*, and *Mycobacterium fortuitum*), demonstrating the power of the assay and positive impact on patient outcomes.

Gonorrhoea is one of the most common sexually transmitted bacterial infections and is caused by *Neisseria gonorrhoeae*. Although there are effective treatments for the disease, there are emerging resistant strains to ceftriaxone and azithromycin, the first-line treatment suggested by WHO, which has gradually become a global health concern. It is critical to have an effective detection system to monitor the epidemiological trend of the resistant strains. To achieve this goal, a method based on multiplex PCR coupled with MALDI-TOF was developed.[24] 24 mutations located in nine genes, penA, 16S rRNA, 23S rRNA, rpsE, porB, gyrA, parC, mtrR, and ponA, were targeted. The 24 mutations were closely associated with resistance to antimicrobial agents. The performance of the MS method was assessed by testing 454 *N. gonorrhoeae* isolates with known minimal inhibitory concentrations (MICs) of six antibiotics, eight non-gonococcal *Neisseria* strains, 214 clinical samples, and three plasmids with a confirmed mosaic penA allele. A total of 446 isolates with known resistance profiles and 178 positive clinical samples were also employed to evaluate the performance of the assay. The results demonstrated that PCR MALDI-TOF accurately detected all 24 resistance-related mutation sites with a limit of detection (LOD) of 10 copies (except for one mutation with 100 copies) per reaction and a specificity of 100%. The method displayed significantly high-throughput potential and one individual could measure 760 samples within 8 hours, providing an effective large-scale epidemiological gonococcal resistant strain surveillance tool.

Multidrug-resistant tuberculosis (MDR-TB) is an important public health burden. To control the spread of drug-resistant strains, it is necessary to have a tool for detection of resistance to different anti-TB drugs. PCR ESI MS was employed to detect resistant genes.[25] Specific genomic regions associated with drug resistance were amplified using multi-locus PCR and the amplicons were analysed by ESI MS. The test detected a number of mutations in genes associated with resistance to isoniazid (INH) (katG, inhA, and its promoter, ahpC promoter), rifampin (RIF) (rpoB), ethambutol (embB), or fluoroquinolones (gyrA). The assay had sensitivities of 100% and 92% for detecting INH and RIF resistance, respectively, and specificities to both RIF and INH exceeding 95%, indicating that the assay is a good tool for the detection of genetic markers of resistance in *M. tuberculosis* isolates.

5.4.4.2 Viral Identification

Qiao and colleagues developed a method based on three-way junction induced exponential RCA (3WJ-eRCA) coupled with MALDI-TOF for SARS-CoV-2 detection.[26] The authors developed an innovative technology to amplify the viral nucleic acids (Figure 5.5). The formation of a three-way junction structure is the key step in the generation of amplification products for MALDI-TOF

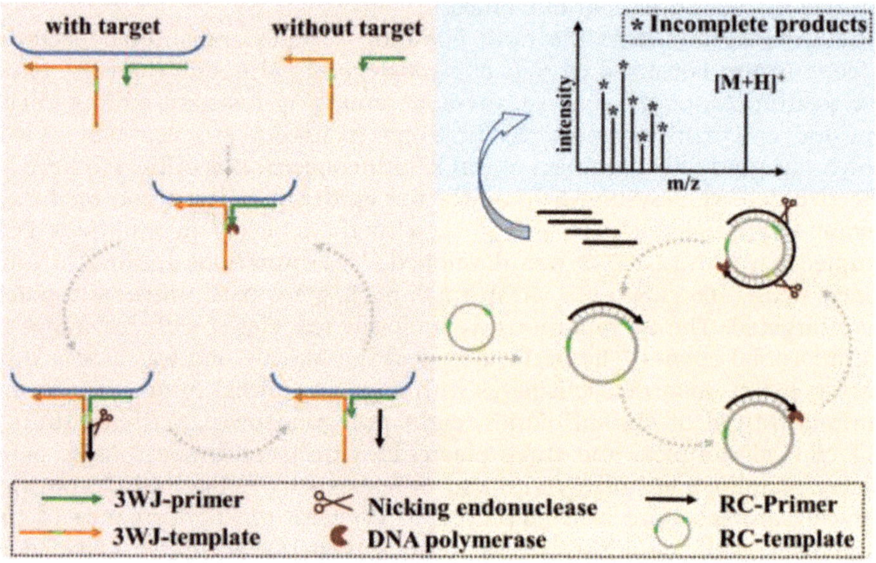

Figure 5.5 The 3WJ-eRCA coupled with MALDI-TOF for SARS-CoV-2 detection. The 3WJ-template is designed to contain a rolling circle primer (RC-primer) sequence at the 5′-end and a complementary sequence to the target sequence of SARS-CoV-2 at the 3′-end, which are separated by the recognition site of the nicking endonuclease. Part of the 3WJ-template sequence is complementary to a 7-base-long sequence at the 3′-end of the 3WJ-primer, which hybridizes the 3WJ-template. The 5′-end of the 3WJ-primer targets the SARS-CoV-2 RNA. When there is a SARS-CoV-2 target, the 3WJ-primer and the 3WJ-template hybridize to viral RNA, respectively; meanwhile, the 3WJ-primer and the 3WJ-template also hybridize through the complementary 7 nucleotides. These interactions allow the formation of a stable viral RNA, 3WJ-primer, and 3WJ-template complex or a three-way junction structure, which allows the generation of a large number of RC-primers through 3WJ-primer extension by Vent (exo-) DNA polymerase, following cleavage by nicking endonuclease. RC-primers are used in the subsequent RCA reaction, where the rolling circle template (RC-template) is composed of a hybridization sequence to RC-primers and nicking sites. Once the RC-primer hybridizes with the RC-template, a cascade of linear RCA reactions occurs, generating a long repeat complementary sequence of the RC-template. The amplified products are detected and analysed by MALDI-TOF MS.

analysis. In the absence of the target RNA, there is no junction structure formation and no amino acid product.

3WJ-eRCA MALDI-TOF was employed to detect SARS-CoV-2. Two genes, N and orf1ab, were targeted. Isothermal amplification of nucleic acids was employed to magnify the genes and the entire amplification process occurred within 30 min. The sensitivity of both N gene and orf1ab reached 10 amol/10 μL. The assays were highly reproducible under many different working conditions. A number of coronaviruses were employed to evaluate the specificity of the assay and the results demonstrate that the N and orf1ab assays are highly specific and detect only SARS-CoV-2.

Saliva is an attractive sample source due to its minimal discomfort and self-collectability. A RT-PCR MALDI-TOF-based method for detection of SARS-CoV-2 in saliva specimens was developed.[27] Five viral targets, three in the N gene (N1, N2, and N3) and two in the ORF1ab gene (ORF1 and Orf1ab), were used for detection. This method demonstrates comparable performance to real-time RT-PCR and is a sensitive and reliable platform for SARS-CoV-2 detection in saliva.

Cytomegalovirus (CMV) infection is associated with significant morbidity and mortality in transplant recipients. Ganciclovir (GCV) is the medication of choice for treatment and prophylaxis of CMV infection after transplantation. Drug resistance develops through a single point mutation in the viral phosphotransferase (UL97) or polymerase (UL54). A primer extension reaction MALDI-TOF method was developed to detect the mutation.[28] The results demonstrate that the assay detects the A594V mutation, encoding GCV resistance, ten days earlier compared to Sanger sequencing and more than 30 days prior to an increase in viral load. The increased sensitivity and rapid turnaround time combined with easy handling and moderate costs indicate that this procedure could make a major contribution to improve transplantation outcomes.

Chronic hepatitis B infection is the most common cause of liver cirrhosis and hepatocellular carcinoma. Antiviral therapy is essential in managing chronic hepatitis B patients. Unfortunately, hepatitis B virus (HBV) variants frequently develop with long-term antiviral therapy and some of these variants exhibit drug resistance, resulting in treatment failure. Timely and accurate identification of HBV drug-resistant variants is critical for treatment of chronic hepatitis B patients. A PCR MALDI TOF assay was developed to detect and investigate 60 HBV variants.[29] A duplex PCR method was used to amplify the region of the reverse transcriptase gene, the precore promoter, and the basal core promoter of the HBV genome. Four multiplex primer extension reactions were used to generate extended nucleic acids, which were detected and analysed by MS. The assay ensures sensitive and accurate detection of HBV variants, providing guidance for patient care.

5.4.4.3 Fungal Identification

Due to the growing immunocompromised population, the incidence of invasive fungal infections is on the rise. Rapid and accurate identification of fungal pathogens is critical for appropriate antifungal drug choice. PCR ESI MS was employed to detect and identify fungi directly from bronchoalveolar lavage (BAL) specimens.[30] The assay was designed to detect and identify over 200 pathogenic fungi. 15 primers targeted against widely conserved fungal genes in the small and large ribosomal subunits, as well as against more narrowly conserved ribosomal genes found in fungal mitochondria and the beta-tubulin gene. After PCR amplification and product purification, the samples were then loaded onto the ESI MS module for detection and analyses.

To evaluate the assay, a core set of fungi was selected for detailed method characterization. The core fungi are medically important and include *Aspergillus fumigatus, Candida dubliniensis*, *Candida glabrata*, *Cryptococcus neoformans*, and a *Mucor* sp. The LODs for the five core fungal species were established. The LODs of these fungal species ranged from less than 1 CFU per mL to 320 CFU per mL with a mean LOD of 19.3 CFU per mL and a median LOD of 3.75 CFU per mL. The specificity of the assay was assessed by challenging the assay with high concentrations of non-target microorganisms: *Haemophilus influenza*, *Chlamydophila pneumonia*, *Legionella pneumophila*, *Pseudomonas aeruginosa*, *Staphylococcus aureus*, *Streptococcus pyogenes*, *Klebsiella pneumonia*, *Moraxella catarrhalis*, *Mycoplasma pneumonia*, influenza viruses, parainfluenza viruses, herpes simplex viruses (HSVs), adenoviruses, respiratory syncytial viruses, coronaviruses, CMVs, *Monocillium humicola*, *Beauveria bassiana*, *Sporobolomyces roseus*, *Oidiodendron echinulatum*, and *Neurospora sitophila*. No cross-reactivity was observed. The overall sensitivity and specificity of the PCR ESI MS were found to be 90.9% and 82.3%, respectively, indicating that PCR ESI MS can be rapid and accurate diagnosis of invasive fungal infections in patients at risk of developing fungal diseases.

The assay and system were also evaluated with regard to their clinical performance. 262 BAL samples were included in this study. The overall sensitivity, specificity, positive predictive value (PPV), negative predictive value (NPV), and positive likelihood ratio + (LR+) for broad levels of fungal detection by PCR ESI MS were 73.0%, 59.5%, 48.2%, 81.1%, and 1.8, respectively. *Aspergillus* spp. represent some of the leading causative agents of opportunistic fungal lung infections among immunocompromised patients. An *Aspergillus* sp. was found to be a frequent colonizing pathogen in the airway in these populations. The development of invasive pulmonary aspergillosis depends on the level of the host immune deficiency. Approximately 57% (20/35) of the isolated *Aspergillus* cases were judged to be related to pulmonary infections. In the 18 cases of probable pulmonary aspergillosis detected by PCR ESI MS, 83% (15/18) were concordant with culture, and half of these cases had increased galactomannan values. The PCR ESI MS system is rapid (6–8 hours) in comparison with culture-based methods, and can be performed directly on uncultured BAL specimens, enabling earlier targeted therapy. The ease and rapidity of fungal diagnosis like PCR ESI MS will have a positive effect on patient outcomes and hospital costs.

Invasive fungal infections are life-threatening complications in patients with hematological malignancies, and adequate early treatment has been shown to be critical for a successful outcome. Rapid identification of non-aspergillus molds, especially molds from the order Mucorales causing mucormycosis, is of particular concern because of their intrinsic resistance to antifungal agents and the aggressive course of the infections. PCR ESI MS was employed to identify molds in BAL in hematological patients with a new pulmonary infiltrate.[31] Among 27 patients, PCR ESI MS was the only positive microbiological test in patients with proven invasive mold infection ($n = 2$) and correctly identified all cases of probable invasive pulmonary aspergillosis ($n = 5$). In patients with a possible invasive mold infection ($n = 5$), PCR ESI MS was positive in three patients. Mucorales was identified with PCR ESI MS in four patients that were all culture negative. The PCR ESI MS results had a clinical impact on antifungal therapy in 12 (44%) of the patients: modification of treatment in 6 (22%) patients and discontinuation in 6 (22%) patients. This study provides proof of concept that routine use of PCR ESI MS for molds in BAL in immunocompromised patients is sensitive and fast and has an impact on clinical decision-making.

5.4.4.4 Mixed Microbial Identification

One of the salient advantages of NAA MS is high-throughput analysis of samples. The system has the potential to process and analyse a large number of samples within one day. In clinical practice, a single infectious disease can be caused by multiple pathogens. Pneumonia may be one of the best examples. There are hundreds of microorganisms that cause pneumonia. The COVID-19 pandemic demonstrates the continuous evolvements of the viruses. The viruses can create new variants in a very short time. To control infections, timely and effective detection and identification of pathogens are essential. NAA MS has the ability to establish a powerful tool for mixed microbial infection identification.

Due to its special physical feature and its exposure to the environment directly, the human respiratory system is and will continue to be affected by microbial infections. Respiratory infections continue to pose threat to public health and is among the top five causes of death worldwide. Many different pathogens, bacteria, viruses, and fungi, are associated with respiratory infections. Different microorganisms may cause similar clinical symptoms; however, treatments for different pathogens vary significantly. Thus, it is important to accurately and rapidly detect respiratory pathogens to provide guidance for precision patient care.

A 24-plex PCR (RT-PCR) MADLI-TOF assay was developed to simultaneously detect and identify 21 common respiratory viruses,[32] including 3 influenza A viruses, 1 influenza B virus, 4 parainfluenza viruses, 1 adenovirus, 4 human coronaviruses, 2 respiratory syncytial viruses, 2 human metapneumoviruses, 2 human bocaviruses, 1 human rhinovirus, and 1 human enterovirus. 102 confirmed positive common respiratory virus samples were employed to evaluate the assay and the test results showed that the assay accurately identified the 21 common respiratory viruses. 332 nasal and throat swabs collected from subjects with suspected viral acute respiratory tract infections were evaluated by the 24-plex PCR (RT-PCR) MADLI-TOF assay. Conventional PCR (RT-PCR) detected 89.5% (297/332) positive samples, while MS detected 92.5% (307/332) positive samples, indicating the improved sensitivity of MS detection. In the era of the coronavirus pandemic, this assay can be extended to include other coronaviruses, such as SARS-Cov, MERS-Cov, SARS-Cov-2, and other potential animal coronaviruses, and can be used in large-scale epidemiological studies of common respiratory virus infections. The results will provide a valuable tool for us to monitor and control the respiratory virus infections.

Meningitis and pneumonia continue to present an enormous public health burden. Among the causative organisms of the illnesses, bacteria are the most common causes of serious disease and deaths. It is critical to accurately and rapidly identify these agents in order to provide effective and precision patient care. A 12-plex PCR MALDI-TOF assay was developed to simultaneously detect and identify 11 key bacterial pathogens related to meningitis and pneumonia.[33] These 11 bacteria are *S. pneumoniae*, *H. influenzae*, *N. meningitidis*, *K. pneumoniae*, *Acinetobacter baumannii*, *P. aeruginosa*, *Staphylococcus aureus*, *M. catarrhalis*, *L. pneumophila*, *M. pneumoniae*, and *Bordetella pertussis*. The assay was highly specific and identified the expected bacteria without cross-reactivity with other pathogens. The sensitivity of the method was 10 copies per reaction. The clinical effectiveness of this assay was evaluated by use of 204 nasopharyngeal swabs from hospitalized children with suspected pneumonia. In total, 81.9% (167/204) of the samples were positive for at least one of the 11 target pathogens. Among the 167 bacteria-positive samples, the rate of multiple infections was 55.7% (93/167), and the most frequent combination was *S. pneumoniae* with *H. influenzae*, representing 46.2% (43/93) two-pathogen mixed infections. Bacterial coinfection is associated with severe disease and high mortality rates, so timely discovery of mixed aetiologies is critical for antimicrobial treatment and clinical outcomes. The MS method is a sensitive and specific molecular detection technique in a multiplex format and with high sample throughput, and will be a powerful tool for pathogen screening and antibiotic selection at an early stage of meningitis and pneumonia.

The role of microbial colonization in disease is complex. Since there is the potential link between the colonization and modification of the microbiome to subsequent infections, a PCR ESI-TOF assay was developed to examine the bacterial and fungal colonization of healthy individuals, and to further understand human microbiome diversity.[34] Nares, oropharynx, and groin specimens from 101 healthy soldiers were tested for the presence of Gram-positive and Gram-negative bacteria (GNB), fungi, and antibiotic resistance genes. The assay detected at least one colonizing organism in each individual: mean number of organisms per subject of 11.8. The mean number of organisms in the nares, groin, and oropharynx was 3.8, 3.8, and 4.2, respectively. The most commonly detected organisms were aerobic Gram-positive bacteria: primarily coagulase-negative *Staphylococcus* (101 subjects: 341 organisms), *Streptococcus pneumoniae* (54 subjects: 57 organisms), *Staphylococcus aureus* (58 subjects: 80 organisms), and *Nocardia asteroides* (45 subjects: 50 organisms). The mecA gene was found in 96 subjects. The most commonly found GNB was *H. influenzae* (20 subjects: 21 organisms) and the most common anaerobe was *Propionibacterium acnes* (59 subjects). *Saccharomyces* species (30 subjects) were the most common fungi detected. PCR ESI-TOF appears to be a useful method for detecting colonized bacterial and fungal organisms.

Sexually transmitted infections (STIs) are still a global public health burden and can be caused by bacteria (*e.g.*, chlamydia, gonorrhoea, and syphilis), viruses (*e.g.*, HIV, HSV, human papillomavirus (HPV), HBV, CMV, and Zika), and parasites (*e.g.*, *Trichomonas vaginalis*). Unfortunately, infections of different sexually transmitted pathogens often present similar clinical symptoms, so it is difficult to distinguish them clinically. Therefore, it is important to develop a sensitive, specific, multitarget, and high-throughput method for detecting various agents responsible for STIs. Multiplex PCR MALDI-TOF was employed to develop a STI MS assay.[35] The assay simultaneously detects and identifies common clinical pathogens associated with STIs, including HSV types 1 and 2 (HSV-1 and HSV-2), *N. gonorrhoeae*, *Chlamydia trachomatis*, *Treponema pallidum*, *T. vaginalis*, *Mycoplasma genitalium*, *Haemophilus ducreyi*, *Mycoplasma hominis*, *Ureaplasma urealyticum*, and *Ureaplasma parvum*. For each microorganism included in the STI-MS assay, two different genes were selected as targets and a human β-globin gene was included as internal control. A 23-plex PCR method was developed to identify the 11 agents.

A panel of 368 sequencing confirmed clinical isolates and samples were employed to evaluate the assay specificity. STI-MS was highly specific and detected only the targeted agents. The sensitivity of the assay ranged from 1.7 to 10.0 copies per reaction.

HPV infection is associated with the development of cervical cancer, head-neck cancer, and other cancers. HPVs pose a significant human health problem since they are among the most common sexually transmitted agents worldwide. 30 HPV types are linked to human cancers. To control HPV transmission and guide patient management, it is essential to have timely and effective tools for detecting and identifying pathogens. A 61-plex PCR MALDI-TOF assay was developed for the 30 HPV types.[36] Two HPV genes, E6 and L1, were targeted for analyses. The assay was highly specific and detected only the corresponding HPV types. The sensitivity of the assay was 10 to 50 copies per reaction. 200 samples from homosexual males (HM) were evaluated by PCR MS and PCR sequencing, respectively. PCR MS detected 129 HPV positive samples, and among them, 51.9% (67/129) of the samples were multiple HPV types, while PCR sequencing detected only one single HPV type in these 129 HPV positive samples. 135 formalin-fixed, paraffin-embedded (FFPE) cervical cancer samples were analysed by PCR MS and commercially available GenoArray (GA) assay, respectively. PCR MS detected 97% (131/135) and GA identified 90.4% (122/135) of the HPV-positive samples. Multiple HPV types were identified in 55.6% (75/135) and 25.2% (34/135) of the cervical cancer samples by PCR MS and GA, respectively. The PCR MS assay allowed the genotyping of more than 1000 samples per day and provided a good tool for large-scale epidemiological investigations of multiple HPV infections and degraded FFPE samples.

Biological threat agents or biothreat agents are among the most aggressive microbes. Early recognition of a biothreat attack is critical to prevent subsequent waves of infections. Because initial symptoms are nonspecific and resemble common respiratory infections, rapid identification of the specific pathogen is essential to emergency management. To deal with this potential problem, a (RT)-PCR ESI MS assay was developed to detect and identify pathogens from human respiratory samples rapidly.[37] The performance characteristics of the assay were evaluated for distinguishing biothreat agents from common bacterial, fungal, and viral respiratory pathogens in BAL fluid specimens from subjects with suspected respiratory infections. In a retrospective case series, 202 BAL fluid specimens were analysed. Samples were processed using standard bacterial, viral, and fungal testing in clinical microbiology laboratories as part of routine care and then were blindly spiked with either water or nucleic acids from biothreat organisms (*Bacillus anthracis*, *Yersinia pestis*, *Francisella tularensis*, *Brucella* spp., *Burkholderia* spp., and *Rickettsia prowazekii*) and tested by (RT)-PCR ESI MS. The sensitivities and specificities of the MS *versus* standard clinical methods were as follows: for mock BT DNA, 98.5% sensitivity and 100% specificity; for bacterial pathogens, 81.8% sensitivity and 73.6% specificity; for viral pathogens, 93.3% sensitivity and 97.3% specificity; and for fungal pathogens, 42.6% sensitivity and 97.8% specificity. Seventy-five BAL specimens that contained 2 or more naturally occurring organisms, as determined by reference testing, were employed to evaluate the assay. The assay correctly detected the pathogens: 53 contained mixtures of only bacteria, 18 contained mixtures of fungal pathogens with Gram-positive or Gram-negative bacteria, 2 contained mixtures of fungal pathogens, 1 contained a mixture of a virus with Gram-negative bacteria, and 1 contained a mixture of a virus with a fungal pathogen. These studies demonstrate that (RT)-PCR ESI MS can detect and identify both biothreat agents and common respiratory pathogens simultaneously.

5.4.4.5 Others

Besides the applications (mostly medically important situations) described earlier, NAA MS can also be applied to any other areas where microbial identifications are necessary, such as food- and water-borne samples, environmental samples, and biothreat samples. In these cases, rapid detection and identification of the associated microorganisms are essential to manage situations timely and effectively.

Food poisoning is illness caused by eating food contaminated with microbes. Identification of pathogens and determination of the sources of contamination are the key steps in preventing the pathogen from spreading. To reach this goal, a PCR ESI MS assay was developed to identify food-borne pathogens.[38] Eight primer pairs, each targeting variable and informative gene segments within mutS and mdh (for Gram-negative enteric bacteria) and prfA and iap (for *Listeria*) where the resultant base compositions provide species, subspecies, and some serotype information, were designed. More than 800 bacterial isolates of *Salmonella enterica*, *Escherichia coli*, and *Shigella* species were employed to evaluate the assay. Overall, 100% of *Salmonella enterica*, 99% of *E. coli*, and 73% of *Shigella* spp. were detected using this assay. This assay has the potential to significantly reduce the time necessary to characterize bacteria in food samples and greatly improve the response time for food-borne bacterial outbreaks.

Acinetobacter species are ubiquitous in soil and water and are an important cause of nosocomial infections. A PCR ESI MS assay was developed to detect and identify bacterial species.[39] Nine primer pairs target the following five genes: trpE, adk, mutY, fumC, and ppa of the *Acinetobacter* genome. Results obtained with the MS were essentially in agreement with those obtained by conventional assay, but in a much shorter time. This rapid analysis allows tracking of transmission for the implementation of appropriate infection control measures in a timely fashion.

Biothreat agents contain a large collection of microbes that can cause severe disease and/or death. Identification of biothreat agents in environmental and clinical specimens is essential to protect citizens in the case of a biological attack. This is a challenge because there are dozens of microbial species that might be used in a biological attack and many have closely related organisms that are harmless. To solve this problem, a PCR ESI MS assay was developed to detect all relevant microbes.[40] The biothreat assay as configured was able to detect all the target organisms and did not misidentify any of the near neighbour organisms as threats. The assay demonstrated a low analytical sensitivity of 40 genome equivalents.

5.5 Conclusions and Perspectives

NAA combined with MS has become a powerful tool for microbial detection, identification, genotyping, and quantification, and AMR determination. Since the first NAA technique, PCR, was developed,[1] the technology has expanded significantly. Theoretically, any nucleic acid product can be detected and analysed by MS. NAA of the samples to be tested has the capacity for improving assay sensitivity. PCR remains a key technique and forms the backbone of molecular diagnostics in clinical microbiology laboratories. While the basic principle of NAA tests has not changed, technologies surrounding this core, including the amplification strategy, amplicon detection, multiplexing of reactions, and automation of the entire process into sample-to-result platforms, have provided a large menu of options for the molecular microbiology laboratory to choose from. Examples of NAA modification include ligation mediated amplification,[4] loop mediated isothermal amplification (LAMP),[5] nucleic acid sequence-based amplification (NASBA),[6] rolling circle amplification (RCA),[7] and transcription-based amplification (TBA).[8] One common feature of these assays is that they employ isothermal synthesis and amplification of nucleic acids, thus avoiding thermocycling and increasing the assay speed compared to standard PCR. One of the most recent NAA techniques is 3WJ-eRCA.[25] MS analysis of RNA is superior to that of DNA since RNA that is transcribed from DNA is more stable. The 2′-OH group on the sugar ring reduces *N*-glycosylic bond fragmentation. NAA technology can be easily and effectively modified to create nucleic acids for subsequent MS detection and analysis.

Because of its speed, easy use, automation, high throughput and low per-sample cost, MS has become a popular instrument for the identification of microorganisms in clinical microbiology laboratories, particularly in large reference laboratories. Ionization is essential for the technology, since MS measurements are based on m/z. Following ionization, the ions are transferred into a mass analyser where they are filtered based on their m/z and detected. In addition to MALDI, several new ionization techniques, including desorption electrospray ionization, direct analysis in real time, and ESI-assisted laser desorption ionization, hold great potential for use in the real-time direct MS-based analyses of pathogens. Furthermore, advances in other MS-related technologies will provide more powerful tools for characterization of pathogens; these technologies include high-speed and high-resolution liquid chromatography, ion mobility spectrometry, and high-capacity linear ion trap and various high-resolution mass analysers, such as Fourier transform ion cyclotron resonance, Fourier transform-Orbitrap, and time-of-flight mass detectors. The combination of these methodologies and instrumentation will certainly enhance the capabilities of MS for the analysis of pathogens. Diversified ionization methods are essential to address the range of samples and sample components requiring analysis. Depending on sample polarity, size, physical makeup, volatility, thermal lability, and other factors, certain ionization techniques work more optimally toward experimental goals than others.

Target identification using MS involves spectrum acquisition, spectrum processing, and spectral matching against the main spectral library. A high-quality mass spectrum database or a library of reference spectra is a cornerstone for MS data analysis in clinical microbiology laboratories. It is essential to have a robust reference library to which NAA MS results are compared. Organisms which are not represented in the library may not be readily identified; however, closely related species or complexes of organisms can be easily separated based on discrimination of nucleotide substitutions.

NAA MS assays are generally qualitative tests, which are well suited for the detection of microbes in clinical specimens. The nucleic acid presence, at any level, is associated with an infection state, which lays the foundation for

NAA MS as a tool for microbial detection, identification, genotyping, and quantification and AMR determination.

This is particularly useful for microorganisms that are hard to grow or unreliable due to the loss of viability during transport. Compared to culture methods, NAA MS demonstrated increased sensitivity and a reduced turnaround time for detection of such microorganisms. Bacterial, viral, fungal, and non-cultivable organisms can be detected following successful amplification of nucleic acid targets. One advantage of NAA MS is the ability to identify organisms directly from patient specimens without the need for culture. This can be especially advantageous for the detection and identification of fastidious organisms or organisms present at a very low concentration in a specimen. An example of this is the diagnosis of prosthetic joint infections (PJI).[41] Recent data have indicated poor recovery of organisms from joints suspected to be harbouring bacteria when routine culture methods are employed. PCR ESI MS is highly sensitive for difficult-to-identify bacteria. MS played an early role in the discovery of the pandemic H1N1 influenza strain.[42] Whereas the pandemic strain was untypeable using routine molecular methods, NAA MS provided rapid and accurate identification, typing, and genetic analysis of the pandemic virus, demonstrating that it is a useful method for rapidly monitoring the circulating influenza virus for emerging patterns of drift.

One advantage of NAA MS is its high throughput. The test can be performed in a 96-, 384-, or 1536 format. MALDI-MS can operate as fast as 10 ms per sample, and a 96-well plate can take less than 10 seconds. The technology provides an unlimited capacity for evaluating a large number of samples. In the field of microbial infections, there is always a huge demand for rapid and high-throughput identification of pathogens. The COVID-19 pandemic by SARS-Cov-2 has created an unsustainable need for a diagnostic test of the virus. Due to Zero-COVID-19 policy in China, some cities with populations of 10 million or more performed COVID-19 tests for every citizen once every day. NAA MS has the potential to meet such a challenge. RT-PCR, which is the current gold standard test, offers sensitive and specific tools for detecting viral RNA. However, this platform often requires a trade-off between speed and throughput. It also faces other logistic issues, such as shortages of molecular grade reagents and consumables, processing plates, pipettes, and RNA extraction kits. RT-PCR MS has the potential to solve the problem, keeping a balance among performance, speed, and throughput, and thus provides guidance for effective infection management.

Microbes, with their large population size, elaborate mechanisms for mutation and genetic exchange, rapid multiplication rates, varied phenotypes, and ability to occupy diverse ecological niches, are consistently evolving and generating huge variants at any moment of time. Microbial genetic polymorphisms have been identified in the genes of microorganisms. The genetic diversity or polymorphic genetic elements may be associated with microbial virulence and disease development. Microbial differences are important and their polymorphisms contribute to differences in associated disease conditions and clinical outcome. It is important to identify microbial polymorphisms and their correlated functions, such as AMR. NAA MS genotyping methods, such as SNP genotyping and MLST, exhibit high throughput and are economical. They are particularly useful for multiplex genotyping and provide cost-effective and efficient methods for high-throughput genotyping. AMR has become a major global health concern and one of the biggest threats to human health. It is challenging to obtain comprehensive AMR profiles using conventional detection methods. NAA MS can overcome the drawbacks of conventional assays and can simultaneously detect 24 mutation sites that are strongly associated with resistance to five antimicrobial agents (spectinomycin, penicillin, cephalosporins, macrolides, and quinolones) located in nine resistance genes in *N. gonorrhoeae*.[24]

Clinically distinct infective syndromes can generally be caused by several different organisms. One syndrome can be caused by many different microbes. It is essential to identify pathogens so that appropriate management can be initiated. Microbial identification can be based on syndromes or groups of microorganisms. There are many such groups, such as respiratory pathogens, gastrointestinal pathogens, urinary pathogens, genital pathogens, pneumonia pathogens, meningitis pathogens, sepsis pathogens, and food-borne pathogens. Each of these groups contains a complex number of microbes. Multiplex NAA MS panels for syndromes or groups of microbes can be established and used to simultaneously detect and identify pathogens related to associated diseases. Such panels have already been established: 12-plex for detecting pathogens causing pneumonia and meningitis,[32] 24-plex for detecting respiratory viruses,[33] and 61-plex for HPV typing.[37] Seven human coronaviruses (HCoVs) have been identified, including HCoV-NL63, HCoV-229E, HCoV-OC43, HCoV-HKU1, SARS-CoV, MERS-CoV, and SARS-CoV-2. A multiplex PCR assay with MS was developed for simultaneous detection and identification of the seven CoVs.[43] Since the most recent human CoVs that caused the epidemic and pandemic were all transmitted from animals, this assay can be extended to include animal CoVs and provides an effective tool for monitoring the epidemiology of CoVs. NAA combined with MS provides a powerful tool for rapid and high-throughput detection and identification of microbes.

Glossary

NAA	Nucleic acid amplification.
Genome	The entire hereditary information about an organism that is encoded by its DNA (or RNA for some viruses).
Bacterial typing	A procedure for identifying types and strains of bacteria.
Metagenome	The global genetic repertoire of an environmental niche that is constituted by diverse organisms such as free-living microorganisms in the wild or the commensals of a particular niche in a mammalian host (from the Greek word 'meta', meaning beyond, and genome).

16S ribosomal RNA gene	A component of the small bacterial and archaeal ribosomal subunits. The gene includes hypervariable regions that contain species-specific signature sequences which are useful for bacterial and archaeal identification at the species level.
MLEE	The characterization of bacterial species by the relative electrophoretic mobility of approximately 15 cellular metabolic enzymes.
MLST	An unambiguous procedure for characterizing isolates of bacterial species using the sequences of internal fragments of (usually) seven housekeeping genes. Approximately 450–500 bp internal fragments of each gene are used, as these can be accurately sequenced on both strands using an automated DNA sequencer.
Pan-genome	The global gene repertoire of a bacterial species that comprises the sum of the core and the dispensable genome (from the Greek word 'pan', meaning whole, and genome).
SNP	DNA sequence variation that occurs when a single nucleotide in the genome differs between members of a species.
Core genome	The pool of genes that are shared by all the strains of the same bacterial species.
Lateral gene transfer	The mechanism by which an individual of one species transfers genetic material (*i.e.*, DNA) to an individual of a different species.
Metagenomics	The study of the genomic repertoire of all the organisms that live in a particular environment and their activities as a collective. The genomic analysis is applied to the entire communities of microorganisms, which bypasses the need to isolate and culture individual microbial species.
Reverse vaccinology	A genomic approach to vaccine development that searches the entire genetic repertoire of a pathogen for protective antigens.

References

1. J. P. Lavigne, P. Espinal, C. Dunyach-Remy, N. Messad, A. Pantel and A. Sotto, *Clin. Chem. Lab. Med.*, 2013, **51**(2), 257.
2. K. B. Mullis, F. A. Faloona, S. Scharf, R. Saiki, G. Horn and H. Erlich, *Cold Spring Harbor Symp. Quant. Biol.*, 1986, **51**, 263.
3. K. B. Mullis and F. A. Faloona, *Meth. Enzymol.*, 1987, **155**, 335.
4. K. B. Mullis, *Sci. Am.*, 1990, **262**, 56.

5. D. Y. Wu and R. B. Wallace, *Genomics*, 1989, **4**, 560.
6. T. Notomi, H. Okayama, H. Masubuchi, T. Yonekawa, K. Watanabe, N. Amino and T. Hase, *Nucleic Acids Res.*, 2000, **28**, E63.
7. J. Compton, *Nature*, 1991, **350**, 91.
8. P. M. Lizardi, X. Huang, Z. Zhu, P. Bray-Ward, D. C. Thomas and D. C. Ward, *Nat. Genet.*, 1998, **19**, 225.
9. D. Y. Kwoh, G. R. Davis, K. M. Whitfield, H. L. Chappelle, L. J. DiMichele and T. R. Gingeras, *Proc. Natl. Acad. Sci. U. S. A.*, 1989, **86**, 1173.
10. C. Jurinke, P. Oeth and D. van den Boom, *Mol. Biotechnol.*, 2004, **26**, 147.
11. I. G. Gut, *Hum. Mutat.*, 2004, **23**, 437.
12. K. Tanaka, H. Waki, Y. Ido, S. Akita, Y. Yoshida and T. Yoshida, *Rapid Commun. Mass Spectrom.*, 1988, **2**, 151.
13. M. Karas and F. Hillenkamp, *Anal. Chem.*, 1988, **60**, 2299.
14. J. B. Fenn, M. Mann, C. K. Meng, S. Wong and C. M. Whitehouse, *Mass Spectrom. Rev.*, 1990, **9**, 37.
15. C. Honisch, Y. Chen, C. Mortimer, C. Arnold, O. Schmidt, D. van den Boom, C. R. Cantor, H. N. Shah and S. E. Gharbia, *PNAS*, 2007, **104**(25), 10649.
16. A. Bizzini and G. Greub, *Clin. Microbiol. Infect.*, 2010, **16**, 1614.
17. C. Benagli, V. Rossi, M. Dolina, M. Tonolla and O. Petrini, *PLoS One*, 2011, **6**, e16424.
18. G. B. Hurst, M. J. Doktycz, A. A. Vass and M. V. Buchanan, *Rapid Commun. Mass Spectrom.*, 1996, **10**, 377.
19. C. Jurinke, D. van den Boom, A. Jacob, K. Tang and H. KÖster, *Anal. Biochem.*, 1996, **237**, 174.
20. E. M. Dunne, E. K. Ong, R. J. Moser, P. M. Siba, S. Phuanukoonnon, A. R. Greenhill, R. M. Robins-Browne, E. K. Mulholland and C. Satzke, *J. Clin. Microbiol.*, 2011, **49**(11), 3756.
21. D. Hou, M. Ju, Y. Wang, D. Zhang, D. Zhu, M. Zhong, C. Zhou, Y. Song and X. Cheng, *Exp. Ther. Med.*, 2020, **20**, 3587.
22. K. Strålin, R. E. Rothman, V. Özenci, K. Barkataki, D. Brealey, N. Dhiman, L. Poling, M. C. Kurz, A. P. Limaye, F. LoVecchio, K. Lowery, L. G. Miller, G. J. Moran, J. S. Overcash, A. Parekh, W. F. Peacock, E. P. Rivers, M. Sims, A. M. Stubbs, M. Sundqvist, M. Ullberg and K. C. Carrollw, *J. Clin. Microbiol.*, 2020, **58**(9), 1.
23. C. L. Brinkman, P. Vergidis, J. R. Uhl, B. S. Pritt, F. R. Cockerill, J. M. Steckelberg, L. M. Baddour, J. J. Maleszewski, W. D. Edwards, R. Sampath and R. Patela, *J. Clin. Microbiol.*, 2013, **51**(7), 2040.
24. Y. Li, L. Xiu, J. Liu, C. Zhang, F. Wang, Y. Yin and J. Peng, *J. Antimicrob. Chemother.*, 2020, **75**, 2817.
25. D. Florea, D. Oţelea, I. D. Olaru and A. Hristea, *GERMS*, 2016, **6**(1), 10.
26. G. Han, Q. Lin, J. Yi, Q. Lyu, Q. Ma and L. Qiao, *Talanta*, 2022, **242**, 123297.

27. M. M. Hernandez, R. Banu, P. Shrestha, A. Patel, F. Chen, L. Cao, S. Fabre, J. Tan, H. Lopez, N. Chiu, B. Shifrin, I. Zapolskaya, V. Flores, P. Y. Lee, S. Castañeda, J. D. Ramírez, J. Jhang, G. Osorio, M. R. Gitman, M. D. Nowak, D. L. Reich, C. Cordon-Cardo, E. M. Sordillo and A. E. Paniz-Mondolfi, *J. Med. Virol.*, 2021, **93**, 5481.
28. S. Zürcher, C. Mooser, A. Urs Lüthi, K. Mühlemann, M. T. Barbani, P. Mohacsi, C. Garzoni, M. Gorgievski-Hrisoho, A. Schaller and L. Flatz, *J. Clin. Virol.*, 2012, **54**(4), 359.
29. J. Luan, J. Yuan, X. Li, S. Jin, L. Yu, M. Liao, H. Zhang, C. Xu, Q. He, B. Wen, X. Zhong, X. Chen, H. L. Y. Chan, J. J. Y. Sung, B. Zhou and C. Ding, *Clin. Chem.*, 2009, **55**(8), 1503.
30. B. Jia, R. Lovari, H. Miller, D. Metzgar, C. Massire, H. Carolan, D. Toleno, F. D'Alessio, R. Rothman, L. B. Blyn and S. X. Zhang, *Diagn. Microbiol. Infect. Dis.*, 2020, **97**(1), 114988.
31. A. KriforsID, V. Özenci, M. Ullberg, M. Ackefors, M. Jädersten, K. Strålin and O. Blennow, *PLoS One*, 2019, **14**(2), e0212812.
32. C. Zhang, Y. Xiao, J. Du, L. Ren, J. Wang, J. Peng and Q. Jina, *J. Clin. Microbiol.*, 2015, **53**(8), 2549.
33. C. Zhang, L. Xiu, Y. Xiao, Z. Xie, L. Ren and J. Peng, *Front. Cell. Infect. Microbiol.*, 2018, **8**, 107.
34. R. Vetor, C. K. Murray, K. Mende, R. Melton-Kreft, K. S. Akers, J. Wenke, T. Spirk, C. Guymon, W. Zera, M. L. Beckius, E. R. Schnaubelt, G. Ehrlich and T. J. Vento, *BMC Infect. Dis.*, 2016, **16**, 338.
35. L. Xiu, C. Zhang, Y. Li, F. Wang and J. Peng, *Infect. Drug Resist.*, 2019, **12**, 2671.
36. J. Peng, L. Gao, J. Guo, T. Wang, L. Wang, Q. Yao, H. Zhu and Q. Jin, *J. Clin. Microbiol.*, 2013, **51**(2), 402.
37. K. Jeng, J. Hardicka, R. Rothmana, S. Yanga, H. Wona, S. Petersona, Y. Hsieha, B. J. Mase, K. C. Carrolla and C. A. Gaydos, *J. Clin. Microbiol.*, 2013, **51**(10), 3300.
38. S. E. Pierce, R. L. Bellb, R. S. Hellberga, C. Cheng, K. Chen, D. M. Williams-Hill, W. B. Martina and M. W. Allard, *Appl. Environ. Microbiol.*, 2012, **78**(23), 8403.
39. J. A. Ecker, C. Massire, T. A. Hall, R. Ranken, T. D. Pennella, C. A. Ivy, L. B. Blyn, S. A. Hofstadler, T. P. Endy, P. T. Scott, L. Lindler, T. Hamilton, C. Gaddy, K. Snow, M. Pe, J. Fishbain, D. Craft, G. Deye, S. Riddell, E. Milstrey, B. Petruccelli, S. Brisse, V. Harpin, A. Schink, D. J. Ecker, R. Sampath and M. W. Eshoo, *J. Clin. Microbiol.*, 2006, **44**(8), 2921.
40. R. Sampath, N. Mulholland, L. B. Blyn, C. Massire, C. A. Whitehouse, N. Waybright, C. Harter, J. Bogan, M. S. Miranda, D. Smith, C. Baldwin, M. Wolcott, D. Norwood, R. Kreft, M. Frinder, R. Lovari, I. Yasuda, H. Matthews, D. Toleno, R. Housley, D. Duncan, F. Li, R. Warren, M. W. Eshoo, T. A. Hall, S. A. Hofstadler and D. J. Ecker, *PLoS One*, 2012, **7**(6), e36528.
41. K. E. Greenwood-Quaintance, J. R. Uhl, A. D. Hanssen, R. Sampath, J. N. Mandrekar and R. Patela, *J. Clin. Microbiol.*, 2014, **52**(2), 642.

42. K. Jeng, C. Massire, T. Zembower, V. Deyde, L. Gubareva, Y. Hsieh, R. Rothman, R. Sampath, S. Penugonda, D. Metzgar, L. Blyn, J. Hardick and C. Gaydos, *J. Clin. Virol.*, 2012, **54**(4), 332.
43. T. Liu, L. Kang, Y. Li, J. Huang, Z. Guo, J. Xu, Y. Hu, Z. Zhai, X. Kang, T. Jiang, H. Li, H. Song, J. Wang, S. Gao, J. Li, X. Zhou, Y. Yuan, B. Zhao, J. Wang and W. Xin, *COVID*, 2021, **2**(1), 5.

CHAPTER 6

Proteomics Analysis for Identification and Antimicrobial Resistance Analysis of Bacteria

DONGXUE ZHANG

Institutes of Biomedical Sciences, Fudan University, Shanghai 200000, China
Email: dxzhang18@fudan.edu.cn

6.1 Introduction

Bacteria are widely present in nature, such as water, soil, atmosphere, food, human and animal body surfaces, and lumens connected to nature, and are closely related to human life. Bacteria can cause infectious diseases and pose a serious threat to public health and social development. For example, bacteria are responsible for 130–175 million cases of urinary tract infections (UTIs) and more than 1.2 million cases of meningitis each year,[1,2] as well as cause bacteremia with extremely high mortality.[3] The emergence of bacteria with antimicrobial resistance (AMR) is a more serious problem. According to the Review on Antimicrobial Resistance, the total loss of gross domestic product (GDP) due to AMR from 2014 to 2050 would be as high as $100.2 trillion, and AMR would result in 10 million deaths each year.[4] The development of techniques and methods for bacterial identification and bacterial AMR analysis is of great importance to address these issues from a long-term development perspective.

Over the past few years, a variety of techniques have been developed and applied for bacterial identification and AMR analysis, such as traditional culture methods, immunoassays, polymerase chain reaction (PCR), next generation sequencing (NGS), and mass spectrometry (MS). Traditional

New Developments in Mass Spectrometry No. 13
Detection and Analysis of Microorganisms by Mass Spectrometry
Edited by Liang Qiao and Jia Yi

Published by the Royal Society of Chemistry, www.rsc.org

methods such as plate culture are time-consuming and it is almost impossible to differentiate bacteria at the species and strain levels using these methods.[5,6] PCR analysis is rapid and sensitive, but has the false-positive problem.[7,8] Matrix-assisted laser desorption/ionization time-of-flight mass spectrometry (MALDI-TOF MS) enables bacterial identification by analysing bacterial ribosomal protein fingerprints, which is a convenient and effective method,[9] and has been commercialized, such as bioMérieux VITEK MS and Bruker MALDI Biotyper.[10,11] However, this convenient MALDI-based method for bacterial identification also has some limitations. It can only be applied to identify bacterial samples with high abundance and purity.[12] In addition, the inherent similarities between bacterial species and a limited number of spectral libraries can result in ineffective discrimination between bacterial species as well as misidentifications.[13]

Proteomics analysis provides comprehensive protein information related to disease initiation and cellular metabolic processes by studying the structure and function of proteins on a large scale, and is an important tool for biomedical analysis, such as early diagnosis of diseases and discovery of biomarkers.[14] Unlike methods that identify bacteria by detecting ribosomal protein fingerprints, proteomics-based analytical methods are capable of obtaining all bacterial protein information without losing any information theoretically. In recent years, proteomics techniques have been developed rapidly, including sample pre-treatment, peptide/protein separation, mass spectrometer performance, isotope labelling for quantification, and bioinformatics data analysis.

Proteomics-based bacterial analysis explores phenotypes beyond the aspects of genomics and provides a wealth of information for the study of bacteria and their behaviours. With the rapid development of proteomics analysis methods, proteomics has been widely applied to all aspects of bacteria research, such as bacterial identification, bacterial differentiation, heredity/variation study, toxin factor study, biomarker discovery, disease diagnosis, pathogenesis study, bacteria–bacteria/host interaction study, immune target exploration, vaccine investigation, AMR determination, AMR mechanistic research, biofilm/persister study, and antimicrobial drug development.

In this chapter, the author provides a comprehensive elaboration of bacterial analysis by the use of various proteomics strategies, with a particular focus on bacterial identification and AMR analysis. The author also prospects the future advances and developments in proteomics-based bacterial analysis. This chapter is mainly divided into the following sections: basic concepts in proteomics, bacterial identification and differentiation by peptide detection, bacterial identification and differentiation by protein detection, bacteria–bacteria/host cell interactions, bacterial AMR analysis, and conclusions and perspectives (Figure 6.1).

6.2 Basic Concepts in Proteomics

The term "proteome" was first proposed by Marc Wilkins to refer to all the proteins expressed by the genome.[15] Proteomics describes the large-scale study

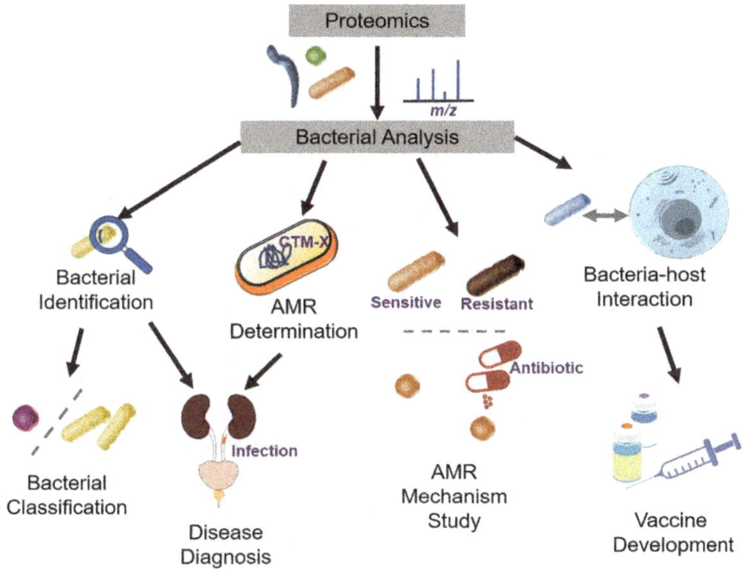

Figure 6.1 The application of proteomics in bacterial analysis.

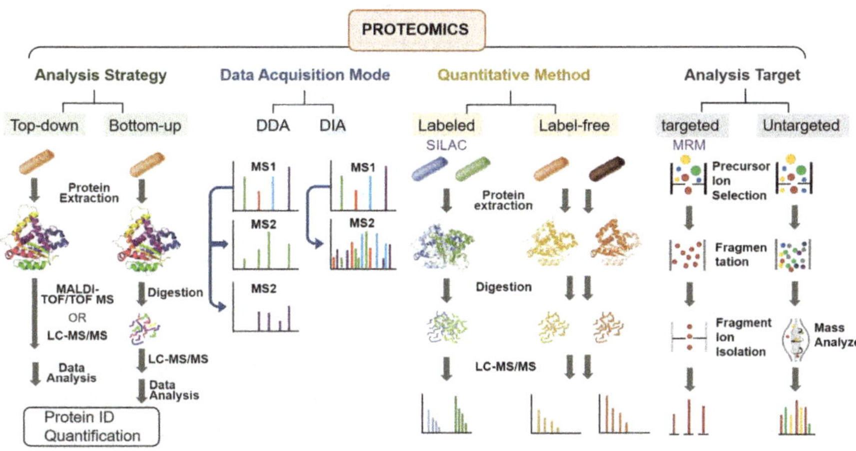

Figure 6.2 The classification of proteomics according to the analysis strategy, data acquisition mode, quantitative method, and analysis target.

of the characteristics of proteins, including protein expression levels, post-translational modifications, and protein–protein interactions.[16] MS is a powerful tool for proteomics analysis. MALDI-TOF MS and liquid chromatography–tandem mass spectrometry (LC–MS/MS) are the two most commonly used MS techniques for proteomics analysis. According to different analysis strategies, analysis targets, data acquisition modes, and quantification methods, proteomics can be divided into different categories (Figure 6.2).

6.2.1 Proteomics Analysis Strategies

Different strategies can be utilized to perform proteomics analysis, mainly including bottom-up proteomics and top-down proteomics. Bottom-up proteomics analysis refers to the characterization of proteins by analysing peptides obtained from proteolytic digestion of proteins.[17] Bottom-up proteomics is also known as shotgun proteomics when it is used to analyse a mixture of proteins.[18] In a typical process of shotgun proteomics analysis, peptide mixtures digested from proteins are separated by online LC, followed by analysis on the basis of electrospray ionization mass spectrometry (ESI-MS).[19] Peptide identification can be achieved by comparing the tandem mass spectra of ion fragments produced by peptide fragmentation with theoretical tandem mass spectra generated from *in silico* digestion of protein databases.[17] Possible proteins are inferred by assigning peptide sequences to proteins and scoring based on the results of peptide identification.

Different from bottom-up proteomics, top-down proteomics does not require prior proteolytic digestion of proteins into peptides and can directly characterize intact proteins. Intact proteins are ionized directly using a soft ionization source such as a MALDI or ESI source, they enter the mass spectrometer, and are then fragmented in tandem mass spectrometry.[20] Protein sequence information can be obtained from the molecular weight of protein fragments. Top-down proteomics is more convenient for protein analysis which avoids the loss of protein information induced by protein digestion, enables accurate protein identification, and characterizes all forms of proteins resulting from post-translational modifications or sequence variations.[21,22] However, protein fractionation, ionization, and fragmentation in the gas phase based on top-down proteomics are difficult. The low proteome coverage and low throughput of top-down proteomics make it lag behind bottom-up proteomics.[23]

6.2.2 Data Acquisition Modes

LC–MS/MS has demonstrably become the most powerful workhorse for the detection of proteins, and its applications have expanded from biological model systems to clinical specimens.[24] For LC–MS/MS-based proteomics analysis, proteomics can be divided into two different data acquisition modes, such as data dependent acquisition (DDA) and data independent acquisition (DIA). DDA is the most primitive and simplest data acquisition mode that can acquire MS1 and fragment information. Proteomics in DDA mode first performs MS1 analysis, followed by precursor ion screening in accordance with pre-set precursor screening conditions. The selected precursor ions are then subjected to fragmentation and MS/MS analysis.[25] This data acquisition mode can reduce the interference of some noise ions to a certain extent. However, DDA preferably acquires high-abundance peptides, resulting in a loss of information for peptides in low abundance.[26] Besides, only the selected precursor ions are fragmented for MS/MS analysis, which results in lower MS/MS coverage. This random sampling process reduces the reproducibility of the analysis, resulting in more missing values for quantification, especially for complex samples.

In contrast to DDA, MS/MS scans are not dependent on MS1 information.[27] DIA divides the *m*/*z* scan range into multiple windows. All precursor ions in each window are fragmented, regardless of the precursor information, and all fragmented ions are further analysed using a high-resolution mass spectrometer.[27] Thus, DIA proteomics is characterized by higher peptide coverage. Besides, compared with DDA which has the inherent problem of biased sampling, DIA proteomics has the characteristics of systematic and unbiased acquisition. DIA has fewer missing values than DDA, has higher reproducibility and accuracy in protein quantification, and outperforms in the identification and quantification of low-abundance proteins.[28,29] DIA is developing rapidly and has great potential for clinical cohort studies and biomarker analysis. However, the MS/MS spectra of DIA are very complex, which results from the fragmentation of multiple peptides together in the same *m*/*z* window, and needs analysis strategies and software tools for data analysis such as deconvolution.[30]

6.2.3 Proteomics Quantitative Methods

Dynamic changes in cellular proteome abundance are very essential to various life processes. Quantitative proteomics is an effective approach for the comprehensive quantification and identification of all proteins expressed in a defined system.[31] Quantitative proteomics strategies are mainly divided into labelled and label-free quantitative proteomics. Commonly used label-free and labelled quantitative proteomics both employ the bottom-up strategy. Labelled quantification compares samples labelled with different mass tags so that detection can be performed based on specific variations in mass.[32] Chemical labelling is one of the most popular labelled quantitative protocols, including stable isotope labelling by amino acids in cell culture (SILAC), isotope-coded affinity tags (ICAT), isobaric tags for relative and absolute quantification (iTRAQ), and tandem mass tags (TMT).[33]

Label-free quantitative proteomics relies on high-resolution and high-precision mass spectrometers. The quantification based on label-free proteomics is realized by comparing the counts of MS/MS spectra or the peak intensities of the same peptides.[34] Compared to label-based quantification, label-free quantitative proteomics is inexpensive without the need for labelling agents and is time-saving.[35] In addition, this quantitative method provides a higher proteome coverage and dynamic range.[36] This method is less sensitive to technical errors and can be used for large-scale protein analysis.

6.2.4 Proteomics Analysis Targets

Depending on the target of the analysis, different categories of analysis methods, such as untargeted proteomics and targeted proteomics, can be selected according to different research requirements. Untargeted proteomics, also known as discovery proteomics, aims to analyse as many proteins as possible in bacterial samples, without targeting any specific protein, to obtain comprehensive protein information.[37] Untargeted proteomics is a powerful tool for exploring new proteins and screening meaningful biomarkers.[38] MS-based label-free quantitative proteomics and SILAC and TMT/iTRAQ quantitative proteomics are all untargeted proteomics.

Targeted proteomics depends on the information from targeted proteins and analyses only specific proteins of interest, with the advantages of better accuracy, sensitivity, and reproducibility.[39] Targeted proteomics techniques mainly include multiple reaction monitoring (MRM) and parallel reaction monitoring (PRM), which can detect low-abundance peptides with higher quantitative accuracy. They are also able to validate and confirm the proteins discovered by untargeted proteomics to account for biological differences. For example, untargeted proteomics is suitable for the early research stage of biomarker discovery, and targeted proteomics is used for the validation and implementation of biomarker discovery.[38]

6.3 Bacterial Identification and Differentiation by Peptide Detection

Proteins are products of gene encoding. Bacteria with different genes also show differences in the expressed proteins. Proteomics based on peptide detection using tandem MS enables rapid characterization of most of the expressed genes of bacteria. By comprehensively detecting bacterial proteins or bacteria-specific proteins, bacteria can be identified and classified at the species level and even the strain level. New proteomics strategies like sample handling and data processing have been developed to identify more bacterial peptides. With the rapid development of proteomics technology, it is also possible to analyse a very small number of bacterial cells.

6.3.1 Bacterial Sample Preparation

For bacterial proteome analysis, sample preparation is a fundamental step prior to MS analysis and directly influences the results of protein identification and quantification. The sample preparation of bacterial proteomics based on peptide detection aims to obtain bacterial peptides, typically including bacterial cell lysis, protein purification and solubilization, and protein reduction, alkylation and digestion.[40] In the process of bacterial cell lysis, bacterial cells suspended in phosphate buffered saline (PBS) are usually resuspended in buffer solutions, such as ammonium bicarbonate buffer with pH 8–9, to provide a vibrant environment for trypsin, and treated with lysis buffers, such as sodium dodecyl sulfate (SDS) solution. Bead beating tubes and homogenizer systems are also used to assist in bacterial cell lysis. The bacterial proteins are then precipitated and extracted using acetone and

TRIzol reagents. The bacterial protein extracts are then incubated with dithiothreitol, iodoacetamide, and trypsin in sequence for disulfide bond reduction and alkylation as well as digestion to obtain bacterial peptides.

Sample preparation is one of the steps that limits the speed of the entire proteomics workflow. In order to improve the sample preparation rate, enhance the digestion efficiency, reduce the degradation and loss of proteins, and reduce the interference of impurities, a variety of novel sample preparation methods have been developed. SDS is a potent surfactant that can disrupt cell membranes and extract proteins which has been commonly used for protein extraction in sample pre-treatment of proteomics. However, this detergent interferes with electrospray ionization and affects the signal of MS. Therefore, SDS needs to be removed prior to MS analysis. The routine procedure for removing SDS is to use acetone sedimentation. Zhou *et al.* developed a novel protocol for SDS removal that applied potassium chloride to realize ion substitution-mediated DS^- precipitation.[41] *Shewanella oneidensis* was selected as the model to perform this method. According to the results obtained from LC–MS/MS, the effective SDS removal rate of this method exceeds 99.9%. In addition, the peptide recovery of this method is also excellent, over 95%, which is superior to those of other standard protocols, such as acetone precipitation and chloroform/methanol precipitation.

Vaezzadeh *et al.* developed a novel bacterial sample pre-treatment method to accelerate digestion efficiency.[42] Pathogenic *Staphylococcus aureus* was selected and enzymatically hydrolysed using conventional digestion methods in combination with microwaves. The whole bacterial proteome was detected and analysed using isoelectric focusing in immobilized pH gradients (IPG-IEF) and LC–MS/MS. With the assistance of microwaves, the efficacy of digestion of bacterial proteins, especially hydrophobic proteins and membrane proteins, was greatly improved. Optimized strategies for proteomics sample preparation, such as contaminant removal strategies and lysis buffer optimization, were also exploited to achieve efficient digestion and high proteomic coverage.[43]

Bacterial protein/peptide fractionation is also of great value in a whole bacterial proteomics analysis. Sodium dodecyl sulfate polyacrylamide gel electrophoresis (SDS–PAGE) is one of the commonly used fractionation strategies in proteomics analysis.[44] Before LC–MS analysis, SDS–PAGE is usually used to separate bacterial protein mixtures prior to bacterial protein digestion. The gel strips of interest are collected and proteins are digested within the gel for MS analysis to obtain protein information. SDS–PAGE can reduce the interference with MS and increase the depth of proteomics coverage.[45] Unlike the protein fractionation method SDS–PAGE, high-pH reverse phase liquid chromatography (RPLC) can realize fractionation at the peptide level, which favours a high-throughput proteome.[44] Wickramasekara *et al.* compared two different protein precipitation protocols, *i.e.*, one-dimensional SDS–PAGE and gel-free MudPIT (multi-dimensional protein identification technique), in proteomics analysis.[46] These two methods were conjugated to LC–MS/MS to analyse proteins of *Burkholderia vietnamiensis*, respectively. More proteins were identified using the gel-based method compared with those identified using the gel-free MudPIT method. By combining these two protocols, a comprehensive coverage of proteomes was achieved and more than 1200 unique proteins were identified.

The chromatographic conditions and mass accuracy and MS/MS frequency of the mass spectrometer are all determining factors for peptide identification. Wöhlbrand *et al.* investigated the effects of LC column length, gradient length, and MS/MS frequency on the identification of bacterial proteins based on shotgun proteomics.[47] They found that more peptides/proteins could be identified under longer gradient conditions. A longer chromatography column and high MS/MS frequency could improve protein coverage. However, efficiency and time cost should be considered. Therefore, optimal parameters of LC and MS need to be set according to the specific characteristics of the sample and the purpose of analysis to acquire the best results.

6.3.2 Bacterial Identification and Differentiation

A simple shotgun proteomics and database search strategy was used to implement bacterial identification rapidly.[48] As shown in Figure 6.3, bacterial proteins were collected, digested into peptides, and analysed using a nano-LC–MS/MS (LTQ Orbitrap XL) system. The obtained peptide fragments

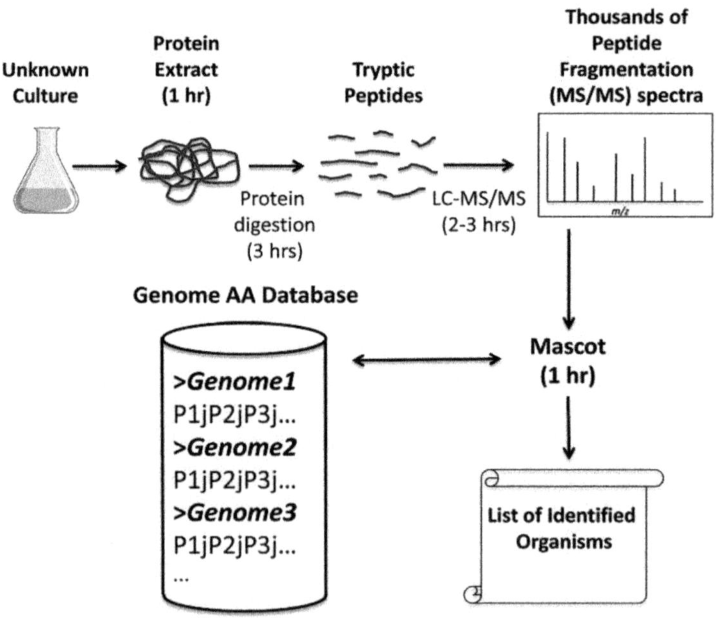

Figure 6.3 Schematic diagram of the bacterial identification workflow based on a simple shotgun proteomics using nano-LC–MS/MS analysis and database search strategy. Reproduced from ref. 48 with permission from Elsevier, Copyright 2013.

were searched against a novel "Genome AA database", represented by tandem proteomes of completed genome sequences. The identified peptides are used to classify bacteria by determining the similarities between the strains being studied and the database bacteria that have been grouped according to their established taxonomic positions, like phylum, class, order, family, genus, and species.[49] By using this protocol, the bacterial species *Escherichia coli* K-12 cells were correctly identified with a probability of more than 99.5%. Besides, *Francisella tularensis* could be distinguished from its closely related species, *Francisella philomiragia* and *Francisella novicida*. Bacterial proteins can also be separated into several fractions for in-depth analysis. Proteins of opportunistic pathogenic *B. vietnamiensis* were extracted into four fractions, including extracellular proteins, intracellular and cell wall proteins, and cell surface proteins, for proteomics analysis.[46] The fraction of interest can be analysed individually to study specific proteins or analysed together for comprehensive bacterial proteomics analysis.

Some proteins or peptides in bacterial cells are specific and can be used for bacterial identification. By utilizing targeted proteomics analysis, proteins of interest can be detected and used as representatives for the determination of bacterial species without the need for detecting all the proteins. For bacteria like *Bacillus*, the highly conserved sequences of chromosomally encoded proteins among closely related species, such as *B. anthracis*, *B. thuringiensis*, and *B. cereus*, make species identification difficult. On the other hand, different plasmids encoding different protein toxins exist in these species. Nguyen *et al.* targeted species-specific tryptic peptides from plasmid-borne proteins and achieved the identification of *Bacillus* at the species level using atmospheric pressure matrix-assisted laser desorption/ionization mass spectrometry (AP-MALDI-MS/MS)-based proteomics.[50] *Bacillus thuringiensis* and *B. anthracis* were identified at the species level, with peptide coverages of 75.4% and 60.3%, respectively.

6.3.3 Data Processing Optimization

For bacterial identification based on proteomics, multiple algorithms are developed to match tandem MS to databases. Pihowski *et al.* introduced a grid-search method named STEPS (Systematic Trial and Error for Parameter Selection) to optimize data filtering criteria and maximize confidence in peptide identification.[51] The usage of datasets derived from *S. oneidensis* cell lysates demonstrated the advantages of this method in terms of the number of true positive identifications. Boulund *et al.* introduced a novel computational approach called TCUP, which compared the generated protein sequence data to a reference database.[52] This TCUP method could automatically discover peptides with unique characteristics to identify bacterial species. Four bacterial species including *E. coli*, *Pseudomonas aeruginosa*, *S. aureus*, and *Streptococcus pneumoniae* were detected by LC–MS/MS and were used to investigate the performance of TCUP. The accuracy of TCUP in peptide classification is about 90.3–98.5% at the species level.

Proteomics-based bacterial identification is typically performed by matching detected peptides to a database. Those bacteria with a sufficient match are thought to be present in the sample. However, there is no standard that dictates how many matches are required to make a reliable identification. Jarman *et al.* proposed a straightforward statistical standard which produced a *p*-value on the results of bacterial or toxin identification.[53] The criteria could be applied regardless of sample quality or data analysis pipeline. The problem of signal erosion could also be overcome. In total 919 LC–MS/MS datasets obtained from 2 toxins and 32 bacterial strains were tested against the criteria and successful identification at the species level with >95% accuracy was achieved.

6.3.4 Identification of Bacteria in Small Numbers

When analysing a small number of bacterial cells, the proteomics for identifying bacteria can be disrupted to some extent. Blankenburg *et al.* developed a new sample preparation method called single-pot, solid-phase enhanced sample preparation (SP3), which is integrated with effective cell disruption for bacterial protein analysis based on proteomics in DDA mode in a limited number of bacterial cells (Figure 6.4A).[54] As shown in

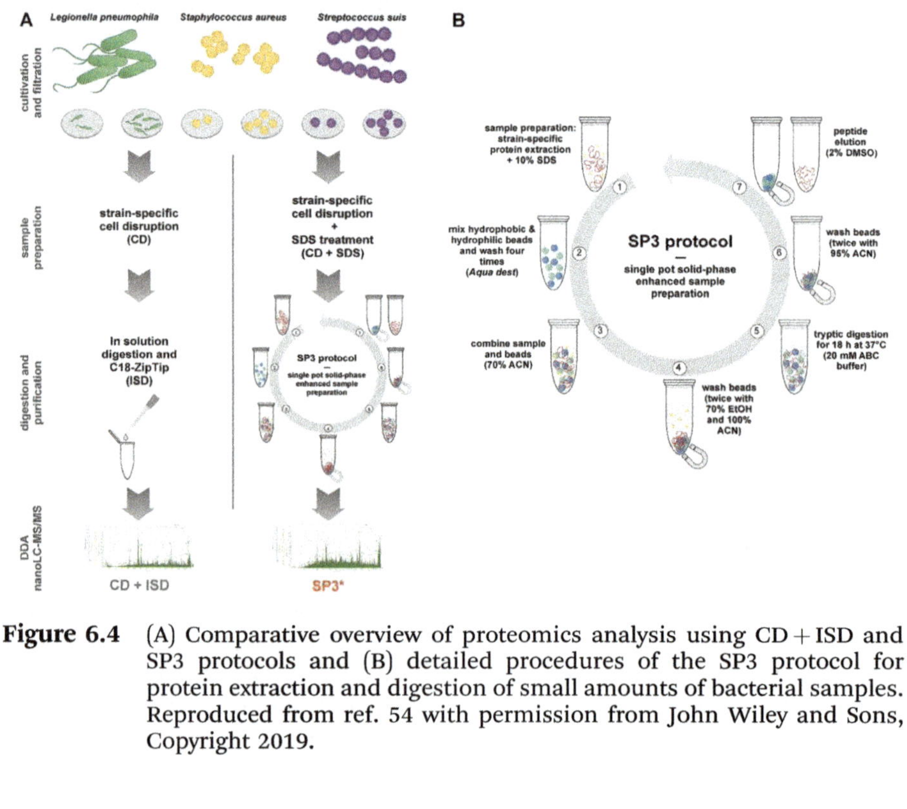

Figure 6.4 (A) Comparative overview of proteomics analysis using CD + ISD and SP3 protocols and (B) detailed procedures of the SP3 protocol for protein extraction and digestion of small amounts of bacterial samples. Reproduced from ref. 54 with permission from John Wiley and Sons, Copyright 2019.

Figure 6.4B, the SP3 process consists of several steps, including mixing of bacterial cells with detergent, a strain-specific enzyme, and magnetic beads, tryptic digestion, magnetic bead washing, and peptide elution. This protein preparation method extremely enhanced the identification of peptides and improved proteome coverage. A strain-specific sample treatment protocol was established using *S. aureus*, *Streptococcus suis*, and *Legionella pneumophila* as model samples. Compared to the strain-specific cell disruption with in-solution digestion (CD + ISD) method, the identified proteins and peptides of *L. pneumophila* increased by 300% and 620%, respectively, when using the SP3 protocol. DDA-based proteomics will be constrained when the sample to be analysed is in low abundance in complex samples, such as when a low proportion of bacterial cells are mixed with human cells. Willems *et al.* developed a hybrid library to perform DIA proteomics analysis by using DDA and *in silico*-predicted spectral libraries.[29] This workflow could be well applied in the identification and quantification of complex *Salmonella*–HeLa cell mixtures and even if bacterial cells were present at fairly low levels in the mixture (1 : 999 dilution).

Commonly used proteomics analysis relies on the bulk culture of bacterial cells, which will lead to heterogeneous responses. It remains difficult to realize single bacterial cell proteomics because of the extremely low abundance of proteins in a single bacterium. Very limited samples, such as a single bacterial colony, can be analysed using modern proteomics protocols while avoiding bulk culture. Gram-negative bacteria like *E. coli* and Gram-positive bacteria like *S. pneumoniae* have been reported to be identified based on single-colony proteomics.[55,56] The proteome of an isolated single colony of *Mycobacteria* was analysed using label-free quantitative proteomics.[57] A total of 7805 unique peptides and 1387 proteins were identified in a single mycobacterial colony.

6.3.5 Identification of Bacteria in Actual Samples

Bacteria can invade the urinary system from outside of the body, causing UTIs, a common disease that poses threats to human health. Rapid and accurate identification of bacteria causing UTIs in urine is of great significance for clinical treatment and therapy. Roux-Dalvai *et al.* developed a method combining DIA-based proteomics with machine learning to identify bacterial species in urine specimens, as shown in Figure 6.5.[58] Bacterial cells inoculated in urine were analysed in DIA mode based on the spectral library established by pure bacterial colonies detected in DDA mode. Then, peptide signatures were defined from these peptides using machine learning classifiers. Signatures in unknown samples were then monitored by PRM for bacterial identification through a prediction algorithm. According to these peptide signatures, bacteria in unknown urine samples could be identified with > 95% accuracy in less than 4 hours by targeted proteomics. Using this method, 15 species of bacteria were identified, accounting for 84% of the

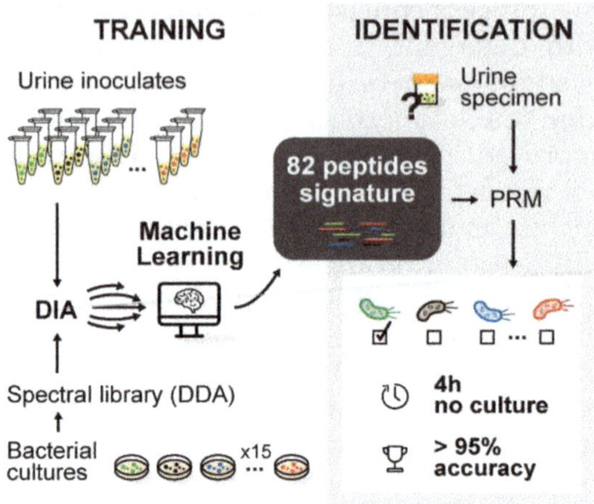

Figure 6.5 A protocol for bacterial species identification in urine specimens using LC–MS/MS and machine learning. Reproduced from ref. 58 with permission from American Society for Biochemistry and Molecular Biology, Inc., Copyright 2019.

bacterial species that cause UTIs. Lozano *et al.* successfully identified 26 microbial isolates from coastal water using phylopeptidomics which is a cutting-edge tandem mass spectrometry proteotyping technique.[59] In addition to common marine bacteria such as *Gammaproteobacteria* and *Alphaproteobacteria*, two isolates belonging to *Balneolaeota* which have not been studied well were identified.

6.3.6 Metaproteomics

Bacterial cells exist and survive in the form of communities. Metaproteomics studies the proteins of microbes from microbial communities in their natural ecosystems and performs protein identification and quantification on a large scale without culture restrictions.[60] Metaproteomics techniques have progressed rapidly and provide a detailed understanding of microbial community composition, interaction mechanisms, and disease mechanisms. Hart *et al.* determined the microbial source classification of the richest proteins in the rumen of dairy cows based on a metaproteomics protocol.[61] *Bacteroidetes*, *Firmicutes*, and *Proteobacteria* were the most abundant taxonomic phyla in the proteome. The gastrointestinal tract has the most abundance of microorganisms, which are closely related to human health. Long *et al.* studied the colorectal cancer (CRC)-associated microbiome using a quantitative metaproteomics method.[62] By comparing the proteomes of microbes in CRC patients and healthy volunteers, 341 proteins were found to have significant differences. Iron intake/transport, oxidative stress, and DNA replication, recombination, and repair were disclosed to play important roles in the progression of CRC.

6.4 Bacterial Identification and Differentiation by Protein Detection

In comparison with proteomics by detecting peptides, proteomics analysis by detecting proteins omits the protein tryptic digestion step, greatly simplifying the experimental procedure and reducing the experimental time. The same gene of bacteria can produce multiple proteoforms due to gene mutations, RNA alternative splicing, and post-translational modification. Studying the proteome at the intact protein level can provide more precise and richer biological information. Accurate identification of proteoforms and proteins can be applied to bacterial identification and differentiation.

6.4.1 Bacterial Identification and Differentiation by MALDI

Demirev *et al.* utilized top-down proteomics to identify intact *Bacillus* spore species based on the identification of individual and unique protein biomarkers using matrix-assisted laser desorption/ionization time-of-flight/time-of-flight mass spectrometry (MALDI-TOF/TOF MS).[63] The intact proteins of *Bacillus* spore species were dissociated in the first TOF analyser, and the corresponding sequence-specific fragments of the major biomarkers were analysed in a reflectron TOF analyser without the prefractionation, digestion, separation, and purification procedures. By comparing the MS/MS spectrum obtained from experiment with the *in silico*-generated MS/MS spectra based on an in-house-developed software package, *Bacillus atrophaeus* and *B. cereus* spores, as well as their mixture, could all be unambiguously identified. The middle-down proteomics strategy is to obtain longer peptides than bottom-up proteomics by using different enzymes, for the analysis and identification of several simultaneous post-translational modifications on longer peptides.[64] Middle-down proteomics can assist in top-down proteomics for bacterial identification. Based on the top-down/middle-down proteomics approach, subtypes of Shiga toxin-producing *E. coli* could be classified, which was consistent with the result of DNA sequencing.[65]

6.4.2 Bacterial Identification and Differentiation by LC–MS/MS

A top-down proteomics platform based on LC–MS/MS was developed to discriminate closely related pathogenic bacterial species by identifying specific proteoforms.[66] *Escherichia coli* was selected to optimize the workflow, including sample preparation, LC separation, MS and MS/MS conditions, and data analysis. By using this platform under the optimized parameters, *Salmonella*, *E. coli*, and *Shigella* were successfully separated at the species level. A multi-dimensional separation platform including size exclusion chromatography (SEC) and RPLC combined with capillary zone electrophoresis (CZE)-tandem mass spectrometry (SEC–RPLC–CZE-MS/MS) was established for proteomics analysis of *E. coli* in a deep top-down strategy

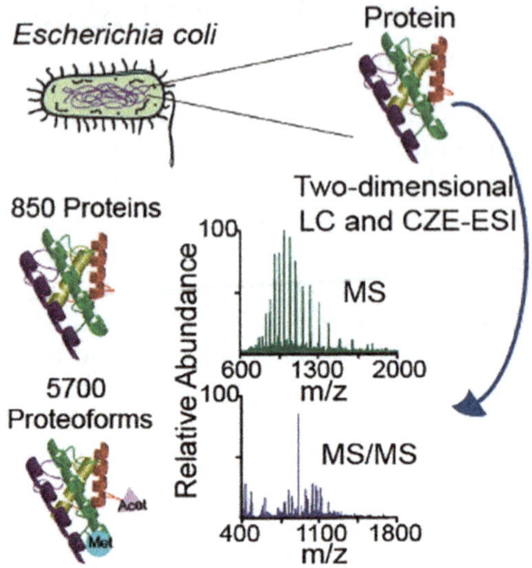

Figure 6.6 The schematic view of deep top-down proteomics for proteoform identification of *E. coli* using SEC–RPLC–CZE-MS/MS. Reproduced from ref. 67 with permission from American Chemical Society, Copyright 2018.

(Figure 6.6).[67] Up to 5700 proteoforms and 805 proteins from *E. coli* were identified by using this SEC–RPLC–CZE-MS/MS platform. The performance of this platform has extremely high peak capacity and could be comparable to the widely used RPLC-MS/MS in terms of proteoform identification quantity and instrument time.

6.5 Bacteria–Bacteria/Host Cell Interactions

6.5.1 Immune Response Study and Vaccine Development

Proteins located on the surfaces of bacteria, secreted out of the bacterial cells, and with immunological functions are all essential proteins in the interaction between bacterial cells and host cells. The identification and study of these proteins plays an important role in elucidating the mechanisms of bacterial colonization and invasion of host cells and contributes to the development of vaccines.

The *Mycobacterium abscessus* complex (MABSC) is capable of causing disease and inducing an immune response. To study immunogenic proteins, Steindor *et al.* utilized a proteomics approach integrated with *in silico* epitope prediction to identify the immune-correlated epitopes of MABSC, which are specific to T helper cells and B cells.[68] Seven clinical complex strains and one standard strain of *M. abscessus* were divided into 14 separated isolates according to the mophotype (smooth and tough). Both the bacterial lysates and the culture supernatant of these isolates were analysed due to protein secretion into the extracellular compartment. Twelve out of the 79 proteins present in each supernatant were identified, and their B-cell and T-cell epitopes were successfully predicted by *in silico* analysis. Distler *et al.* focused on the proteins secreted by bacteria into the supernatant to understand the mechanism of the interaction between pathogen-derived effector proteins and the cell membrane of host cells, as well as the mechanism of virulence factor action.[69] They designed nanosponges (NSs) coated with membranes isolated from macrophages or red blood cells. As shown in Figure 6.7, bacterial virulence factors from the bacterial supernatant were captured by NSs, digested by trypsin, and further analysed by LC–MS/MS-based quantitative proteomics to determine host cell-specific virulence factors.

Proteomics could be combined with expression library screening which consisted of standard recombinant DNA and immunochemistry techniques to rapidly identify microbial immunoproteomes.[70] By using this method, 207 proteins of the *E. coli* (strain O157:H7) immunome were identified in bovine reservoirs in three weeks. This proteomics-based expression library screening method has great potential for discovering bacterial proteome targets for disease diagnosis and treatment.

In addition to studying immune responses, proteomics is also used in vaccine development. Proteins localized on bacterial surfaces can be considered as immunogens and contribute to vaccine development. Five kinds of bacterial subfractionation methods were utilized to obtain bacterial outer membranes for further proteomics studies.[71] The method combining French press cell lysis with treatment with chaotropic reagents was discovered to be applicable to a lot of Gram-negative bacteria such as *E. coli*, *P. aeruginosa*, and *Yersinia pseudotuberculosis*. The outer membrane proteins of *Helicobacter pylori* were digested and analysed using a simple system comprised of two-dimensional electrophoresis and MALDI-TOF MS to identify proteins in the outer membrane fraction of bacteria to generate vaccine candidates.[72] Bacteria can attach to host cells through a special protein–carbohydrate

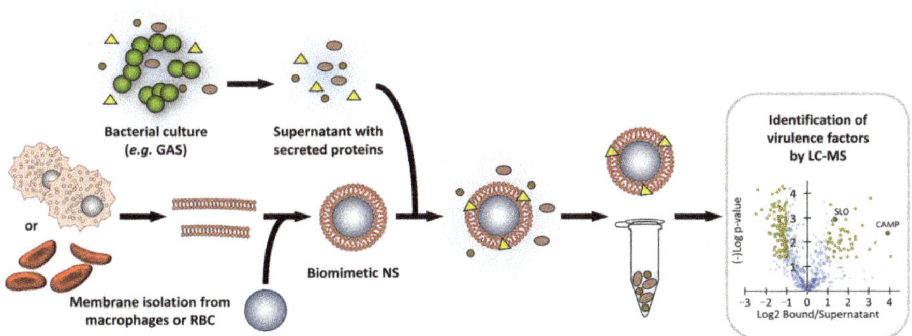

Figure 6.7 Scheme of proteomic identification of proteins secreted into the supernatant based on the system composed of biomimetic NSs and LC–MS/MS. Reproduced from ref. 69 with permission from American Chemical Society, Copyright 2017.

interaction. Thomas *et al.* utilized a carbohydrate-containing crosslinking probe to select the surface adhesion proteins of *H. pylori*.[73] After SDS–PAGE and enzymatic digestion, approximately 300 fmol purified adhesion proteins were successfully identified using MALDI-TOF MS. Besides the surface-exposed proteins, the immunoproteome and secretome can also contribute to vaccine development.[74] Bumann *et al.* aimed to analyse secreted proteins from *H. pylori* to develop a vaccine.[75] By using two-dimensional gel electrophoresis followed by LC–MS/MS, 26 secreted proteins were identified from bacterial supernatants that could contribute to the vaccine development.

6.5.2 Interactions in Bacterial Communities

Proteomics is a powerful means for understanding the behaviours of bacteria in microbial communities. With the richness of the diversity of species in the community, the proteome of the colonist that dominates changes greatly, indicating that interactions between species have an influence on bacterial metabolism.[76] Three cystic fibrosis-related pathogens, including *Burkholderia cepacia*, *P. aeruginosa*, and *S. aureus*, were mixed together for further interaction analysis based on proteomics.[77] By comparing the proteomes of the microbial community with or without the impact of ceftazidime, proteins associated with bacterial interactions, responses to antibiotic treatment, and metabolic shifts were identified and found to have significant changes, illustrating that an interaction between bacterial species occurs on the basis of SOS response and metabolite competition. To study the effects of interspecies interactions on biofilm development, Herschend *et al.* applied metaproteomics analysis to study protein abundance in biofilms of single bacterial species, including *Stenotrophomonas rhizophila*, *Xanthomonas retroflexus*, *Microbacterium oxydans*, and *Paenibacillus amylolyticus*, as well as in the biofilms of their communities, to investigate interspecies interactions.[78] Cross-feeding specific amino acids between species promoted the development of a community. Competition also existed in the community, such as the opposite regulation of fermentation and nitrogen pathways between *X. retroflexus* and *P. amylolyticus*.

6.6 Bacterial AMR Analysis

AMR testing plays an important role in the use of clinical antibiotics for treatment of bacterial infectious diseases. Phenotype-based methods for AMR testing make it difficult to discriminate bacteria at the species and strain levels. Due to the low correlation between resistance genes and phenotypes, sequencing-based approaches can lead to false-positive results in resistance analysis. MS-based proteomics analysis overcomes these disadvantages and can be well applied to AMR analysis. Bacterial AMR analysis is usually accomplished by directly detecting AMR determinants. In addition, proteomics analysis provides abundant protein molecular information, and is a powerful technique to illustrate the biological pathways of biological systems and guide the understanding of AMR. Comparative proteomics methods, such as comparing differently expressed proteins between antibiotic-resistant and antibiotic-sensitive strains and comparing bacterial cells treated with and without antibiotics, are commonly used to identify the essential proteins that promote the progression of AMR and analyse the mechanism of AMR development.

6.6.1 AMR Determinant Detection

Chen *et al.* combined genomics and proteomics by using whole gene sequencing (WGS) and LC–MS/MS-based shotgun proteomics to investigate the AMR determinants of four *Campylobacter jejuni* isolates and further detect their AMR.[79] Compared to the genomic approach in which three isolates were identified with the β-lactamase gene, $bla_{OXA\text{-}61}$, one isolate was identified as having this resistance gene by the proteomics-based method, which is consistent with the results of phenotypic testing. Therefore, the proteomics-based approach has higher accuracy in AMR detection. To thoroughly complete bacterial AMR phenotypic prediction in an unbiased manner, Blumenscheit *et al.* performed bacterial analysis using untargeted LC–MS/MS-based proteomics in DIA mode, and developed a software tool named rawDIAtect for AMR-related peptide detection.[80] A total of 29 isolates from 7 species with different types of AMR, such as carbapenemase-producing, extended-spectrum β-lactamase (ESBL)-producing, and aminoglycoside acetyltransferase-carrying isolates, were selected as model samples to demonstrate the performance of this protocol. Eleven AMR determinants representing 13 protein isoforms were detected and identified. The sensitivity and specificity of this AMR detection method were as high as 98% and 100%, respectively.

6.6.2 Bacterial AMR Mechanistic Study by Strain Comparison

Understanding the mechanism of bacterial AMR emergence is of great importance. Comparative proteomics analysis has great potential to link AMR phenotypes to biomolecules and is a powerful approach to elucidate the mechanism of AMR.[81] Both different types and abundances of expressed proteins can be used to account for AMR. Comparing the proteomes between antibiotic-resistant and antibiotic-sensitive strains, as well as between strains with different degrees of antibiotic resistance, is one of the ways to study AMR-related proteins.

Monteiro *et al.* utilized classical proteomics based on two-dimensional gel electrophoresis LC–MS/MS to study the protein expression variations between ESBL-producing *E. coli* (ESBL-*E. coli*) and sensitive *E. coli*.[82] Resistant proteins such as β-lactamase CTX-M-1 and TEM, as well as proteins associated with ROS, were highly expressed. Thirty-nine proteins with significant variations were identified in the rifampicin-resistant strain of *Brucella melitensis* compared to its sensitive strain.[83] Some up-regulated proteins were enriched in the metabolic pathways of the tricarboxylic acid (TCA) cycle, arginine biosynthesis, and fatty acid synthesis and metabolism, which are closely related to bacterial survival and metabolism. Four multidrug-resistant *E. coli* strains present in slaughterhouses were studied based on comparative proteomics.[84] Proteins associated with bacterial pathogenesis, bacterial resistance, and potential antibacterial targets were determined in all four strains.

6.6.3 Bacterial AMR Mechanistic Study by Antibiotic Stimulation

The comparison of proteomes between bacterial strains with or without antibiotic treatment is another method for AMR research. Studies on the AMR mechanism of Gram-negative pathogens such as *P. aeruginosa* and *E. coli* could all be accomplished by comparing the bacterial proteomes between antibiotic stimulation and normal growth.[85] Sharma *et al.* analysed the fluctuations in the proteome of carbapenemase-producing *K. pneumoniae* under the influence of meropenem.[86] Proteins that underwent significant variations during meropenem treatment were functionally assigned. Proteins involved in DNA/RNA modifications and carbapenems cleavage were overexpressed. The study of bacterial survival mechanisms of antimicrobial peptides could also be realized by comparing sensitive and resistant bacterial strains under the influence of antimicrobial peptides based on quantitative proteomics.[87] Ma *et al.* researched on the antibacterial mechanism of daptomycin against *S. aureus*.[88] By using iTRAQ quantitative proteomics analysis, 51 proteins of *S. aureus*, which participate in catalytic activity, nucleotide binding, protein binding, and other biological activities, were found to show variations when the bacterial cells were affected by daptomycin at sub-inhibitory concentrations. Moreover, 7 cell membrane proteins showed significant changes, which were closely correlated with the decrease in bacterial membrane potential, due to the antibacterial activity of daptomycin.

Besides, antibiotics at different concentrations and at different stimulating times can also result in the different expression levels of bacterial proteins. Low concentrations of ciprofloxacin led to up-regulation of proteins associated with iron and polyamine uptake, as well as DNA repair in *P. aeruginosa*. When treated with ciprofloxacin at high concentrations, proteins associated with the mexCD–OprJ efflux pump were overexpressed in *P. aeruginosa*.[82] The proteome variations of *Mycobacterium tuberculosis* were studied at different times (24 h, 48 h, 4 days, and 6 days) on treatment with double minimum inhibitory concentration (MIC) antibiotics (isoniazid, rifampin, moxifloxacin, mefloquine, and bedaquiline) based on quantitative shotgun proteomics.[89] Some proteins encoded by genes, such as LpqY and Rv1006, were found to contribute to metabolic escape pathways to prolong bacterial survival under the impact of antibiotics at bactericidal concentrations.

Proteomics can also be used for the development of new antibiotics and the study of bactericidal mechanisms. Karanja *et al.* investigated the bactericidal mechanism of newly synthesized alkynyl isoquinoline compounds using proteomics techinques.[90] These antimicrobial compounds showed a stronger clearance ability against methicillin-resistant *S. aureus* than vancomycin and would not induce resistance. Proteomics analysis showed that the cell wall and nucleic acid biosynthesis of *S. aureus* were disturbed when treated with these bactericidal compounds. Proteomics integrated with machine learning can identify proteins that can be used to predict signatures of antibiotics and classify antibiotics based on their different modes of action, which is fundamental for new drug development.[91]

6.6.4 Multi-omics-based Bacterial AMR Analysis

Proteomics can be combined with other omics such as genomics, transcriptomics, and metabolomics, to thoroughly analyse bacterial AMR. Malmström *et al.* integrated WGS, shotgun MS, and DIA-MS to analyse bacterial strains.[92] WGS was used to build a database for constructing a spectral library on the basis of shotgun MS analysis. Based on the spectral library, bacterial peptides could be quantified and differentially expressed proteins could be identified between non-invasive and invasive *Streptococcus pyogenes* strains. Proteins with significantly increased abundance in invasive strains were enriched for the protein functions arginine deiminase pathway, *S. pyogenes* virulome and sucrose metabolism. Cheng *et al.* utilized iTRAQ-based quantitative proteomics to analyse the proteome of levofloxacin-resistant *Vibrio alginolyticus*.[93] Important metabolic pathways such as the TCA cycle showed significant down-regulation. The authors also used transcriptomics to validate gene expression in the TCA cycle by using quantitative reverse transcription polymerase chain reaction (RT-PCR). By combining proteomics and metabolomics, Li *et al.* compared the changes in proteins and metabolites between *E. coli* treated with and without ciprofloxacin.[94] Ciprofloxacin was found to inhibit DNA replication and enhance the expression of DNA gyrase and DNA topoisomerase.

6.6.5 Bacterial Biofilm Study

Biofilms are bacterial colonies with a self-produced matrix that provide a good physiological environment for bacteria and protect them from antibiotics.[95] Biofilms are closely related to bacterial AMR. Studying the proteome of biofilms facilitates the study of bacterial AMR and the discovery of sterilization methods. A DIA-based proteomics technique was used to identify 2385 proteins from a wild-type strain of *Aeromonas hydrophila* and its cob-deleted mutant.[96] A total of 385 increased proteins and 46 decreased proteins were found in this mutant, which could produce more biofilms than the wild-type strain. Metabolic pathways like the ribosome, bacterial secretion system, protein export, RNA degradation, β-lactam resistance, and oxidative phosphorylation pathways were enriched based on these proteins with significant variations.

Proteome comparison between biofilm and planktonic forms of bacterial cells based on comparative proteomics is widely used for biofilm analysis. Erdmann utilized DIA technique-based proteomics to analyse clinical *P. aeruginosa* isolates in planktonic and biofilm forms.[97] Firstly, the *P. aeruginosa* genome database was selected as the reference for the construction of a spectral library. According to the spectral library, 6232 peptides were identified, corresponding to 1996 proteins of 27 clinical *P. aeruginosa* strains. By using principal component analysis (PCA), biofilm and planktonic bacteria could be distinctively divided into two groups, and biofilm proteome profiles between the strains were more different than planktonic groups. The differently expressed proteins between planktonic and biofilm forms of *M. tuberculosis* (a virulent type strain) were also studied based on the proteomics protocol.[98] A total of 47 different proteins were found in the biofilm of *M. tuberculosis* that are involved in cell wall and cell processes, conserved hypotheticals, and intermediary metabolism and respiration.

Apart from the aforementioned research by label-free quantitative proteomics analysis, labelled quantitative methods for proteomics are also used in biofilm studies. Chua *et al.* utilized pulsed-SILAC quantitative proteomics, to quantify the newly expressed proteins of the biofilm subunit of colistin-resistant *P. aeruginosa* (Figure 6.8).[99] Colistin-resistant bacteria migrated to the top region of the biofilm killed by colistin, and then formed new resistant

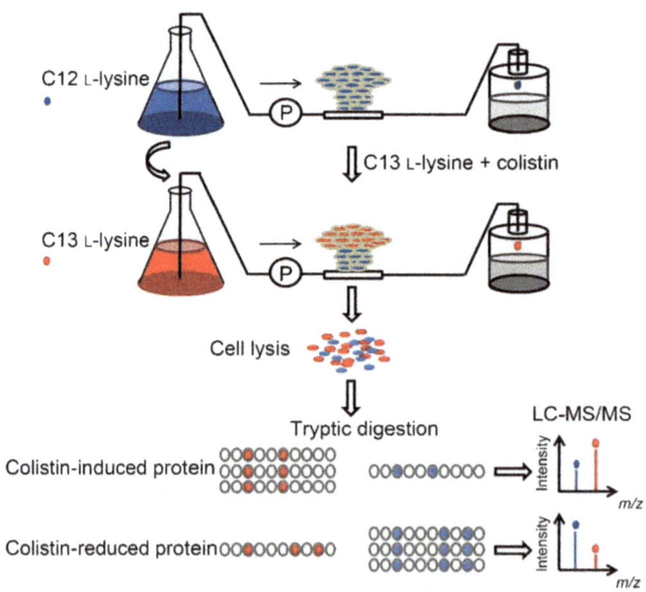

Figure 6.8 Workflow of colistin-resistant and colistin-sensitive subpopulations based on pulsed-SILAC proteomics analysis. Reproduced from ref. 99 with permission from Springer Nature, Copyright 2016.

sub-populations through quorum sensing. The researchers found that protein PilF, which resembles type IV pili, and protein LasB, which is related to quorum sensing in colistin-related subpopulations, were up-regulated.

Biofilms are heterogeneous and have different spatial protein expressions. The spatial heterogeneity of *P. aeruginosa* biofilms was studied by combining laser ablation sample transfer (LAST) which enables sampling at specific sites and label-free proteomics based on LC–MS/MS.[100] Although *P. aeruginosa* is an aerobic bacterium, a higher abundance of proteins related to adenosine triphosphate (ATP) synthase were identified and quantified in the anoxic region of the biofilm than in the oxic region. Besides, proteins associated with arginine and polyamine metabolism, as well as virulence and environmental stress mediation, were all in higher abundance in the anoxic region.

6.6.6 Bacterial Filamentation Study

Bacterial cells tend to change their morphology, like undergoing filamentation, when they are disturbed by antibiotics. Morphological change is a feedback strategy of bacterial cells existing in stressful environments and is closely related to the production of daughter cells with stronger AMR.[95,101] On the basis of this phenomenon, Zhang *et al.* integrated microfluidic chips with proteomics based on MALDI-TOF MS to study the differential proteins during bacterial morphological changes when bacterial cells were stimulated with antibiotics, as shown in Figure 6.9.[102] The clinically isolated carbapenem-resistant *K. pneumoniae* strain was selected as a model to demonstrate this method. By combining transcriptomics analysis, three proteins, including

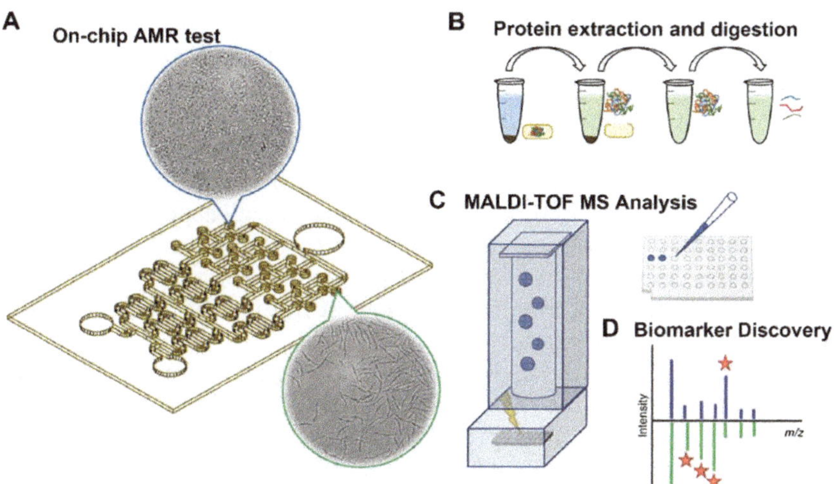

Figure 6.9 Workflow of studying the protein mutation during bacterial morphological changes under the impact of antibiotics based on microfluidic chips and MALDI-TOF MS. Reproduced from ref. 102 with permission from American Chemical Society, Copyright 2019.

fimbrial subunit type 3, penicillin-binding protein activator LpoB, and 30S ribosomal protein S14, were found to be closely related to bacterial AMR.

Culture shaking speed can also induce bacterial filamentation. *Pseudomonas putida* cultured at a low shaking speed preferably undergo filamentation and exhibit more resistance to heat and saline stressors compared to culturing at a high shaking speed. By using LC–MS/MS based proteomics, the expression of the heat protein ibpA and recombinase RecA was found to be induced at a lower culture shaking speed.[103]

6.7 Conclusions and Perspectives

MS-based proteomics can obtain abundant protein information that can be well applied to bacterial analysis. In this chapter, the applications of proteomics techniques in bacterial analysis were comprehensively described, including proteomics analysis strategies, data acquisition modes and quantitative approaches, proteomics analysis for bacterial identification and classification, proteomics analysis to study bacteria–bacteria/host cell interactions, and proteomics applications in bacterial AMR diagnosis and AMR mechanistic research.

MS-based proteomics is an analytical technique that includes sample preparation, protein/peptide purification and separation, MS analysis, and data processing. The optimized method of bacterial protein extraction and digestion simplifies the experimental procedure and enhances the data acquisition efficiency of protein/peptide analysis, laying a good foundation for MS analysis. The mass spectrometer as a detection instrument is at the heart of proteome analysis. The rapid development of mass spectrometers constantly refreshes the ability of protein detection. For example, the emergence of ion mobility mass spectrometers has ushered in a new era of four-dimensional proteomics. On the basis of traditional three-dimensional parameters, including retention time, *m*/*z*, and ion intensity, the addition of mobility separation greatly improves the scanning speed and detection sensitivity of the mass spectrometer. Data processing algorithms are designed to identify more proteins/peptides with higher accuracy and lower error rates. Therefore, all aspects of proteomics analysis (sample pretreatment, mass spectrometry, data processing, *etc.*) strive to promote the development of bacterial proteomics towards precise identification and quantification, even for the smallest number of bacterial cells and complex bacterial compositions.

Rapid and accurate bacterial identification is essential for the diagnosis of clinical bacterial infectious diseases, and is the basis for subsequent treatment and medication. Using proteomics techniques to identify bacterial species enables more accurate bacterial identification by obtaining a variety of protein information. However, this proteomics-based method, especially when using bottom-up proteomics, is somewhat complex and time-consuming compared to the ribosomal protein fingerprinting method based on MALDI-TOF MS, which already has satisfactorily high enough accuracy and covers the vast majority of bacteria in clinics.

Proteomics provides protein information that reflects gene expression and function in bacterial cells. Comparative proteomics is a powerful tool for discovering and screening biomarkers. By comparing the proteomes of drug-sensitive and drug-resistant bacteria, or non-invasive and invasive bacteria, or non-pathogenic and pathogenic bacteria, biomarkers of bacterial AMR, or invasive bacterial diseases, or infectious diseases can be discovered and screened, which can be further applied to the development of new drugs. Comparative proteomics also has specific advantages in mechanistic studies. Proteins with significant variations related to bacterial AMR or infectious diseases can be functionally annotated and enriched on metabolic pathways to demonstrate the emergence and progression of bacterial AMR and infectious diseases.

Proteomics can be combined with a lot of omics techniques, including genomics, transcriptomics, metabolomics, and phenomics. Information can be complemented between the various omics. The combination of these omics, *i.e.*, multi-omics, can provide more abundant biomolecular information and provide a comprehensive understanding of bacteria and their physiological activity and pathogenicity. Metaproteomics is an effective tool for studying bacterial communities: it studies the states of bacteria that are closer to natural conditions and reflects the real situation. For example, it is often used to study the relationship between the state of flora in the human gut or human oral cavity and diseases.

Overall, compared to culture-based and nucleic acid-based methods, MS-based proteomics has greater potential for obtaining more information in bacterial analysis, especially in the study of molecular mechanisms of bacteria-related diseases. In the future, MS-based proteomics techniques will be further improved and combined with artificial intelligence to cover more bacteria-associated research fields, such as early diagnosis of bacterial infectious diseases, disease assessment, and single bacterial proteome detection.

Abbreviations

AMR	Antimicrobial resistance
AP-MALDI-MS/MS	Atmospheric pressure matrix-assisted laser desorption/ionization mass spectrometry
ATP	Adenosine triphosphate
CD + ISD	Strain-specific cell disruption with in-solution digestion
CRC	Colorectal cancer
CZE	Capillary zone electrophoresis
DDA	Data dependent acquisition
DIA	Data independent acquisition
ESI-MS	Electrospray ionization mass spectrometry
ESBL	Extended-spectrum β-lactamase
ESBL-*E. coli*	Extended-spectrum β-lactamase-producing *Escherichia coli*
GDP	Gross domestic product
ICAT	Isotope-coded affinity tag

IPG-IEF	Isoelectric focusing in immobilized pH gradients
iTRAQ	Isobaric tags for relative and absolute quantification
LAST	Laser ablation sample transfer
LC–MS/MS	Liquid chromatography–tandem mass spectrometry
MABSC	*Mycobacterium abscessus* complex
MALDI-TOF MS	Matrix-assisted laser desorption/ionization time-of-flight mass spectrometry
MALDI-TOF/TOF MS	Matrix-assisted laser desorption/ionization time-of-flight/time-of-flight mass spectrometry
MIC	Minimum inhibitory concentration
MRM	Multiple reaction monitoring
MS	Mass spectrometry
MudPIT	Multi-dimensional protein identification techniques
NGS	Next generation sequencing
NSs	Nanosponges
PBS	Phosphate buffered saline
PCA	Principal component analysis
PCR	Polymerase chain reaction
PRM	Parallel reaction monitoring
RPLC	Reverse phase liquid chromatography
RT-PCR	Reverse transcription polymerase chain reaction
SDS	Sodium dodecyl sulfate
SDS–PAGE	Sodium dodecyl sulfate–polyacrylamide gel electrophoresis
SEC	Size exclusion chromatography
SEC–RPLC–CZE-MS/MS	Size exclusion chromatography and reverse phase liquid chromatography combined with capillary zone electrophoresis tandem mass spectrometry
SILAC	Stable isotope labelling by amino acids in cell culture
SP3	Single-pot, solid-phase enhanced sample preparation
STEPS	Systematic Trial and Error for Parameter Selection
TCA	Tricarboxylic acid
TMT	Tandem mass tag
UTIs	Urinary tract infections
WGS	Whole gene sequencing

Acknowledgements

This work was supported by the China Postdoctoral Science Foundation (2022M720806).

References

1. T. A. Russo and J. R. Johnson, *Microbes Infect.*, 2003, **5**, 449–456.
2. J. Paireau, A. Chen, H. Broutin, B. Grenfell and N. E. Basta, *Lancet Global Health*, 2016, **4**, e370–e377.

3. G. R. Corey, *Clin. Infect. Dis.*, 2009, **48**(Suppl. 4), S254–S259.
4. J. O'Neill, The Review on Antimicrobial Resistance, 2014.
5. F. Yu, Y. Li, M. Li, L. Tang and J. J. He, *Biosens. Bioelectron.*, 2017, **89**, 880–885.
6. K. S. Gracias and J. L. McKillip, *Can. J. Microbiol.*, 2004, **50**, 883–890.
7. Z. A. Khan, M. F. Siddiqui and S. Park, *Diagnostics*, 2019, **9**, 49.
8. D. Zhang, Y. Zhang, F. Yin, Q. Qin, H. Bi, B. Liu and L. Qiao, *Analyst*, 2021, **146**, 515–520.
9. D. Ziegler, J. F. Pothier, J. Ardley, R. K. Fossou, V. Pflüger, S. de Meyer, G. Vogel, M. Tonolla, J. Howieson, W. Reeve and X. Perret, *Appl. Microbiol. Biotechnol.*, 2015, **99**, 5547–5562.
10. B. W. Buchan, K. M. Riebe and N. A. Ledeboer, *J. Clin. Microbiol.*, 2012, **50**, 346–352.
11. A. M. Huang, D. Newton, A. Kunapuli, T. N. Gandhi, L. L. Washer, J. Isip, C. D. Collins and J. L. Nagel, *Clin. Infect. Dis.*, 2013, **57**, 1237–1245.
12. D. Zhang, H. Bi, B. Liu and L. Qiao, *Anal. Chem.*, 2018, **90**, 5512–5520.
13. J. Rychert, *J. Infectiology Epidemiol.*, 2019, **2**, 1–5.
14. K. Chandramouli and P. Y. Qian, *Hum. Genomics Proteomics*, 2009, **2009**, 239204.
15. M. R. Wilkins, J. C. Sanchez, A. A. Gooley, R. D. Appel, I. Humphery-Smith, D. F. Hochstrasser and K. L. Williams, *Biotechnol. Genet. Eng. Rev.*, 1996, **13**, 19–50.
16. B. Domon and R. Aebersold, *Science*, 2006, **312**, 212–217.
17. Y. Zhang, B. R. Fonslow, B. Shan, M.-C. Baek and J. R. Yates III, *Chem. Rev.*, 2013, **113**, 2343–2394.
18. P. Chaurand, D. B. Friedman and R. M. Caprioli, in *The Molecular Basis of Cancer*, ed. J. Mendelsohn, P. M. Howley, M. A. Israel, J. W. Gray and C. B. Thompson, W.B. Saunders, Philadelphia, 3rd edn, 2008, pp. 293–307.
19. V. S. Priyadharshini and L. M. Teran, in *Precision Medicine for Investigators, Practitioners and Providers*, ed. J. Faintuch and S. Faintuch, Academic Press, 2020, pp. 255–261.
20. A. D. Catherman, O. S. Skinner and N. L. Kelleher, *Biochem. Biophys. Res. Commun.*, 2014, **445**, 683–693.
21. K. A. Brown, J. A. Melby, D. S. Roberts and Y. Ge, *Expert Rev. Proteomics*, 2020, **17**, 719–733.
22. L. M. Smith and N. L. Kelleher, *Nat. Methods*, 2013, **10**, 186–187.
23. A. D. Catherman, O. S. Skinner and N. L. Kelleher, *Biochem. Biophys. Res. Commun.*, 2014, **445**, 683–693.
24. F. Zhang, W. Ge, G. Ruan, X. Cai and T. Guo, *Proteomics*, 2020, **20**, e1900276.
25. L. Krasny and P. H. Huang, *Mol. Omics*, 2021, **17**, 29–42.
26. D. L. Tabb, L. Vega-Montoto, P. A. Rudnick, A. M. Variyath, A. J. Ham, D. M. Bunk, L. E. Kilpatrick, D. D. Billheimer, R. K. Blackman, H. L. Cardasis, S. A. Carr, K. R. Clauser, J. D. Jaffe, K. A. Kowalski, T. A. Neubert, F. E. Regnier, B. Schilling, T. J. Tegeler, M. Wang, P. Wang,

J. R. Whiteaker, L. J. Zimmerman, S. J. Fisher, B. W. Gibson, C. R. Kinsinger, M. Mesri, H. Rodriguez, S. E. Stein, P. Tempst, A. G. Paulovich, D. C. Liebler and C. Spiegelman, *J. Proteome Res.*, 2010, **9**, 761–776.
27. J. D. Egertson, A. Kuehn, G. E. Merrihew, N. W. Bateman, B. X. MacLean, Y. S. Ting, J. D. Canterbury, D. M. Marsh, M. Kellmann, V. Zabrouskov, C. C. Wu and M. J. MacCoss, *Nat. Methods*, 2013, **10**, 744–746.
28. J. Li, L. S. Smith and H. J. Zhu, *Drug Discovery Today: Technol.*, 2021, **39**, 49–56.
29. P. Willems, U. Fels, A. Staes, K. Gevaert and P. Van Damme, *J. Proteome Res.*, 2021, **20**, 1165–1177.
30. A. Doerr, *Nat. Methods*, 2015, **12**, 35–35.
31. Y. Yakkioui, Y. Temel, E. Chevet and L. Negroni, in *Meth. Enzymol.*, ed. A. K. Shukla, Academic Press, 2017, vol. 586, pp. 229–246.
32. L. Zhang and J. E. Elias, *Methods Mol. Biol.*, 2017, **1550**, 185–198.
33. V. J. Patel, K. Thalassinos, S. E. Slade, J. B. Connolly, A. Crombie, J. C. Murrell and J. H. Scrivens, *J. Proteome Res.*, 2009, **8**, 3752–3759.
34. S. Anand, M. Samuel, C. S. Ang, S. Keerthikumar and S. Mathivanan, *Methods Mol. Biol.*, 2017, **1549**, 31–43.
35. D. A. Megger, L. L. Pott, M. Ahrens, J. Padden, T. Bracht, K. Kuhlmann, M. Eisenacher, H. E. Meyer and B. Sitek, *Biochim. Biophys. Acta*, 2014, **1844**, 967–976.
36. Z. Li, R. M. Adams, K. Chourey, G. B. Hurst, R. L. Hettich and C. Pan, *J. Proteome Res.*, 2012, **11**, 1582–1590.
37. J. G. Meyer and B. Schilling, *Expert Rev. Proteomics*, 2017, **14**, 419–429.
38. C. A. Sobsey, S. Ibrahim, V. R. Richard, V. Gaspar, G. Mitsa, V. Lacasse, R. P. Zahedi, G. Batist and C. H. Borchers, *Proteomics*, 2020, **20**, e1900029.
39. S. Saleh, A. Staes, S. Deborggraeve and K. Gevaert, *Proteomics*, 2019, **19**, e1800435.
40. N. Hansmeier, S. Sharma and T.-C. Chao, in *Proteomics in Systems Biology: Methods and Protocols*, ed. J. Geddes-McAlister, Springer US, New York, NY, 2022, pp. 63–70.
41. J.-Y. Zhou, G. P. Dann, T. Shi, L. Wang, X. Gao, D. Su, C. D. Nicora, A. K. Shukla, R. J. Moore, T. Liu, D. G. Camp II, R. D. Smith and W.-J. Qian, *Anal. Chem.*, 2012, **84**, 2862–2867.
42. A. R. Vaezzadeh, J. M. P. Deshusses, P. Waridel, P. Francois, C. G. Zimmermann-Ivol, P. Lescuyer, J. Schrenzel and D. F. Hochstrasser, *J. Microbiol. Meth.*, 2010, **80**, 56–62.
43. M. Baniasad, Y. Kim, M. Shaffer, A. Sabag-Daigle, I. Leleiwi, R. A. Daly, B. M. M. Ahmer, K. C. Wrighton and V. H. Wysocki, *Anal. Bioanal. Chem.*, 2022, **414**, 2317–2331.
44. X. Yin, Y. Zhang, X. Liu, C. Chen, H. Lu, H. Shen and P. Yang, *Analyst*, 2015, **140**, 1314–1322.
45. V. Gautier, E. Mouton-Barbosa, D. Bouyssié, N. Delcourt, M. Beau, J.-P. Girard, C. Cayrol, O. Burlet-Schiltz, B. Monsarrat and A. Gonzalez de Peredo, *Mol. Cell. Proteomics*, 2012, **11**, 527–539.

46. S. Wickramasekara, J. Neilson, N. Patel, L. Breci, A. Hilderbrand, R. M. Maier and V. Wysocki, *J. Biomed. Biotechnol.*, 2011, **2011**, 701928.
47. L. Wöhlbrand, R. Rabus, B. Blasius and C. Feenders, *J. Mol. Microbiol. Biotechnol.*, 2017, **27**, 199–212.
48. D. M. Tracz, S. J. McCorrister, P. M. Chong, D. M. Lee, C. R. Corbett and G. R. Westmacott, *J. Microbiol. Meth.*, 2013, **94**, 54–57.
49. J. P. Dworzanski, S. V. Deshpande, R. Chen, R. E. Jabbour, A. P. Snyder, C. H. Wick and L. Li, *J. Proteome Res.*, 2006, **5**, 76–87.
50. J. Nguyen and S. C. Russell, *J. Am. Soc. Mass Spectrom.*, 2010, **21**, 993–1001.
51. P. D. Piehowski, V. A. Petyuk, J. D. Sandoval, K. E. Burnum, G. R. Kiebel, M. E. Monroe, G. A. Anderson, D. G. Camp 2nd and R. D. Smith, *Proteomics*, 2013, **13**, 766–770.
52. F. Boulund, R. Karlsson, L. Gonzales-Siles, A. Johnning, N. Karami, O. Al-Bayati, C. Åhrén, E. R. B. Moore and E. Kristiansson, *Mol. Cell. Proteomics*, 2017, **16**, 1052–1063.
53. K. H. Jarman, N. C. Heller, S. C. Jenson, J. R. Hutchison, B. L. D. Kaiser, S. H. Payne, D. S. Wunschel and E. D. Merkley, *J. Proteome Res.*, 2018, **17**, 3075–3085.
54. S. Blankenburg, C. Hentschker, A. Nagel, P. Hildebrandt, S. Michalik, D. Dittmar, K. Surmann and U. Völker, *Proteomics*, 2019, **19**, e1900192.
55. J. Zhou, L. Zhang, H. Chuan, A. Sloan, R. Tsang and K. Cheng, *Methods in Molecular Biology*, 2019, pp. 113–121.
56. S. Fortuin, J. Iradukunda, A. J. Nel, J. M. Blackburn and N. C. Soares, *MethodsX*, 2021, **8**, 101277.
57. J. Iradukunda, T. Ganief, J. M. Blackburn and N. C. Soares, *Methods Mol. Biol.*, 2021, **2259**, 181–189.
58. F. Roux-Dalvai, C. Gotti, M. Leclercq, M. C. Hélie, M. Boissinot, T. N. Arrey, C. Dauly, F. Fournier, I. Kelly, J. Marcoux, J. Bestman-Smith, M. G. Bergeron and A. Droit, *Mol. Cell. Proteomics*, 2019, **18**, 2492–2505.
59. C. Lozano, M. Kielbasa, J. C. Gaillard, G. Miotello, O. Pible and J. Armengaud, *Microorganisms*, 2022, **10**, 719.
60. B. J. Kunath, G. Minniti, M. Skaugen, L. H. Hagen, G. Vaaje-Kolstad, V. G. H. Eijsink, P. B. Pope and M. Arntzen, *Adv. Exp. Med. Biol.*, 2019, **1073**, 187–215.
61. E. H. Hart, C. J. Creevey, T. Hitch and A. H. Kingston-Smith, *Sci. Rep.*, 2018, **8**, 10504.
62. S. Long, Y. Yang, C. Shen, Y. Wang, A. Deng, Q. Qin and L. Qiao, *npj Biofilms Microbi.*, 2020, **6**, 14.
63. P. A. Demirev, A. B. Feldman, P. Kowalski and J. S. Lin, *Anal. Chem.*, 2005, **77**, 7455–7461.
64. P. B. Pandeswari and V. Sabareesh, *RSC Adv.*, 2019, **9**, 313–344.
65. C. K. Fagerquist, W. J. Zaragoza, B. G. Lee, J. C. Yambao and B. Quiñones, *Clin. Mass Spectrom.*, 2019, **11**, 27–36.
66. M. Dupré, M. Duchateau, C. Malosse, D. Borges-Lima, V. Calvaresi, I. Podglajen, D. Clermont, M. Rey and J. Chamot-Rooke, *J. Proteome Res.*, 2021, **20**, 202–211.

67. E. N. McCool, R. A. Lubeckyj, X. Shen, D. Chen, Q. Kou, X. Liu and L. Sun, *Anal. Chem.*, 2018, **90**, 5529–5533.
68. M. Steindor, V. Nkwouano, A. Stefanski, K. Stuehler, T. R. Ioerger, D. Bogumil, M. Jacobsen, C. R. Mackenzie and R. Kalscheuer, *Microbes Infect.*, 2019, **21**, 154–162.
69. U. Distler and S. Tenzer, *ACS Nano*, 2017, **11**, 11768–11772.
70. I. T. Kudva, B. Krastins, H. Sheng, R. W. Griffin, D. A. Sarracino, P. I. Tarr, C. J. Hovde, S. B. Calderwood and M. John, *Mol. Cell. Proteomics*, 2006, **5**, 1514–1519.
71. M. Thein, G. Sauer, N. Paramasivam, I. Grin and D. Linke, *J. Proteome Res.*, 2010, **9**, 6135–6147.
72. D. N. Chakravarti, M. J. Fiske, L. D. Fletcher and R. J. Zagursky, *Vaccine*, 2000, **19**, 601–612.
73. T. Larsson, J. Bergström, C. Nilsson and K. A. Karlsson, *FEBS Lett.*, 2000, **469**, 155–158.
74. D. Bumann, P. R. Jungblut and T. F. Meyer, *Proteomics*, 2004, **4**, 2843–2848.
75. D. Bumann, S. Aksu, M. Wendland, K. Janek, U. Zimny-Arndt, N. Sabarth, T. F. Meyer and P. R. Jungblut, *Infect. Immun.*, 2002, **70**, 3396–3403.
76. R. S. Mueller, V. J. Denef, L. H. Kalnejais, K. B. Suttle, B. C. Thomas, P. Wilmes, R. L. Smith, D. K. Nordstrom, R. B. McCleskey, M. B. Shah, N. C. Verberkmoes, R. L. Hettich and J. F. Banfield, *Mol. Syst. Biol.*, 2010, **6**, 374.
77. S. Kluge, M. Hoffmann, D. Benndorf, E. Rapp and U. Reichl, *Proteomics*, 2012, **12**, 1893–1901.
78. J. Herschend, Z. B. V. Damholt, A. M. Marquard, B. Svensson, S. J. Sørensen, P. Hägglund and M. Burmølle, *Sci. Rep.*, 2017, 7, 16483.
79. C.-Y. Chen, C. G. Clark, S. Langner, D. A. Boyd, A. Bharat, S. J. McCorrister, A. G. McArthur, M. R. Graham, G. R. Westmacott and G. Van Domselaar, *Proteom. Clin. Appl.*, 2020, **14**, 1800182.
80. C. Blumenscheit, Y. Pfeifer, G. Werner, C. John, A. Schneider, P. Lasch and J. Doellinger, *Anal. Chem.*, 2021, **93**, 14599–14608.
81. R. Burchmore, *Expert Rev. Proteomics*, 2014, **11**, 1–3.
82. R. Monteiro, M. Hébraud, I. Chafsey, P. Poeta and G. Igrejas, *J. Proteomics*, 2016, **145**, 167–176.
83. V. Sandalakis, A. Psaroulaki, P.-J. De Bock, A. Christidou, K. Gevaert, G. Tsiotis and Y. Tselentis, *J. Proteome Res.*, 2012, **11**, 2374–2385.
84. S. Ramos, N. Silva, M. Hébraud, H. M. Santos, J. D. Nunes-Miranda, L. Pinto, J. E. Pereira, J. L. Capelo, P. Poeta and G. Igrejas, *OMICS*, 2016, **20**, 362–374.
85. S. L. Keasey, B. M. Zhao, R. G. Ulrich and M. J. Suh, *Methods Mol. Biol.*, 2021, **2296**, 381–392.
86. D. Sharma, A. Garg, M. Kumar and A. U Khan, *J. Proteomics*, 2019, **200**, 102–110.
87. S. J. Balboa and L. M. Hicks, *Meth. Enzymol.*, 2022, **663**, 259–271.

88. W. Ma, D. Zhang, G. Li, J. Liu, G. He, P. Zhang, L. Yang, H. Zhu, N. Xu and S. Liang, *J. Proteomics*, 2017, **150**, 242–251.
89. L. Danelishvili, N. Shulzhenko, J. J. J. Chinison, L. Babrak, J. Hu, A. Morgun, G. Burrows and L. E. Bermudez, *Antimicrob. Agents Chemother.*, 2017, **61**, 00430–00417.
90. C. W. Karanja, N. Naganna, N. S. Abutaleb, N. Dayal, K. I. Onyedibe, U. Aryal, M. N. Seleem and H. O. Sintim, *Molecules*, 2022, **27**, 5085.
91. Y. Yu, A. O'Rourke, Y. H. Lin, H. Singh, R. V. Eguez, S. Beyhan and K. E. Nelson, *ACS Infect. Dis.*, 2020, **6**, 2120–2129.
92. L. Malmström, A. Bakochi, G. Svensson, O. Kilsgård, H. Lantz, A. C. Petersson, S. Hauri, C. Karlsson and J. Malmström, *J. Proteomics*, 2015, **129**, 98–107.
93. Z. X. Cheng, M. J. Yang, B. Peng, X. X. Peng, X. M. Lin and H. Li, *J. Proteomics*, 2018, **181**, 83–91.
94. W. X. Li, S. Zhang, X. Y. Wang, J. Yu, Z. Q. Li, W. X. Lin and X. M. Lin, *J. Proteomics*, 2018, **179**, 61–70.
95. S. Banerjee, K. Lo, N. Ojkic, R. Stephens, N. F. Scherer and A. R. Dinner, *Nat. Phys.*, 2021, **17**, 403–409.
96. G. Wang, Y. Wang, L. Zhang, Q. Cai, Y. Lin, L. Lin and X. Lin, *J. Proteomics*, 2020, **225**, 103848.
97. J. Erdmann, J. G. Thöming, S. Pohl, A. Pich, C. Lenz and S. Häussler, *Cells*, 2019, **8**, 1129.
98. C. Wang, Q. Zhang, Y. Wang, X. Tang, Y. An, S. Li, H. Xu, Y. Li, W. Luan, X. Wang, M. Liu and L. Yu, *Electrophoresis*, 2019, **40**, 2736–2746.
99. S. L. Chua, J. K. H. Yam, P. Hao, S. S. Adav, M. M. Salido, Y. Liu, M. Givskov, S. K. Sze, T. Tolker-Nielsen and L. Yang, *Nat. Commun.*, 2016, 7, 10750.
100. A. C. Pulukkody, Y. P. Yung, F. Donnarumma, K. K. Murray, R. P. Carlson and L. Hanley, *PLoS One*, 2021, **16**, e0250911.
101. J. Bos, Q. Zhang, S. Vyawahare, E. Rogers, S. M. Rosenberg and R. H. Austin, *Proc. Natl. Acad. Sci. U. S. A.*, 2015, **112**, 178–183.
102. D. Zhang, Y. Yang, Q. Qin, J. Xu, B. Wang, J. Chen, B. Liu, W. Zhang and L. Qiao, *Anal. Chem.*, 2019, **91**, 2352–2359.
103. A. Crabbé, B. Leroy, R. Wattiez, A. Aertsen, N. Leys, P. Cornelis and R. Van Houdt, *BMC Microbiol.*, 2012, **12**, 282.

CHAPTER 7

Metabolomics as a Tool for Phenotypic Characterization and Strain Improvement

MARVIN NATHANAEL IMAN,[†] RIFQI AHMAD RIYANTO[†] AND SASTIA PRAMA PUTRI*

Department of Biotechnology, Graduate School of Engineering, Osaka University, Osaka 565-0871, Japan
*Email: sastia_putri@bio.eng.osaka-u.ac.jp

7.1 Introduction

Metabolomics is an interdisciplinary field of study that examines the complete set of metabolites in a target organism by using advanced analytical methods. It offers insight into variations in metabolite profiles and can uncover complex biological changes through the use of statistical analysis techniques. In the post-genomic era, due to its ability to exhaustively and quantitatively profile metabolites in a biological system, metabolomics has become an indispensable tool for numerous scientific disciplines.[1]

The key strengths of metabolomics analysis are its prediction power, sensitivity, and specificity, as well as its versatility. Metabolomics analysis has great prediction power as it analyses metabolites, which are the end products of cellular processes and are considered to be the most accurate representation of the functional state of a cell. As such, metabolomics allows for the identification of patterns and changes in metabolic pathways, which

[†]These authors contributed equally.

New Developments in Mass Spectrometry No. 13
Detection and Analysis of Microorganisms by Mass Spectrometry
Edited by Liang Qiao and Jia Yi

Published by the Royal Society of Chemistry, www.rsc.org

can then be used to make predictions about the functional state of a cell. Metabolomics analyses are also sensitive and specific as they allow the detection of low-abundance metabolites with high accuracy.[1] Additionally, metabolomics analysis is versatile as it can analyse various sample types and can be applied to various research fields, including medical research, synthetic biology, medicine, and predictive modelling of plants, animals, and microbial systems.[2–6]

Recent advancements in mass spectrometry (MS)-based metabolomics have opened new opportunities for improving the bioproductivity of microorganisms in the biomanufacturing industry. The ability to comprehensively analyse the metabolites within a cell has established metabolomics as a versatile and powerful tool for metabolic engineering. Unlike the traditional method of modifying single genes through trial-and-error, metabolomics provides a systematic understanding of cellular metabolism and helps to identify the rate-limiting steps in synthetic biological systems.

In this chapter, the authors provide a comprehensive discussion on microbial metabolomics, with an emphasis on phenotypic characterization and strain improvement. The authors also discuss prospective advances and breakthroughs in phenotypic characterization and strain enhancement based on metabolomics. This chapter focuses on the following topics: MS-based metabolomics methodologies, microbial metabolomics application for strain improvement, case studies, current challenges, and future outlook.

7.2 Methodologies of MS-based Metabolomics

The general framework of a metabolomics study includes five major components: the generation of a working hypothesis, selection of a metabolomics approach, data acquisition method (for determining and optimizing analytical conditions), sample preparation protocol (for sampling, quenching, metabolite extraction, and storage), and data analysis method (for both statistical analysis and visualization).

The first step in a metabolomics study is to pose a relevant biological question and create an appropriate working hypothesis accordingly. To build a robust working hypothesis, researchers must collect as much background information as possible on the organism of interest. Not only will this aid in data interpretation, but this is also essential for determining how many samples and what kinds of samples are necessary, which eventually leads to an effective experimental design. Based on this information, researchers can then decide on the type of analysis, the relevant analytical tools, and the process for preparing biological samples. After acquiring metabolome data, data mining can be performed using statistical methods such as univariate and multivariate analyses. It is then followed by data interpretation, the generation of new hypotheses based on the newly acquired information, and the testing of these hypotheses using an improved experimental design, repeating the cycle.[7]

7.2.1 Analysis Strategies

It is important for researchers to choose the appropriate type of analysis method in advance. Metabolomics has three major approaches: targeted analysis (specific and quantitative measurement of a small set of known metabolites), metabolic profiling (untargeted, high-throughput analysis of a large number of metabolites, including unknowns), and metabolic fingerprinting (rapid, exhaustive analysis of biochemical fingerprints to discriminate between distinct groups, without the need for metabolite identification).[8]

In metabolomics analysis, there is often a trade-off between data quality and metabolite coverage. High data quality requires detailed and specific targeting methods for sensitive quantification, accurate identification, and high accuracy measurement, but may result in limited metabolite coverage. On the other hand, broader methods that allow for more comprehensive metabolite coverage often sacrifice data quality for the sake of wide coverage. The optimal balance between data quality and metabolite coverage will depend on the specific research question and the desired outcome.[9]

In recent years, analysis strategies involving dynamic metabolomics techniques have become increasingly popular. Dynamic metabolomics is a powerful tool for characterizing the metabolic changes that occur in response to various stimuli in an organism. This method involves monitoring the levels of metabolites in a sample over a period of time, allowing researchers to observe metabolic shifts that are indicative of changes in cellular processes. By analysing the temporal changes in the levels of various metabolites, it is possible to determine which metabolic pathways are active or inactive, and how they respond to different stimuli. Data generated from these techniques can then be used to create models of metabolic networks, which provide insights into how metabolic pathways are regulated and how they respond to changes in the environment.[10]

7.2.2 Analytical Platforms

The ultimate objective of metabolomics analysis is to exhaustively profile all metabolites present in biological samples. Unfortunately, due to the wide range of physical and chemical characteristics of metabolites (*e.g.*, polarity, molecular weight, and solubility), there is currently no technology that allows for the simultaneous analysis of all metabolites. As such, narrowing the number of target metabolites through the use of specific analytical instruments is often required. When needed, whole metabolome data can subsequently be acquired through the combination of data from several different analytical instruments targeting different metabolite groups. In the next sections, we discuss the characteristics, development, and most recent applications of the analytical equipment typically employed in microbial metabolomics.

7.2.2.1 GC/MS

Gas chromatography–mass spectrometry (GC/MS) separates complex mixtures of compounds into individual components using gas chromatography, and then detecting and identifying each component based on its unique mass spectrum using MS. One of the key advantages of GC/MS is its high reproducibility, making it a reliable tool for quantitative analyses. The high degree of reproducibility in GC/MS is due to several factors, such as the precise control of column temperature, the use of highly uniform and stable chromatography columns, and the ability to generate highly accurate and consistent mass spectra. The use of the retention index (RI) in conjunction with MS is also a vital factor contributing to the high reproducibility of GC/MS. The RI is a measure of the relative retention time of a compound on the chromatography column and is calculated based on the retention times of a set of reference compounds. By using the RI of a target compound in conjunction with its mass spectrum, GC/MS can provide highly specific and reproducible identification of the target compound.

One important way to improve the detection and quantification of target compounds in GC/MS is through the use of derivatization techniques. Derivatization involves chemically modifying the target compounds to enhance their chromatographic and mass spectral properties. This process can improve the detection sensitivity of the target compounds, or even allow for the detection of compounds that would otherwise be undetectable by GC/MS. Common derivatization techniques include silylation, acylation, and alkylation. Common derivatization reagents include methoxyamine hydrochloride and *N*-methyl-*N*-(trimethylsilyl)trifluoroacetamide.[11]

One of the latest advancements in the GC system is the use and further development of three-dimensional comprehensive GC, exploiting a thermal-flow modulation-hybrid GC × GC × GC–MS system. This was developed as an approach to address the challenges associated with the gas phase analysis of exceedingly complex samples.[12]

7.2.2.2 LC/MS

In contrast to GC/MS, liquid chromatography–mass spectrometry (LC/MS) separates complex mixtures of compounds into individual components using LC. By selecting the appropriate column and mobile phases, a broad spectrum of metabolites ranging from low to high molecular weight and hydrophobic to hydrophilic metabolites can be analysed. A common ionization technique employed in LC/MS-based techniques is electrospray ionization (ESI), which can ionize a wide range of compounds.[13] However, ESI lacks quantitative capabilities due to the presence of the matrix effect (ionization suppression and enhancement due to the co-eluted matrix), which was initially observed by Kebarle *et al.*[14] This matrix effect can be avoided in a number of ways, one of which is by labelling molecules with stable isotopes.[15] Unfortunately, unlike the use of the RI in GC/MS-based analyses, there is no reliable method for standardizing retention time information in LC/MS metabolomics. Therefore, it is vital to create an analytical method that can guarantee a constant retention time throughout the entirety of the chromatogram.

7.2.2.3 CE/MS

Capillary electrophoresis–mass spectrometry (CE/MS) is an effective analytical separation technique for profiling polar and charged metabolites, particularly for compound classes such as amino acids, nucleotides, small organic acids, and sugar phosphates. While CE/MS has unique advantages in specific cases, it is still largely considered to be technically challenging and less reproducible in comparison to GC/MS and LC/MS, which leads to the limited use of CE/MS in metabolomics studies.[16] The limited use of CE/MS may also be attributed to the lack of standard operating procedures and workflows despite new developments in sample throughput and quality control (QC).[17] One of the latest CE/MS applications in microbial metabolomics is regarding the effects of *N*-acetyl-D-glucosamine (GlcNAc) on the intracellular metabolism of nitrogen-containing compounds in yeast species.[18]

7.2.2.4 SFC/MS

Lipids have diverse and crucial roles in a variety of biological events, making them intriguing targets for metabolomics.[19] Bamba *et al.* established a novel method for lipid analysis utilizing supercritical fluid chromatography–mass spectrometry (SFC/MS).[20] SFC, which has characteristics of both GC and HPLC, is a separation technique with high resolution and high throughput that uses a supercritical fluid (a non-condensable fluid over liquid gas at the critical point) as the mobile phase. Because of the supercritical fluid's high diffusion coefficient, SFC has a significantly greater separation capacity than HPLC. Additionally, due to the fact that the polarity of supercritical carbon dioxide is analogous to that of hexane, SFC can also be used to analyse hydrophobic metabolites. Furthermore, the use of a polar organic solvent (such as methanol) as a modifier can significantly modify the polarity of the mobile phase. As a result, SFC/MS can be utilized for the high-throughput and high-resolution investigation of a large array of metabolites, including lipids.[21] An example of the use of CE/MS in microbial analysis is the recently published study that uses online supercritical fluid extraction–supercritical fluid chromatography–triple-quadrupole mass spectrometry (SFE–SFC–QqQ-MS) to achieve high-throughput analysis of microbial colonies without the need for pre-treatments through the use of an automated metabolite extraction process.[22]

7.2.2.5 MALDI-MS

Matrix-assisted laser desorption/ionization, often known as MALDI, utilizes the application of matrix uniformly distributed throughout the sample to aid sample ionization. During the analysis, nitrogen laser pulse (*i.e.*, UV) irradiation is applied to the surface of the mixture to ionize the matrix together with the sample. The matrix serves three basic functions: facilitating the ionization of target compounds by conveying laser energy, inhibiting the photodecomposition of targets, and dissociating the molecular ion or enhancing selective binding by inhibiting or enhancing fragmentation. In contrast to GC/MS or LC/MS, MALDI-MS does not involve chromatography-based separation which results in its limited selectivity.

Due to its disadvantages such as poor quantitative performance, low reproducibility, and matrix interference with ionization, the use of MALDI-MS is often limited to the identification of molecules with high molecular weight, such as proteins. An example of the use of MALDI-MS in microbial metabolomics is the imaging analysis of *Pseudomonas putida* biofilm development.[23]

7.2.2.6 DI-MS

In direct infusion mass spectrometry (DI-MS), samples are injected directly into the ESI source using a syringe pump or nanospray chip without chromatographic separation. Due to its rapid analysis, DI-MS has a very high throughput. Its quantitative performance, however, is inferior to that of LC/MS due to its strong matrix effect. DI-MS also lacks retention time information for metabolites, reducing its molecular selectivity compared to LC/MS and MALDI-MS. FT-ICR and Orbitrap MS have been used to address this issue by applying MS/MS and accurate mass analysis to DI-MS metabolomics.[24] In addition to this, the spectral stitching method has previously been applied to improve the dynamic range and mass accuracy in direct infusion nanoelectrospray Fourier transform ion cyclotron resonance mass spectrometry (DI-nESI FT-ICRMS). This method includes the stitching of a number of selected ion monitoring (SIM) windows using advanced algorithms.[25]

An automated method called flow injection analysis/mass spectrometry (FIA/MS), which uses high-performance liquid chromatography (HPLC) yet does not require a column, is becoming increasingly popular. For instance, Fuhrer and colleagues developed an analysis technique based on FIA/MS capable of analysing over 1400 samples of *Escherichia coli* cellular extract per day.[26] To expand its metabolite coverage this method can be coupled with chromatography.

As there is currently no instrument capable of measuring all types of metabolites, users should select the most suitable approach for their study, taking into consideration aspects such as resolution, sensitivity, throughput, and expenses.

7.2.3 Sample Preparation Strategies

Sample preparation is one of the most crucial aspects of a successful metabolomics study. The goal of sample preparation is to obtain a representative and homogeneous sample, suitable for the subsequent analysis, allowing for the accurate and reproducible quantification of metabolites. Proper sample preparation is crucial as it directly impacts the quality and accuracy of the resulting data. Incorrect or poorly executed sample preparation can result in loss of metabolites, contamination, and artifacts that can skew the results.

All analytical procedures, including metabolome analyses, need to be validated through a number of parameters before they can be applied to biological studies. These parameters include accuracy, precision, repeatability, intermediate precision, specificity, and detection limit. In a metabolomics study, the use of internal standards and QC samples plays a crucial role in ensuring the quality and accuracy of the obtained data.

QC samples, such as pooled biological samples and reference materials, are used to monitor the performance of the analytical workflow. By analysing a set of QC samples at various stages of the experimental process, it becomes possible to assess the accuracy, precision, and reproducibility of the results. The use of QC samples helps to detect and correct systematic errors that may arise during sample preparation, instrumentation, or data processing, and ensures that the data obtained are reliable and accurate.[27]

The use of internal standards is a common practice to normalize and quantify metabolites in a sample. Internal standards are compounds that are added to a sample before analysis, and serve as references for instrument calibration. By using internal standards, it becomes possible to accurately quantify the concentrations of metabolites in a sample, even if there is variation in the instrument performance or sample preparation. Internal standards help to normalize the data and to reduce the impact of systematic errors that can occur during the experimental process.[28] An ideal internal standard compound should fulfil the following criteria: it should be extremely rare for it to exist in biological samples; it should not interfere with the metabolite signals that are obtained from the samples; and it should not be affected by the ion suppression that occurs during the electrospray ionization (ESI) process.

Experimental procedures in microbial metabolomics generally include biomass cultivation, fast sampling, metabolic activity and endogenous enzymatic arrest (quenching), metabolite extraction, and intracellular reactant (metabolite) measurement.[29] Widely used techniques for quantifying extracellular metabolites in cell-free supernatants include low-temperature filtering and centrifugation. Liquid batch culture, continuous culture in a chemostat, and batch culture on a filter support (also known as filter culture) are three methods that are frequently used for cultivating microbial cells for the purpose of conducting metabolomic research in the field of microbial metabolomics. Liquid batch culture is frequently utilized due to its simplicity and convenience, although chemostats offer significant advantages for controlling culture conditions and ensuring reproducibility. The primary benefits of filter culture are the simplicity of manipulating the cellular nutritional environment and the convenience of sampling. Sampling metabolites from liquid batch or chemostat cultures involves comparable challenges: collecting a specific volume of culture fluid; separating cells from the surrounding medium without affecting their metabolome; and extracting metabolites from the isolated cells.[30]

A rapid quenching method should be validated, particularly in the investigation of central metabolism-related metabolites and cofactors including glucose phosphates, NAD(P)H, and ATP. As the pool size of these metabolites is dynamically altered by external stimuli, the enzymes responsible for metabolism must be inhibited as promptly as possible, as the inter-conversion of metabolites might occur in seconds. There are numerous techniques for the rapid collection of microbiological samples, one of the simplest being manual sampling of culture flasks using pipettes or syringes, followed by quick spraying into the flask containing quenching solution. Using syringes prefilled with quenching solution is an alternative to this procedure.[31] Recently, a study demonstrated direct injection of living *E. coli* cells for real-time metabolome profiling using a high-resolution mass spectrometer (time of flight) which enables automated monitoring of about 300 compounds in 15–30 s cycles over several hours.[32]

The selection of a suitable quenching and extraction process is greatly dependent on the size, density, and membrane composition of the organism. An appropriate quenching process should be able to immediately halt cellular metabolic activity without causing considerable cell membrane damage, as this may result in the release of intracellular metabolites from cells. It has been reported that the use of pre-cooled organic solvents such as cold methanol and cold ethanol is an effective approach for quenching in eukaryotes. However, when exposed to the commonly used cold methanol procedure, prokaryotic bacteria such as *E. coli* tend to leak internal metabolites.[33] Furthermore, neither buffering nor increasing the ionic strength of the quenching solution was found to be effective in preventing this.[34] This problem with leakage is most likely a result of their cell wall and membrane architecture and composition, which are known to be distinct from those of eukaryotic cells. Therefore, rapid filtering remains the preferred method for quenching prokaryotes, even when rapid sampling is required. After quenching has completely stopped the metabolism, the metabolites must be collected from the cells. The objective here is to permanently inactivate endogenous enzymes and permeabilize the cells in order to achieve the highest yield of cellular metabolites.[33] The most commonly used method is organic solvent extraction using methanol and chloroform. Physical stresses such as high temperature and freeze–thaw can also be used.[35] Ideally, the extraction procedure should be able to extract a wide range of metabolites without altering these metabolites. As such, physical or chemical modification as well as degradation should be minimized.

7.2.4 Data Analysis Methods

Metabolomics analyses generate massive amounts of data which require proper handling and consideration. The goal of metabolomics data analysis is to extract biologically relevant information from the generated data.

Multivariate analysis is a group of methods that are used to analyse complex data sets. In metabolomics, two commonly used methods are principal component analysis (PCA) and partial least squares discriminant analysis (PLS-DA). PCA is a dimension reduction technique that transforms a large number of variables into a smaller number of variables through the use of principal components. PCA helps in visualizing patterns and relationships between metabolites and in identifying outliers or unusual observations in the data. PLS-DA, on the other hand, is a supervised classification method that can be used to separate different groups of samples. PLS-DA is particularly useful for identifying differentially expressed metabolites between groups, and for making predictions based on the metabolic profile.

Metabolite set enrichment analysis (MSEA) is another important method in metabolomics data analysis, which assesses the enrichment of predefined metabolite sets in the data. MSEA combines gene set analysis, which assesses the enrichment of gene sets in gene expression data, and metabolomics data analysis, to provide a systems biology perspective. MSEA is used to identify biological pathways and networks that are affected by a particular condition, such as disease or treatment, and to uncover potential metabolic biomarkers. MSEA can provide insights into the functional significance of changes in the metabolome and help to integrate the results from metabolomics studies with other omics data.[36] Commonly used platforms to perform MSEA analysis are MetaboAnalyst,[37] MBRole,[38] and MMEASE.[39]

7.3 Microbial Metabolomics Application for Strain Improvement

As the utilization of biological systems for industrial purposes becomes more prevalent, finding reliable ways to improve the bioproduction capabilities of chassis organisms is of great importance. Traditional methods for improving strain bioproductivity have often been restricted to the modification of singular genes without considering the metabolic system of the chassis organism.[40] However, recent advancements in omics technologies, such as MS-based metabolomics, now allow for a more holistic approach to metabolic engineering.[41]

By utilizing techniques such as principal component analysis (PCA) and partial least squares-discriminant analysis (PLS-DA), metabolomics enables quantification of metabolites throughout a bioproduction system. This leads to a better understanding of cellular metabolome states and genetic influences of organisms on production phenotype and makes MS-based metabolomics a versatile and powerful tool for improving strains.[1] Figure 7.1 illustrates an overview of the typical workflow of metabolomics that results in inputs for strain improvement.[42]

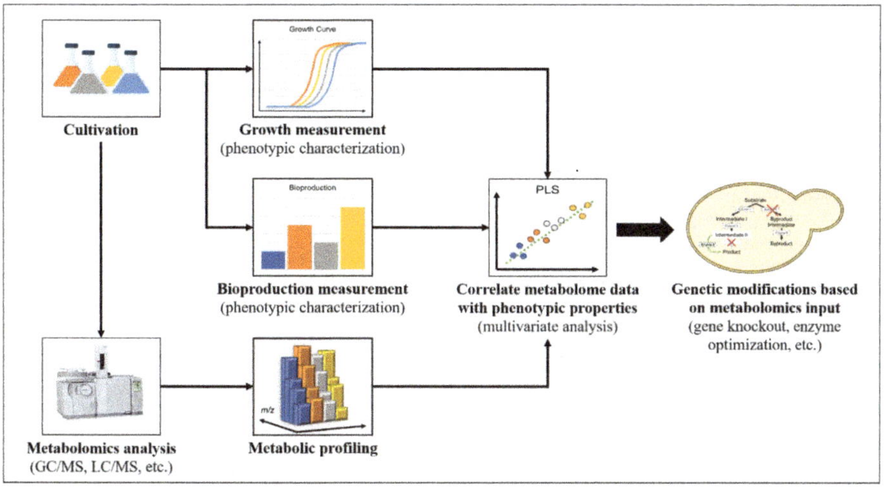

Figure 7.1 Overview of the workflow of microbial metabolomics application for strain improvement. Despite the many studies on its role in understanding cellular mechanisms and changes, there are relatively few on its application in identifying rate-limiting steps in synthetic biological systems for actual improvement of host microorganisms. Reproduced from ref. 42, https://doi.org/10.3389/fmolb.2022.1057709, under the terms of CC BY 4.0 license, http://creativecommons.org/licenses/by/4.0/.

7.3.1 Strategies to Improve Strains Using Metabolomics Data

The primary objective of utilizing metabolomics in strain improvement is to optimize bioproduct production. By analysing intermediates and end products simultaneously, metabolomics can pinpoint bottlenecks or rate-limiting steps in the bioproduction system, such as suboptimal enzyme activity and the presence of competing pathways, as shown in Figure 7.2.[42]

This allows for targeted improvements to the system. An accumulation of intermediates is often a sign of suboptimal enzyme performance, which can be improved by employing techniques such as gene overexpression through the construction of a ribosomal binding site (RBS) library and adjusting promoter strength.[43] Competing pathways divert important intermediate metabolites away from the desired products and lead to decreased bioproduction. Identification of these competing pathways allows for the knockout of genes in those pathways to redirect flux towards the desired product.[44] Cell growth is another important factor to consider, and in some cases, identifying genes that confer stress tolerance can enhance bioproductivity, especially in situations where high concentrations of the desired end product or a side product can inhibit growth.[45]

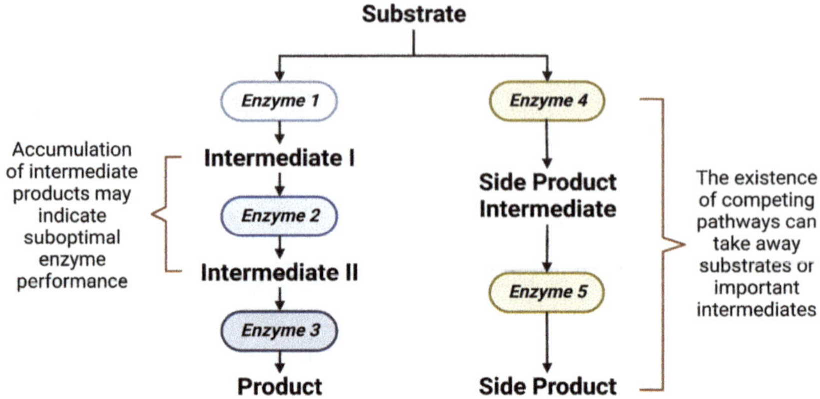

Figure 7.2 Rate-limiting steps or bottlenecks in a bioproduction system may include suboptimal enzyme activity or the presence of competing pathways. Reproduced from ref. 42, https://doi.org/10.3389/fmolb.2022.1057709, under the terms of CC BY 4.0 license, http://creativecommons.org/licenses/by/4.0/.

7.3.2 Metabolomics-driven Strain Improvement in Various Organisms

The ability to generate meaningful optimization inputs based on metabolomics data is heavily dependent on the presence of genetic and pathway information about the host organism. For some hosts a complete metabolic map is available, allowing for the applications of advanced techniques such as genome-scale metabolic modelling (GSM)[46,47] and multiomics-based techniques.[48] On the other hand, interpreting metabolomics data collected from other host organisms where genetic information is limited can be a more challenging task. This section provides examples of metabolomics-driven strain improvement studies in a diverse range of common and less common hosts.

Escherichia coli and *Saccharomyces cerevisiae* are the two most commonly used host organisms to produce heterologous metabolites. As such, these two canonical hosts have been extensively characterized over the years, making them ideal for metabolomics-driven strain improvement. In a recent study, Webb *et al.* demonstrated a multiomics-based framework integrating metabolomics, transcriptomics, proteomics, and lipidomics to improve the styrene bioproduction in engineered *E. coli.*[48] They leveraged on the wealth of characterization data available for *E. coli* hosts such as the library of transcription factors[49] and an integrated multiomics database, MORF,[50] to develop a multiomics-based production strain improvement (MOBpsi) framework to improve the biomanufacturing of toxic chemicals in microbial hosts.[48] Similarly, Strucko *et al.* leveraged on the wealth of characterization methods available for *S. cerevisiae*, including a genome-scale metabolic model[47] to optimize glycerol utilization, which is crucial in improving glycerol-based bioprocesses, using a combination of metabolomics, transcriptomics, and proteomics techniques.[46]

Metabolomics-driven strain improvement has also been applied to photosynthetic cyanobacterial hosts, which have great potential for bioenergy production. These organisms are capable of harnessing sunlight to produce energy-rich compounds through photosynthesis, making them attractive candidates for bioenergy production. However, the genetic and metabolic pathways of these organisms are not as well characterized as those of canonical hosts such as *E. coli* and *S. cerevisiae*, making the use of metabolomics-driven strain improvement more challenging. Nevertheless, researchers have been able to use metabolomics to improve the efficiency of photosynthetic bioproduction in cyanobacterial hosts by identifying metabolic pathways that can be targeted to improve the overall growth of the organism. For example, in the *Synechocystis* sp. PCC 6803, researchers have used metabolomics to identify metabolic bottlenecks that limit the bioproduction efficiency of succinate, and then used this information to design genetic modifications that improve the organism's performance.[51] Additionally, in other cyanobacteria such as *Synechococcus elongatus* PCC 7942, metabolomics-driven strain improvement has been used to increase the production of biofuels such as 1-butanol.[52] Overall, while the genetic and metabolic pathways of photosynthetic cyanobacterial hosts are less characterized than those of canonical hosts, metabolomics-driven strain improvement is still a powerful tool for optimizing the performance of these organisms for bioenergy production.

The use of metabolomics-driven metabolic engineering in hosts other than *E. coli*, *S. cerevisiae*, and cyanobacteria is less common. However, these non-canonical hosts are still utilized because of their innate ability to produce industrially relevant compounds, even though it may be more challenging to gain insight from the obtained metabolomics data due to the lack of genetic information and characterization. Some notable examples include the strain improvements in *Corynebacterium glutamicum*, an important bacterium in industry, particularly for large-scale production of amino acids, such as L-glutamate[53] and in the fungus *Aureobasidium pullulans*, a host commonly used in industry for the production of various enzymes and compounds, including polymalic acid (PMA) and its monomer L-malic acid (MA).[54] Overall, while the genetic and metabolic pathways of other non-canonical hosts are less characterized than those of canonical hosts, metabolomics-driven strain improvement is still a powerful tool for optimizing the performance of these organisms for various industrial applications.

7.4 Case Studies

In this section, a case study of using metabolomics-driven strain improvement strategies to improve the bioproduction of 1-propanol in *Escherichia coli* is presented. Ohtake and colleagues (2022) demonstrated the GC/MS and ion pair LC–MS/MS-based metabolomics strategies to systematically identify targets for phenotype improvement.[41]

In a previous study, the yield of 1-propanol in the production system *E. coli* SYN3 was improved by the utilization of glucose *via* the heterologous citramalate pathway in tandem with the native threonine pathway,[55] as illustrated in Figure 7.3.

While the overexpression of the threonine pathway beyond the basal levels in the production system *E. coli* SYN12 was expected to improve the 1-propanol yield, no significant change in 1-propanol titer was observed. Ohtake *et al.* investigated this phenomenon using a PCA multivariate analysis of metabolomics data obtained using GC/MS and ion-pair LC–MS/MS (Figure 7.4).[41]

In the PCA analysis of GC/MS data, SYN12 was observed to group on the negative axis of PC1, which correlated with increased norvaline and 2-aminobutyrate levels based on the loading plot. The accumulation of norvaline and 2-aminobutyrate reveals the existence of competing pathways which take up key intermediate metabolites away from 1-propanol production. On the other hand, PCA analysis of ion-pair LC–MS/MS data showed SYN12 to be grouped on the positive axis of PC1, which correlated with increased 2-ketobutyrate levels. The accumulation of 2-ketobutyrate, an intermediate metabolite in 1-propanol biosynthesis pathways, suggests suboptimal enzyme activity in the conversion of this metabolite into 1-propanol.[41]

To redirect the flux towards 1-propanol, Ohtake *et al.* integrated deletion of the *avtA* gene, the role of which was to convert 2-ketobutyrate into 2-aminobutyrate. Interestingly, this deletion did not improve the 1-propanol yield and further worsened the accumulation of 2-ketobutyrate. This suggests that the vital bottleneck in this production system was the suboptimal enzyme performance in converting 2-ketobutyrate into 1-propanol.[41]

Enzymes responsible for converting 2-ketobutyrate into 1-propanol in the 1-propanol biosynthesis pathway are KivD and YqhD. To optimize the activity levels of these enzymes, Ohtake *et al.* constructed two RBS libraries to modulate the expression levels of *KivD* and *YqhD*. They found that the optimization of *YqhD* expression resulted in the significant increase in YqhD enzyme activity, glucose consumption rate, and 1-propanol yield (Figure 7.5).[41]

This study showed the use of metabolomics data to identify the existence of competing pathways and suboptimal enzyme activity and further demonstrated how the generated input could lead to strain improvement.

7.5 Future Outlook

Despite certain obstacles, like low throughput and scalability issues, metabolomics-based strain improvement is gaining popularity as a reliable strategy for metabolic engineering projects. Recently, this strategy has been frequently integrated into the comprehensive Design-Build-Test-Learn (DBTL) methodology of synthetic biology, particularly in the testing and learning stages.

With continuous advances in metabolomics technology, the potential for metabolomics-driven strain improvement strategies will likely increase. For example, advancements in single-cell metabolomics[56] and real-time

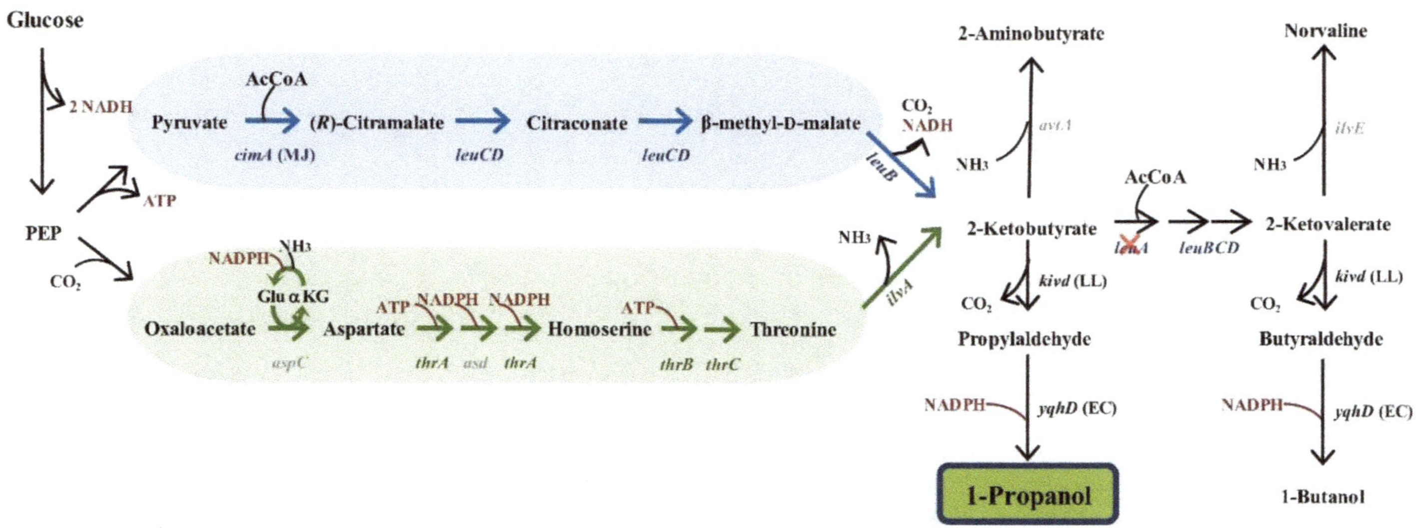

Figure 7.3 The keto-acid pathway for 1-propanol production featuring the combination of a heterologous citramalate pathway and a native threonine pathway. Reproduced from ref. 41, https://doi.org/10.3389/fmicb.2022.871624, under the terms of CC-BY 4.0 licence, http://creativecommons.org/licenses/by/4.0/.

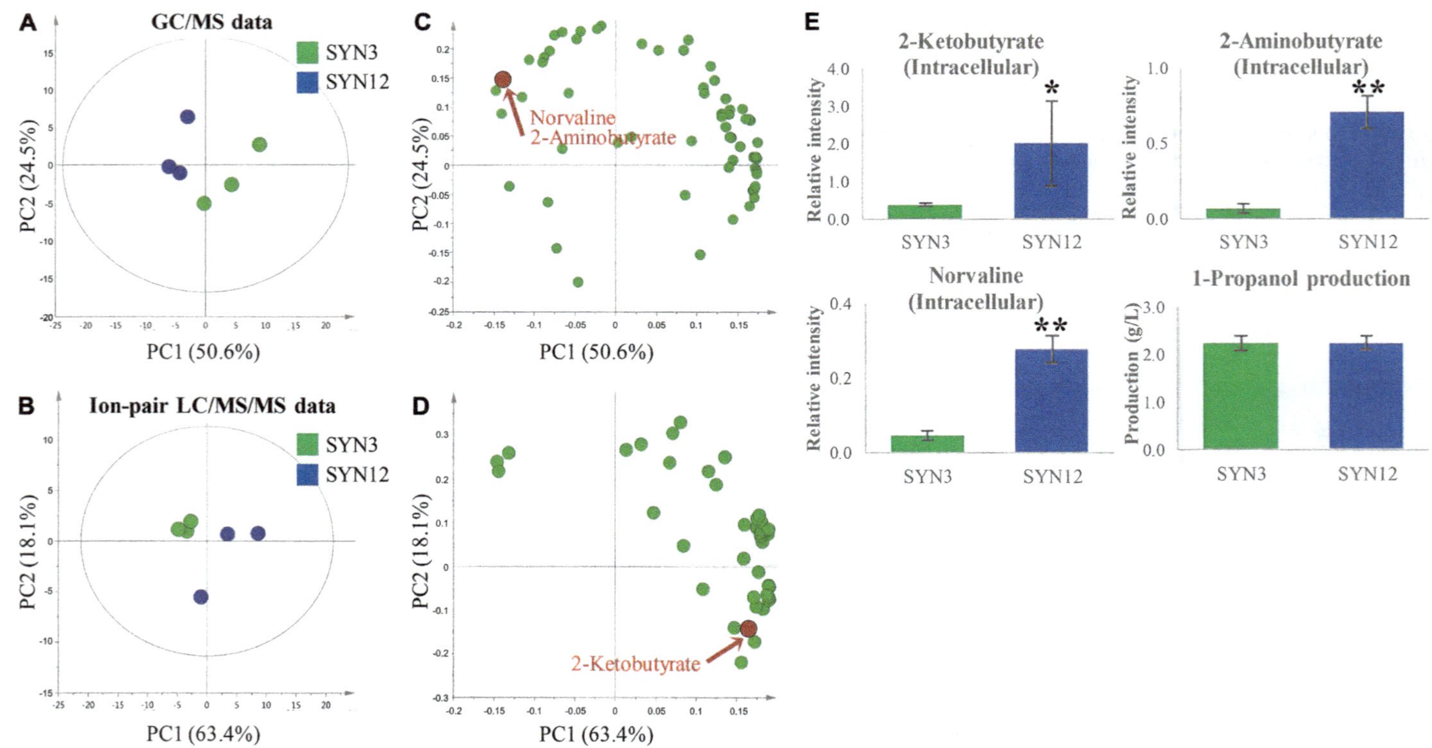

Figure 7.4 PCA analysis of metabolomics data comparing SYN3 and SYN12, obtained using: (A) GC/MS and (B) ion-pair LC–MS/MS. (C) Loading plot of GC/MS data. (D) Loading plot of ion-pair LC–MS/MS data. (E) Relative intensity levels of key intermediate metabolites. Reproduced from ref. 41, https://doi.org/10.3389/fmicb.2022.871624, under the terms of CC-BY 4.0 licence, http://creativecommons.org/licenses/by/4.0/.

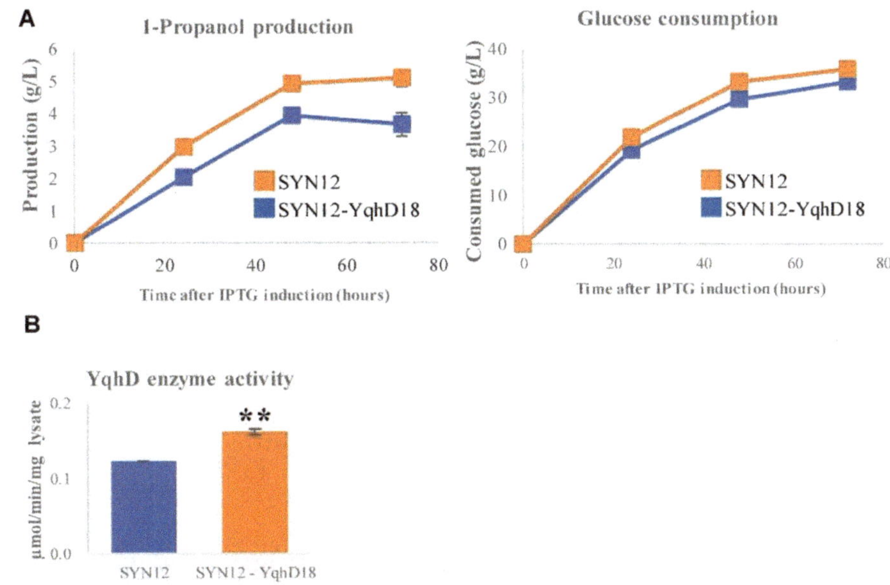

Figure 7.5 Optimization of YqhD expression resulted in an increase in (A) 1-propanol yield and glucose consumption and (B) YqhD enzyme activity. Reproduced from ref. 41, https://doi.org/10.3389/fmicb.2022.871624, under the terms of CC-BY 4.0 licence, http://creativecommons.org/licenses/by/4.0/.

analysis[57] can provide a more accurate understanding of the metabolic state of strains. The combination of MS and imaging techniques[58,59] can also provide spatial information on metabolite distribution. Additionally, ongoing research is focused on developing analytical methods to identify more metabolites, particularly those with fast turnover rates.[60]

Abbreviations

ATP	Adenosine triphosphate
CE/MS	Capillary electrophoresis–mass spectrometry
DI-MS	Direct infusion mass spectrometry
ESI-MS	Electrospray ionization-mass spectrometry
GC/MS	Gas chromatography–mass spectrometry
GSM	Genome-scale metabolic modelling
HPLC	High-performance liquid chromatography
IPTG	Isopropyl β-D-1-thiogalactopyranoside
LC–MS	Liquid chromatography–mass spectrometry
LC–MS/MS	Liquid chromatography–tandem mass spectrometry
MA	L-Malic acid
MS	Mass spectrometry
PCA	Principal component analysis

PLS-DA	Partial least squares-discriminant analysis
PMA	Polymalic acid
RI	Retention index
RBS	Ribosomal binding site
SFC/MS	Supercritical fluid chromatography–mass spectrometry
TOF	Time of flight
UPLC	Ultra-performance liquid chromatography
QqQ-MS	Triple-quadrupole mass spectrometer

Acknowledgements

The authors would like to thank Prof. Eiichiro Fukusaki for his conceptual guidance. This work was supported by JSPS Grants-in-Aid for Scientific Research grant number 20308851 to Sastia Prama Putri.

References

1. S. P. Putri, Y. Nakayama, F. Matsuda, T. Uchikata, S. Kobayashi, A. Matsubara and E. Fukusaki, *J. Biosci. Bioeng.*, 2013, **115**, 579–589.
2. X. W. Zhang, Q. H. Li, Z. Di Xu and J. J. Dou, *RSC Adv.*, 2020, **10**, 3092–3104.
3. K. A. Hollywood, K. Schmidt, E. Takano and R. Breitling, *Curr. Opin. Biotechnol.*, 2018, **54**, 114–120.
4. A. Villate, M. San Nicolas, M. Gallastegi, P. A. Aulas, M. Olivares, A. Usobiaga, N. Etxebarria and O. Aizpurua-Olaizola, *Plant Sci.*, 2021, **303**, 110789.
5. L. Fontanesi, *Anim. Front.*, 2016, **6**, 73–79.
6. H. Mallick, E. A. Franzosa, L. J. Mclver, S. Banerjee, A. Sirota-Madi, A. D. Kostic, C. B. Clish, H. Vlamakis, R. J. Xavier and C. Huttenhower, *Nat. Commun.*, 2019, **10**, 1–11.
7. W. Weckwerth, *Methods in molecular biology: Metabolomics methods and protocols*, Humana Press, 2007.
8. V. Shulaev, *Briefings Bioinf.*, 2006, 7, 128–139.
9. R. N. Trethewey, *Curr. Opin. Plant Biol.*, 2004, **7**, 196–201.
10. L. Li, H. Hoefsloot, A. A. de Graaf, E. Acar and A. K. Smilde, *BMC Bioinf.*, 2022, **23**, 1–22.
11. W. Pongsuwan, E. Fukusaki, T. Bamba, T. Yonetani, T. Yamahara and A. Kobayashi, *J. Agric. Food Chem.*, 2007, **55**, 231–236.
12. T. J. Trinklein, S. Schöneich, P. E. Sudol, C. G. Warren, D. V. Gough and R. E. Synovec, *J. Chromatogr. A*, 2020, **1634**, 461654.
13. C. M. Whitehouse, R. N. Dreyer, M. Yamashita and J. B. Fenn, *Anal. Chem.*, 1985, **57**, 675–679.
14. M. G. Ikonoraou, A. T. Blades and P. Kebarle, *Anal. Chem.*, 1990, **62**, 957–967.
15. Y. Izumi, A. Okazawa, T. Bamba, A. Kobayashi and E. Fukusaki, *Anal. Chim. Acta*, 2009, **648**, 215–225.
16. J. M. Büscher, D. Czernik, J. C. Ewald, U. Sauer and N. Zamboni, *Anal. Chem.*, 2009, **81**, 2135–2143.

17. W. Zhang and R. Ramautar, *Electrophoresis*, 2021, **42**, 381–401.
18. K. Inokuma, M. Matsuda, D. Sasaki, T. Hasunuma and A. Kondo, *Microb. Cell Fact.*, 2018, **17**, 1–11.
19. C. N. Serhan, *Prostaglandins Other Lipid Mediators*, 2005, 77, 4–14.
20. T. Bamba, N. Shimonishi, A. Matsubara, K. Hirata, Y. Nakazawa, A. Kobayashi and E. Fukusaki, *J. Biosci. Bioeng.*, 2008, **105**, 460–469.
21. T. Bamba, J. W. Lee, A. Matsubara and E. Fukusaki, *J. Chromatogr. A*, 2012, **1250**, 212–219.
22. M. Takekana, T. Yoshida, E. Yoshida, S. Ono, S. Horie, C. J. Vavricka, M. Hiratani, K. Tsuge, J. Ishii, Y. Hayakawa, A. Kondo and T. Hasunuma, *J. Chromatogr. B: Biomed. Sci. Appl.*, 2023, **1215**, 123588.
23. R. Pitchapa, S. Dissook, S. P. Putri, E. Fukusaki and S. Shimma, *Metabolites*, 2022, **12**, 1148.
24. R. J. M. Weber, A. D. Southam, U. Sommer and M. R. Viant, *Anal. Chem.*, 2011, **83**, 3737–3743.
25. A. D. Southam, T. G. Payne, H. J. Cooper, T. N. Arvanitis and M. R. Viant, *Anal. Chem.*, 2007, **79**, 4595–4602.
26. T. Fuhrer, D. Heer, B. Begemann and N. Zamboni, *Anal. Chem.*, 2011, **83**, 7074–7080.
27. W. B. Dunn, I. D. Wilson, A. W. Nicholls and D. Broadhurst, *Bioanalysis*, 2012, **4**, 2249–2264.
28. M. Sysi-Aho, M. Katajamaa, L. Yetukuri and M. Orešič, *BMC Bioinf.*, 2007, **8**, 1–17.
29. M. R. Mashego, K. Rumbold, M. De Mey, E. Vandamme, W. Soetaert and J. J. Heijnen, *Biotechnol. Lett.*, 2007, **29**, 1–16.
30. C. A. Crutchfield, W. Lu, E. Melamud and J. D. Rabinowitz, *Mass spectrometry-based metabolomics of yeast*, Elsevier Inc., 2nd edn, 2010, vol. 470.
31. C. J. Bolten, P. Kiefer, F. Letisse, J. C. Portais and C. Wittmann, *Anal. Chem.*, 2007, **79**, 3843–3849.
32. H. Link, T. Fuhrer, L. Gerosa, N. Zamboni and U. Sauer, *Nat. Methods*, 2015, **12**, 1091–1097.
33. C. L. Winder, W. B. Dunn, S. Schuler, D. Broadhurst, R. Jarvis, G. M. Stephens and R. Goodacre, *Anal. Chem.*, 2008, **80**, 2939–2948.
34. H. Taymaz-Nikerel, M. de Mey, C. Ras, A. ten Pierick, R. M. Seifar, J. C. van Dam, J. J. Heijnen and W. M. van Gulik, *Anal. Biochem.*, 2009, **386**, 9–19.
35. S. P. Putri, S. Yamamoto, H. Tsugawa and E. Fukusaki, *J. Biosci. Bioeng.*, 2013, **116**, 9–16.
36. J. Xia and D. S. Wishart, *Nucleic Acids Res.*, 2010, **38**, 71–77.
37. J. Xia and D. S. Wishart, *Nat. Protoc.*, 2011, **6**, 743–760.
38. M. Chagoyen and F. Pazos, *Bioinformatics*, 2011, **27**, 730–731.
39. Q. Yang, B. Li, S. Chen, J. Tang, Y. Li, Y. Li, S. Zhang, C. Shi, Y. Zhang, M. Mou, W. Xue and F. Zhu, *J. Proteomics*, 2021, **232**, 104023.
40. P. Opgenorth, Z. Costello, T. Okada, G. Goyal, Y. Chen, J. Gin, V. Benites, M. De Raad, T. R. Northen, K. Deng, S. Deutsch, E. E. K. Baidoo,

C. J. Petzold, N. J. Hillson, H. Garcia Martin and H. R. Beller, *ACS Synth. Biol.*, 2019, **8**, 1337–1351.
41. T. Ohtake, N. Kawase, S. Pontrelli, K. Nitta, W. A. Laviña, C. R. Shen, S. P. Putri, J. C. Liao and E. Fukusaki, *Front. Microbiol.*, 2022, **13**, 1–8.
42. M. N. Iman, E. Herawati, E. Fukusaki and S. P. Putri, *Front. Mol. Biosci.*, 2022, **9**, 1–7.
43. T. Siegl, B. Tokovenko, M. Myronovskyi and A. Luzhetskyy, *Metab. Eng.*, 2013, **19**, 98–106.
44. K. Nitta, W. A. Laviña, S. Pontrelli, J. C. Liao, S. P. Putri and E. Fukusaki, *J. Biosci. Bioeng.*, 2019, **127**, 301–308.
45. E. Ohta, Y. Nakayama, Y. Mukai, T. Bamba and E. Fukusaki, *J. Biosci. Bioeng.*, 2016, **121**, 399–405.
46. T. Strucko, K. Zirngibl, F. Pereira, E. Kafkia, E. T. Mohamed, M. Rettel, F. Stein, A. M. Feist, P. Jouhten, K. R. Patil and J. Forster, *Metab. Eng.*, 2018, **47**, 73–82.
47. J. Förster, I. Famili, P. Fu, B. Palsson and J. Nielsen, *Genome Res.*, 2003, **13**, 244–253.
48. J. P. Webb, A. C. Paiva, L. Rossoni, A. Alstrom-Moore, V. Springthorpe, S. Vaud, V. Yeh, D. P. Minde, S. Langer, H. Walker, A. Hounslow, D. R. Nielsen, T. Larson, K. Lilley, G. Stephens, G. H. Thomas, B. B. Bonev, D. J. Kelly, A. Conradie and J. Green, *Metab. Eng.*, 2022, **72**, 133–149.
49. L. Liang, R. Liu, K. E. O. Foster, A. Choudhury, S. Cook, J. C. Cameron, W. V. Srubar and R. T. Gill, *Metab. Eng.*, 2020, **57**, 74–84.
50. V. Springthorpe, R. Leaman, D. Sifouna, J. Bennett and G. Thomas, *Access Microbiol.*, 2020, **2**, 10.
51. T. Hasunuma, M. Matsuda, Y. Kato, C. J. Vavricka and A. Kondo, *Metab. Eng.*, 2018, **48**, 109–120.
52. A. M. Fathima, D. Chuang, W. A. Laviña, J. Liao, S. P. Putri and E. Fukusaki, *Biotechnol. Biofuels*, 2018, **11**, 1–12.
53. H. Kawaguchi, K. Yoshihara, K. Y. Hara, T. Hasunuma, C. Ogino and A. Kondo, *Microb. Cell Fact.*, 2018, **17**, 1–16.
54. J. Feng, J. Yang, W. Yang, J. Chen, M. Jiang and X. Zou, *Biotechnol. Biofuels*, 2018, **11**, 1–13.
55. C. R. Shen and J. C. Liao, *Metab. Eng.*, 2013, **17**, 12–22.
56. R. Hu, Y. Li, Y. Yang and M. Liu, *Mass Spec. Rev.*, 2023, **42**, 67–94.
57. M. T. Judge, Y. We, F. Tayyari, A. Hattori, J. Glushka, T. Ito, J. Arnold and A. S. Edison, *Front. Mol. Biosci.*, 2019, **6**, 1–16.
58. G. Schleyer, N. Shahaf, C. Ziv, Y. Dong, R. A. Meoded, E. J. N. Helfrich, D. Schatz, S. Rosenwasser, I. Rogachev, A. Aharoni, J. Piel and A. Vardi, *Nat. Microbiol.*, 2019, **4**, 527–538.
59. M. J. Taylor, J. K. Lukowski and C. R. Anderton, *J. Am. Soc. Mass Spectrom.*, 2021, **32**, 872–894.
60. F. Mousavi, B. Bojko and J. Pawliszyn, *Microbial Metabolomics*, 2019, vol. 1859.

CHAPTER 8

Integrated Analysis for Identification, Phenotyping, and Antimicrobial Susceptibility Testing (AST) of Bacteria Using Mass Spectrometry, Machine Learning, and Multi-omics Analysis

RUTAN ZHANG,*[a] BRIAN J. WERTH[b] AND LIBIN XU[a]

[a] Department of Medicinal Chemistry, University of Washington, Seattle, WA, USA; [b] Department of Pharmacy, University of Washington, Seattle, WA, USA
*Email: rtzhang@uw.edu

8.1 Introduction

The global mortality attributable to antimicrobial resistance (AMR) is increasing and is projected to reach 10 million per year by 2050.[1] Prompt identification and antimicrobial susceptibility profiling of infecting bacterial pathogens is key to the successful treatment of bacterial infections and to making informed public health policy for antimicrobial use. Given the high rate of antimicrobial use, international travel, and looming climate-related mass-migration, conflict, and sanitation challenges, the rapid and low-cost

New Developments in Mass Spectrometry No. 13
Detection and Analysis of Microorganisms by Mass Spectrometry
Edited by Liang Qiao and Jia Yi

Published by the Royal Society of Chemistry, www.rsc.org

identification of antimicrobial-resistant pathogens is one of the most urgent public health priorities. Historically, the identification and susceptibility testing of bacteria was done on the basis of a battery of biochemical assays performed and interpreted over a few days. This laborious but low-tech approach has been increasingly replaced with rapid molecular and mass spectrometry (MS) techniques in wealthy health systems, where they have been shown to significantly reduce the time to identify pathogens, and thus reduce the time to appropriate antimicrobial therapy, which improves patient outcomes. MS-based methods of identifying antimicrobial-resistant bacteria are uniquely promising, and while they have revolutionized clinical microbiology laboratory practice, they are not yet used to their full potential.

Since Woese and Fox defined 16S rRNA genes as molecular and evolutionary markers to distinguish members of the domain Archaea from those belonging to Bacteria in 1977, 16S rRNA gene sequencing has become the gold standard for microbial identification.[2,3] However, this genotypic approach remains relatively laborious and expensive. MS-based techniques have been proven to be highly efficient and accurate in identifying bacteria on the species level, mainly by ribosomal protein profiling.[3] To date, the MS-based approach has been widely used for fast bacterial typing by researchers working in clinical laboratories,[4–9] food fermentation and production,[10] food and water hygiene,[11–15] agricultural biotechnology and animal science,[15–17] aquatic and marine science,[18,19] natural product and drug discovery,[20] and environmental science.[21]

Matrix-assisted laser desorption/ionization (MALDI) time-of-flight (TOF) MS can achieve fast and high-throughput analysis of samples without or with little sample pretreatment. This technique has been widely used in MS imaging of targeted metabolites, lipids, polysaccharides, peptides, nucleic acids, and proteins in organisms.[22–24] Liquid chromatography–tandem MS (LC–MS/MS) allows for two-dimensional separation of complex samples, providing an extended dynamic range and numerous detectable molecules due to the high peak capacity, typically hundreds to thousands of proteins, metabolites, and lipids in every single LC–MS/MS experiment.[25]

Metabolites (including both non-polar lipids and polar metabolites), which are non-genetically encoded small molecules with molecular weights <1000 Da, directly reflect the real-time holistic state of biological systems.[26,27] Metabolomics is defined as a comprehensive approach to qualitatively and quantitatively study all the metabolites of a sample in a given state.[28,29] Metabolomics has gained its popularity in extensive research fields, such as microbiology, human health, diagnostic biomarker discovery, cancer research, drug discovery, and food analysis.[30–38] The study of characteristic small molecule profiles for microbial identification could date back to 1970s by Anhalt and Fenselau.[39] In their study, they, for the first time, demonstrated the proof-of-concept of taxonomic classification by small molecules, specifically phospholipids and ubiquinones, using combined pyrolysis–MS. Foschi *et al.* found that variations in lactobacillus metabolome obtained by ^{1}H-NMR analysis were correlated with taxonomy.[40]

Voorhees *et al.* differentiated ten bacterial strains at the species level based on fatty acid profiling using MALDI-MS.[41] In the meantime, more and more matrices have been developed for the LDI of small molecules.[42,43]

The focus of this chapter is on the recent developments of MS-based techniques as well as bioinformatics as tools for subspecies (*i.e.*, strain level) typing and profiling.

8.2 Approaches and Applications

8.2.1 MS-based Biotyping and Phenotyping

Two MS techniques, MALDI-TOF MS and LC–MS/MS, have been widely and systematically studied in typing and profiling bacteria. MALDI-TOF MS-based identification platforms have gained massive popularity due to their low cost, easy sample preparation, fast analysis time, and multiple commercially available databases, such as Biotyper (Bruker Daltonics), SARAMIS (bioMérieux), and Andromas (Andromas SAS).[44] Both the Bruker and bioMérieux platforms have been approved for clinical use in both Europe and the USA.[45]

As shown in Figure 8.1, in comparison with traditional culturing methods, the time-to-result (TTR) of the MALDI-TOF MS identification system could be drastically shortened from days to hours by nonculture-based enrichment techniques.[6,12,46] The development of nano/biosensors (ligands including but not limited to antibodies, aptamers, bacteriophages or their tail proteins, antimicrobial peptides, host cell receptors, nucleic acid probes, and ionic liquids) could further shorten the enrichment step from hours to minutes as well as decrease the bacterial colony-forming unit (CFU) limit of detection by forming analyte–ligand complexes.[12,47,48]

Both the Bruker BioTyper and bioMérieux VITEK MS systems have demonstrated excellent microbial typing accuracy consistent with conventional

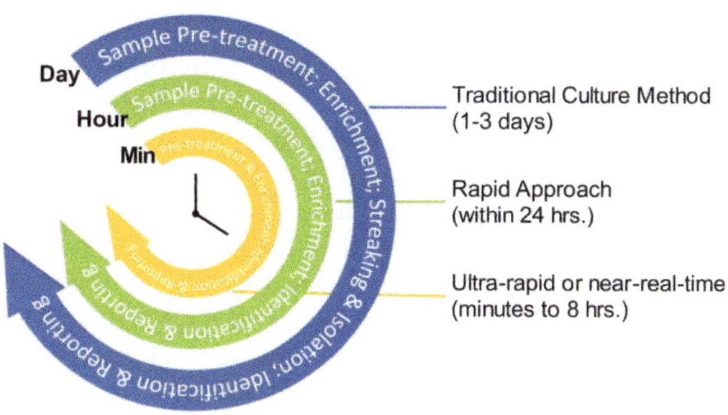

Figure 8.1 Typical workflow for foodborne and clinical pathogen identification depends on the speed of the assay.

methods at the species level. However, there are many microorganisms that cannot be identified with current library-based approaches, which might result in misidentifications or low confidence identities. For instance, *Shigella* spp. cannot usually be discriminated from *Escherichia coli*;[49] *Bacillus anthracis* is systematically confused with other *Bacillus* species;[49] *Campylobacter sputorum* subspecies *sputorum* DSM 10535 and *Campylobacter sputorum* subspecies *bubulus* DSM 5363 cannot be distinguished by MALDI-TOF MS; however, these organisms also share 100% 16S rRNA sequence.[50] A recent study also concluded that MALDI-TOF MS is not useful for the routine assignment of *Salmonella* serovars.[25] Many investigators have attempted to increase the discriminatory power of library-based approaches by optimizing sample preparation, developing new LDI matrices or culture conditions, and improving database searching algorithms.[3,49]

To achieve more accurate strain-level identification without further increasing profile data richness and quality, bioinformatics-enabled approaches have been systematically studied, which use genome/protein sequence databases to identify unique biomarker proteins or PCR products of particular marker genes.[3] In these cases, tandem mass spectrometers, such as MALDI-TOF/TOF and LC–MS/MS systems, are commonly needed to facilitate the identification of characteristic protein or peptide profiles. Meanwhile, beyond identification, more information about microbial characteristics, *i.e.*, phenotype profiles, could be derived from MALDI-TOF or LC–MS spectra.

Antimicrobial susceptibility testing (AST) provides useful information in guiding clinical decision-making and is commonly performed by minimum inhibitory concentration (MIC) measurement. However, in some instances, MALDI-TOF MS can be used to detect or deduce certain resistance phenotypes by detecting key drug-resistance proteins such as beta-lactamases.[4,51,52] Identification of methicillin-resistant *Staphylococcus aureus* (MRSA), vancomycin-resistant enterococci (VRE), and strains that carry various beta-lactamases are examples of those with the most obvious clinical significance.[6,53–56]

However, many drug-resistant strains evolved from single nucleotide polymorphism (SNP) mutations; therefore, it is very difficult to differentiate them from susceptible wild-type strains solely using proteomic fingerprints. In such scenarios, in addition to the identification of DNA-PCR products by MALDI-TOF or electro-spray ionization MS (ESI-MS),[3,57,58] metabolic level alteration tends to have higher resolution and provide supplementary information in recognizing AMR. We differentiated three clinical extended-spectrum beta-lactamase (ESBL)-positive *E. coli* from four susceptible strains using 11 metabolites by Monte Carlo cross validation (MCCV) using MetaboAnalyst 3.0.[59] The alterations of lipid profiles of 3 beta-lactam-resistant MRSA strains derived from N315 were observed using an LC–MS-based method. Interestingly, for each strain, the variation of its lipid profiles could be monitored after exposure to 6 different antibiotics.[60] A bioinformatics workflow, IDBac, managed to distinguish 301 environmental bacterial subgroups based on natural product spectra (0.2–2 kDa).[20] In general, metabolites could facilitate microbial identification, especially at the strain level; in the meantime, metabolic profiles could be further explored for the AMR mechanism.

8.2.2 Databases, Software, and Classification Algorithms

Some big challenges for MALDI-TOF MS-based bacterial typing are the quality, mass accuracy, and reproducibility of MS spectra due to the heterogeneity of matrix crystallization.[51] Both commercial and free open-source packages have been developed to eliminate or offset mass shifts. ClinProTools, a commercial software package for Bruker instruments, can automate this workflow of importing data, spectral recalibration and average spectral calculation, peak picking and peak calculation, and peak statistical calculation.[6] MALDIquant,[61] a free R package, could import Bruker data and has gained popularity with versatile compatibility with a series of R packages, such as MassSpecWavelet for peak finding, *speaq* for peak alignment, and LIBSVM for support vector machine (SVM) analysis.[6] An open-source Python application, SPeDE, can help pre-process mass spectra for dereplication.[62,63] Another challenge is the difficulty of identifying polymicrobial samples, which has been partly solved by using biomarkers, similarity coefficients, and linear regression.[64,65] BacteriaMS-Mixture, an R package developed by Yang *et al.*, could facilitate the reliable characterization of bacterial mixtures up to 6 strains.[66]

For library-based bacterial characterization and typing, there are two types of data processing approaches: unsupervised methods, such as cosine similarity coefficients, clustering analysis, and principal component analysis (PCA), which generally perform strain categorization and/or differentiation; and supervised learning methods, such as partial least squares-discriminant analysis (PLS-DA), which aim to identify bacteria from a spectral input.[47,64,67,68] Commercial software packages like SIMCA, SPSS, MATLAB, and R packages, such as *ropls*, are available for conducting PLS-DA analysis.[68–70] In addition to commercial MALDI-TOF spectral libraries, small-scale free databases targeting specific research fields are also available, *e.g.*, SpectraBank,[71] RKI,[72] DWL,[73] DRIAMS,[74] and more general proteomics repositories, such as well-known MassIVE,[75,76] which also contains MALDI-TOF datasets.

To take full advantage of computational science, machine learning (ML) was introduced for bacterial recognition. Initial attempts, such as the Manchester Metropolitan University Search Engine (MUSE™), were dominated by neural networks.[77,78] Most recently, a lot of studies have applied multiple mainstream ML algorithms to bacterial typing, including support vector machines (SVMs), artificial neural networks (ANNs), random forests (RFs), *k*-nearest neighbor algorithms (KNN), naïve Bayes classifiers (NB), genetic algorithms (GAs), multiple logistic regression (MLR), and decision tree (DT) induction, typically resulting in extremely high accuracies of more than 95% for most species.[46,74,76,77,79–82] Briefly, to filter non-characteristic protein peaks, ML algorithms typically use a set of known strains as the training set to map the different categories into multidimensional space such that they are maximally separated. This trained model can then map each unknown strain into the space to assign it an identity.[6] Moreover, by training the peak list from known organisms with specific phenotypes, such as AMR, a single spectrum with several different classifiers could provide AST information, clonality, and presence or absence of a beta-lactamase.[6,76] Most strikingly, ML has been proven to be able to discriminate bacteria simply by analyzing the images of colonies at the species level.[79,83]

Bioinformatics-enabled bacterial typing, *i.e.*, the so-called "shotgun" proteomics method, mostly relies on publicly available proteome resources, such as the National Center for Biotechnology Information (NCBI) and UniProtKB, along with proteomics-related software for protein identification, such as MASCOT, Comet, MaxQuant, SEQUEST, and X!Tandem.[3,4,84–88] LC–MS/MS-based bottom-up proteomics have been widely used for the identification of taxon-specific protein/peptide markers with higher taxonomical resolution than MALDI-TOF MS-based typing.[25,86,88,89] Similarity coefficients, discriminant analysis, and NCBI BLAST (basic local alignment search tool) were mostly used for subspecies classification.[25,87,88,90] A big advantage of bioinformatics-enabled typing is its superb compatibility with downstream bioinformatic and pathway analyses. The most commonly used open-source software, databases, and web tools for bottom-up proteomics, metabolomics, lipidomics, and pathway analysis are summarized in Table 8.1.

8.2.3 Multi-omics-based Pathway and Meta-analysis

Beyond identification, metabolic pathway analysis has been explored with MALDI-TOF MS spectra, mainly by the occurrence of beta-lactam-resistant identifiers.[6] In our previous research, we also demonstrated the integration of LC–MS/MS and MALDI-TOF MS platforms for the study of the drug-resistance mechanism of several clinical *E. coli* strains.[59] As shown in Figure 8.2, MS-based multi-omics are able to elucidate the variation of a bacterial phenotype, such as antimicrobial susceptibility to a specific antibiotic, based on the alteration of the bacterial genome, proteome, and metabolome. However, MALDI-TOF MS-based SNP genotyping is only commercially available with the Sequenom MassARRAY platform, and its throughput can be severely impeded by the number of individual reactions, which has limited the widespread use of this technique.[57]

Another popular application for this multi-omics-based phenotyping is the study of gut microbiota. Grégory *et al.* detailed the culturomics- and metaproteomics-based methodology.[49] Overall, the MALDI-TOF MS-based approach could be supplementary to the conventional 16S rDNA sequencing-based metagenomics, with a dramatic extension of the microbial repertoire using multiple culture conditions.

Most recently, integrative multi-omics data have been successfully applied to large-scale meta-analysis. Through multi-omics-based meta-analyses, it is possible to identify disease-related gene signatures, biomarkers, and

Table 8.1 Frequently used open-source software, databases, and web tools for microbial phenotyping.

Database/platform	Description	URL
UniProt	Protein sequence and functional information	https://www.uniprot.org/
Comet	Protein identification	https://uwpr.github.io/Comet/
MaxQuant	Protein identification	https://www.maxquant.org/
Lipid Maps	Lipid identification	https://www.lipidmaps.org/
MassBank	Metabolite identification	https://mona.fiehnlab.ucdavis.edu/
HMDB	Metabolite identification	https://hmdb.ca/
Metabolomics Workbench	Metabolite identification	https://www.metabolomicsworkbench.org/
STRING	Protein–protein interaction and functional enrichment	https://string-db.org/
MetaboAnalyst	Metabolomics data analysis	https://www.metaboanalyst.ca/
KEGG	Pathway database collection	https://www.genome.jp/kegg/
BioCyc	Pathway database collection	https://biocyc.org/
DAVID	Functional annotation tool collection	https://david.ncifcrf.gov/home.jsp

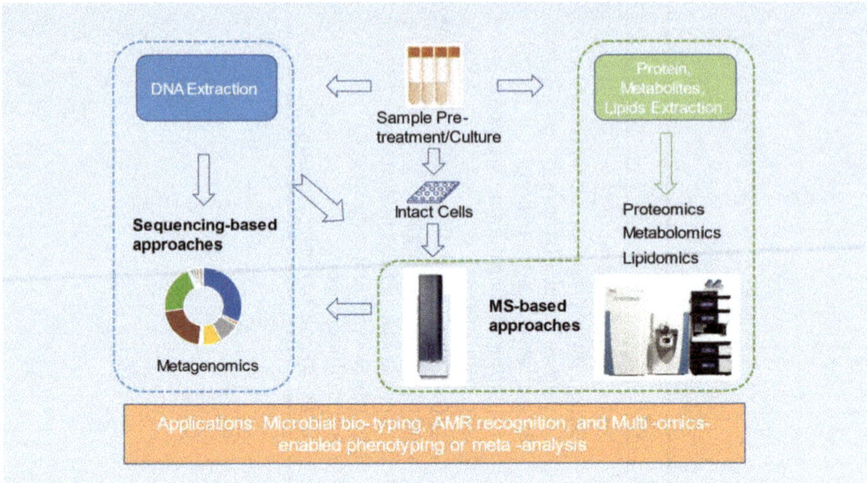

Figure 8.2 Schematic diagram of MS-based multi-omics platforms for microbial bio-typing, pathway enrichment, and meta-analysis.

pathways, providing new insight into a phenotype of interest.[91] Wang *et al.* found that butyrate, homocysteine, and palmitate were the microbial metabolites best correlated with chronic obstructive pulmonary disease (COPD)-associated host genes by meta-analysis and further validated the hypothesis by a pilot multi-omics study of 18 COPD patients and 10 controls whereby the airway metagenome, metabolome, and host transcriptome were simultaneously characterized.[92] Furthermore, genome-wide association studies (GWAS) could also be conducted to identify SNPs that regulate phenotypic traits.[93]

8.3 Conclusions and Future Potential

Although most metabolites are present in samples of the same microbial species, the subtle alteration of metabolite quantities can be captured and analyzed by ML methods for strain-level identification. Rationalization of this approach is due in part to characteristic/phenotypic changes caused by genetic mutations, which may affect the downstream regulation of many genes. These small molecules could function as potential biomarkers to facilitate strain typing, especially for drug-resistant SNP variants. With a single MALDI-TOF spectrum, it is possible to obtain both proteomic and metabolic fingerprints for accurate strain-level identification. However, to fully automate this process, metabolic data need to be standardized with consistent LDI matrices, and then incorporated into database libraries, which is laborious and would require collaboration among research labs and leading companies in bacterial typing. Recent studies have focused on only a few of the most common pathogens, like *E. coli*, *S. aureus*, and *Enterococcus faecalis*. Large-scale studies are still absent, which was partly caused by the difficulty in finding appropriate matrices without interfering with low-mass range metabolites. Moreover, ML methods also need to be optimized for metabolite-based bacterial typing, especially for background/noise peak filtering, peak statistical evaluation, and characteristic peak screening. Lastly, the availability of publicly available MALDI-TOF spectral databases is still limited, which will inevitably impede the training of ML predictors for more accurate large-scale species-specific typing.

The foundation of integration of MS-based multi-omics for bacterial phenotyping is well established and has been beneficial to biology scientists; however, these tools have been adopted more slowly in clinical settings due to the need for a consistent streamlined approach for all samples. Commercially available software packages, such as iPathwayGuide, allow for meta-analyses across multi-omics data. In clinical settings, the mechanistic study of beta-lactam by MALDI-TOF MS has been widely accepted, and some AST assays have even been used for routine diagnosis. However, there is no single method that is approved for phenotyping all strains. For many SNP variants without protein determinants of resistance, metabolic analysis could provide higher resolution by simply identifying the hydrolysis of a beta-lactam antibiotic or profiling small molecule biomarkers. In comparison with MALDI-TOF MS, high-resolution LC–MS/MS tends to have more protein or metabolite peaks detected, which is suitable for deep phenotyping and pathway analysis. Even so, the LC–MS/MS-based bacterial typing method is still time-consuming and less competitive than the MALDI-TOF MS-based approach, especially in a clinical laboratory. Thus, MALDI-TOF MS is expected to continue dominating MS-based bacterial typing and profiling for the years to come. Despite the advantages of MS-based approaches, traditional AST assays would not be replaced entirely anytime soon due to the high costs of MS equipment and relatively sophisticated operating conditions, which could drastically limit their application in low-resource countries with the highest need for AMR interventions.

Abbreviations

MALDI-TOF MS	Matrix-assisted laser desorption/ionization time-of-flight mass spectrometry
LC–MS	Liquid chromatography–mass spectrometry
ESI	Electro-spray ionization
AST	Antimicrobial/antibiotic susceptibility testing
AMR	Antimicrobial resistance
CFU	Colony-forming unit
SNP	Single nucleotide polymorphism
ML	Machine learning
ESBL	Extended-spectrum beta-lactamase
MRSA	Methicillin-resistant *Staphylococcus aureus*
VRE	Vancomycin-resistant enterococci
GWAS	Genome-wide association study

References

1. J. O'Neill, Tackling drug-resistant infections globally: final report and recommendations, Government of the United Kingdom, 2016.
2. J. J. Faith, J. F. Colombel and J. I. Gordon, Identifying strains that contribute to complex diseases through the study of microbial inheritance, *Proc. Natl. Acad. Sci. U. S. A.*, 2015, **112**(3), 633–640.
3. T. R. Sandrin, J. E. Goldstein and S. Schumaker, MALDI TOF MS profiling of bacteria at the strain level: A review, *Mass Spectrom. Rev.*, 2013, **32**(3), 188–217.
4. P. J. Hart, *et al.*, A method for the detection of antibiotic resistance markers in clinical strains of Escherichia coli using MALDI mass spectrometry, *J. Microbiol. Methods*, 2015, **111**, 1–8.
5. E. Torres-Sangiao, C. L. Rodriguez and C. Garcia-Riestra, Application and Perspectives of MALDI-TOF Mass Spectrometry in Clinical Microbiology Laboratories, *Microorganisms*, 2021, **9**(7), 1539.
6. M. L. DeMarco and B. A. Ford, Beyond Identification: Emerging and Future Uses for MALDI-TOF Mass Spectrometry in the Clinical Microbiology Laboratory, *Clin. Lab. Med.*, 2013, **33**(3), 611–628.
7. M. De-la-Fuente, *et al.*, Detection of Polyclonality among Clinical Isolates from Prosthetic Joint Infections, *J. Clin. Microbiol.*, 2015, **53**(12), 3766–3772.
8. A. AbdulWahab, *et al.*, Discrepancy in MALDI-TOF MS identification of uncommon Gram-negative bacteria from lower respiratory secretions in patients with cystic fibrosis, *Infect. Drug Resist.*, 2015, **8**, 83–88.
9. L. Grenga, O. Pible and J. Armengaud, Pathogen proteotyping: A rapidly developing application of mass spectrometry to address clinical concerns, *Clin. Mass Spectrom.*, 2019, **14**, 9–17.
10. C. H. Huang, *et al.*, Identification and Classification for the Lactobacillus casei Group, *Front. Microbiol.*, 2018, **9**, 1974.
11. F. Giacometti, *et al.*, Application of MALDI-TOF MS for the subtyping of Arcobacter butzleri strains and comparison with their MLST and PFGE types, *Int. J. Food Microbiol.*, 2018, **277**, 50–57.
12. A. K. Bhunia, One day to one hour: how quickly can foodborne pathogens be detected?, *Future Microbiol.*, 2014, **9**(8), 935–946.
13. A. Verma and K. Ambatipudi, Challenges and opportunities of bovine milk analysis by mass spectrometry, *Clin. Proteomics*, 2016, **13**, 8.
14. C. C. Kern, R. F. Vogel and J. Behr, Differentiation of Lactobacillus brevis strains using Matrix-Assisted-Laser-Desorption-Ionization-Time-of-Flight Mass Spectrometry with respect to their beer spoilage potential, *Food Microbiol.*, 2014, **40**, 18–24.
15. J. M. Membre and S. Guillou, Latest developments in foodborne pathogen risk assessment, *Curr. Opin. Food Sci.*, 2016, **8**, 120–126.
16. J. Salplachta, *et al.*, Characterization of Dickeya and Pectobacterium species by capillary electrophoretic techniques and MALDI-TOF MS, *Anal. Bioanal. Chem.*, 2015, **407**(25), 7625–7635.

17. M. H. Josefsen, *et al.*, Monitoring Campylobacter in the poultry production chain - From culture to genes and beyond, *J. Microbiol. Methods*, 2015, **112**, 118–125.
18. N. T. Popovic, *et al.*, Differentiation of environmental aquatic bacterial isolates by MALDI-TOF MS, *Environ. Res.*, 2017, **152**, 7–16.
19. E. M. Hartmann, *et al.*, Proteomics meets blue biotechnology: A wealth of novelties and opportunities, *Mar. Genomics*, 2014, **17**, 35–42.
20. M. S. Costa, *et al.*, Minimizing Taxonomic and Natural Product Redundancy in Microbial Libraries Using MALDI-TOF MS and the Bioinformatics Pipeline IDBac, *J. Nat. Prod.*, 2019, **82**(8), 2167–2173.
21. I. C. Santos, Z. L. Hildenbrand and K. A. Schug, Applications of MALDI-TOF MS in environmental microbiology, *Analyst*, 2016, **141**(10), 2827–2837.
22. H. L. Liu, *et al.*, A hybrid ionic liquid-matrix material, TiO_2-Si-NH_3+CHC-, as a novel matrix for the analysis of small molecules by MALDI-TOF MS, *Int. J. Mass Spectrom.*, 2015, **376**, 85–89.
23. B. Unnikrishnan, *et al.*, Functional gold nanoparticles coupled with laser desorption ionization mass spectrometry for bioanalysis, *Anal. Methods*, 2016, **8**(46), 8123–8133.
24. E. W. Y. Ng, M. Y. M. Wong and T. C. W. Poon, Advances in MALDI Mass Spectrometry in Clinical Diagnostic Applications, *Chemical Diagnostics: From Bench to Bedside*, 2014, vol. 336, pp. 139–175.
25. S. H. Chen, *et al.*, Identification of Salmonella Taxon-Specific Peptide Markers to the Serovar Level by Mass Spectrometry, *Anal. Chem.*, 2019, **91**(7), 4388–4395.
26. M. J. van der Werf, *et al.*, Microbial metabolomics: Toward a platform with full metabolome coverage, *Anal. Biochem.*, 2007, **370**(1), 17–25.
27. F. Mozzi, *et al.*, Metabolomics as a tool for the comprehensive understanding of fermented and functional foods with lactic acid bacteria, *Food Res. Int.*, 2013, **54**(1), 1152–1161.
28. O. Fiehn, Metabolomics—the link between genotypes and phenotypes, in *Functional Genomics*, ed. C. Town, Springer Netherlands, Dordrecht, 2002, pp. 155–171.
29. I. Bertini, X. Y. Hu and C. Luchinat, Global metabolomics characterization of bacteria: pre-analytical treatments and profiling, *Metabolomics*, 2014, **10**(2), 241–249.
30. V. Mapelli, L. Olsson and J. Nielsen, Metabolic footprinting in microbiology: methods and applications in functional genomics and biotechnology, *Trends Biotechnol.*, 2008, **26**(9), 490–497.
31. A. Hirayama, *et al.*, Quantitative Metabolome Profiling of Colon and Stomach Cancer Microenvironment by Capillary Electrophoresis Time-of-Flight Mass Spectrometry, *Cancer Res.*, 2009, **69**(11), 4918–4925.
32. N. Psychogios, *et al.*, The Human Serum Metabolome, *PLoS One*, 2011, **6**(2), e16957.
33. A. Sreekumar, *et al.*, Metabolomic profiles delineate potential role for sarcosine in prostate cancer progression, *Nature*, 2009, **457**(7231), 910–914.

34. J. R. Denery, *et al.*, Metabolomics-Based Discovery of Diagnostic Biomarkers for Onchocerciasis, *PLoS Neglected Trop. Dis.*, 2010, **4**(10), e834.
35. X. J. Wang, *et al.*, Urine Metabolomics Analysis for Biomarker Discovery and Detection of Jaundice Syndrome in Patients With Liver Disease, *Mol. Cell. Proteomics*, 2012, **11**(8), 370–380.
36. T. Frank, *et al.*, Metabolite profiling of barley: Influence of the malting process, *Food Chem.*, 2011, **124**(3), 948–957.
37. R. D. Beger, A Review of Applications of Metabolomics in Cancer, *Metabolites*, 2013, **3**(3), 552–574.
38. J. Rabinowitz, *et al.*, Metabolomics in drug target discovery, in *Cold Spring Harbor symposia on quantitative biology*, Cold Spring Harbor Laboratory Press, 2011.
39. J. P. Anhalt and C. Fenselau, Identification of Bacteria Using Mass-Spectrometry, *Anal. Chem.*, 1975, **47**(2), 219–225.
40. C. Foschi, *et al.*, Novel approaches for the taxonomic and metabolic characterization of lactobacilli: Integration of 16S rRNA gene sequencing with MALDI-TOF MS and ^{1}H-NMR, *PLoS One*, 2017, **12**(2), e0172483.
41. K. J. Voorhees, *et al.*, Modified MALDI MS fatty acid profiling for bacterial identification, *J. Mass Spectrom.*, 2013, **48**(7), 850–855.
42. M. H. Lu, *et al.*, Nanomaterials as Assisted Matrix of Laser Desorption/Ionization Time-of-Flight Mass Spectrometry for the Analysis of Small Molecules, *Nanomaterials*, 2017, 7(4), 87.
43. R. Zenobi, Single-Cell Metabolomics: Analytical and Biological Perspectives, *Science*, 2013, **342**(6163), 1243259.
44. T. Luzzatto-Knaan, A. V. Melnik and P. C. Dorrestein, Mass spectrometry tools and workflows for revealing microbial chemistry, *Analyst*, 2015, **140**(15), 4949–4966.
45. J. Rohn, First mass spectrometry diagnostic approved in US, *Nat. Biotechnol.*, 2013, **31**, 862.
46. A. Freiwald and S. Sauer, Phylogenetic classification and identification of bacteria by mass spectrometry, *Nat. Protoc.*, 2009, **4**(5), 732–742.
47. Y. D. Zhu, *et al.*, Sensitive and fast identification of bacteria in blood samples by immunoaffinity mass spectrometry for quick BSI diagnosis, *Chem. Sci.*, 2016, 7(5), 2987–2995.
48. M. L. Bhaisare, *et al.*, Rapid and direct MALDI-MS identification of pathogenic bacteria from blood using ionic liquid-modified magnetic nanoparticles (Fe_3O_4@SiO_2), *J. Mater. Chem. B*, 2014, **2**(29), 4671 4683.
49. D. Gregory, *et al.*, How mass spectrometric approaches applied to bacterial identification have revolutionized the study of human gut microbiota, *Expert Rev. Proteomics*, 2018, **15**(3), 217–229.
50. P. Schumann and T. Maier, MALDI-TOF Mass Spectrometry Applied to Classification and Identification of Bacteria, *New Approaches to Prokaryotic Systematics*, 2014, vol. 41, pp. 275–306.
51. S. Schubert and M. Kostrzewa, MALDI-TOF Mass Spectrometry in the Clinical Microbiology Laboratory; Beyond Identification, *Current and*

Emerging Technologies for the Diagnosis of Microbial Infections, 2015, vol. 42, pp. 501–524.
52. K. Sparbier, *et al.*, Matrix-Assisted Laser Desorption Ionization-Time of Flight Mass Spectrometry-Based Functional Assay for Rapid Detection of Resistance against beta-Lactam Antibiotics, *J. Clin. Microbiol.*, 2012, **50**(3), 927–937.
53. M. Wolters, *et al.*, MALDI-TOF MS fingerprinting allows for discrimination of major methicillin-resistant *Staphylococcus aureus* lineages, *Int. J. Med. Microbiol.*, 2011, **301**(1), 64–68.
54. P. M. Griffin, *et al.*, Use of Matrix-Assisted Laser Desorption Ionization-Time of Flight Mass Spectrometry To Identify Vancomycin-Resistant Enterococci and Investigate the Epidemiology of an Outbreak, *J. Clin. Microbiol.*, 2012, **50**(9), 2918–2931.
55. I. Wybo, *et al.*, Differentiation of cfiA-Negative and cfiA-Positive Bacteroides fragilis Isolates by Matrix-Assisted Laser Desorption Ionization-Time of Flight Mass Spectrometry, *J. Clin. Microbiol.*, 2011, **49**(5), 1961–1964.
56. K. A. Jackson, *et al.*, Optimisation of intact cell MALDI method for fingerprinting of methicillin-resistant *Staphylococcus aureus*, *J. Microbiol. Methods*, 2005, **62**(3), 273–284.
57. M. W. Syrmis, *et al.*, Comparison of a multiplexed MassARRAY system with real-time allele-specific PCR technology for genotyping of methicillin-resistant *Staphylococcus aureus*, *Clin. Microbiol. Infect.*, 2011, **17**(12), 1804–1810.
58. S. Sauer and M. Kliem, Mass spectrometry tools for the classification and identification of bacteria, *Nat. Rev. Microbiol.*, 2010, **8**(1), 74–82.
59. R. T. Zhang, *et al.*, TiO_2-Assisted Laser Desorption/Ionization Mass Spectrometry for Rapid Profiling of Candidate Metabolite Biomarkers from Antimicrobial-Resistant Bacteria, *Anal. Chem.*, 2018, **90**(6), 3863–3870.
60. R. T. Zhang, *et al.*, Synergy Between Beta-Lactams and Lipo-, Glyco-, and Lipoglycopeptides, Is Independent of the Seesaw Effect in Methicillin-Resistant *Staphylococcus aureus*, *Front. Mol. Biosci.*, 2021, **8**, 688357.
61. S. Gibb and K. Strimmer, MALDIquant: a versatile R package for the analysis of mass spectrometry data, *Bioinformatics*, 2012, **28**(17), 2270–2271.
62. C. Dumolin, *et al.*, Introducing SPeDE: High-Throughput Dereplication and Accurate Determination of Microbial Diversity from Matrix-Assisted Laser Desorption-Ionization Time of Flight Mass Spectrometry Data, *mSystems*, 2019, **4**(5), e00437–19.
63. C. Dumolin, *et al.*, Network Analysis Based on Unique Spectral Features Enables an Efficient Selection of Genomically Diverse Operational Isolation Units, *Microorganisms*, 2021, **9**(2), 416.
64. L. Zhang, S. Smart and T. R. Sandrin, Biomarker- and similarity coefficient-based approaches to bacterial mixture characterization using matrix-assisted laser desorption ionization time-of-flight mass spectrometry (MALDI-TOF MS), *Sci. Rep.*, 2015, **5**, 15834.

65. P. Mahe, *et al.*, Automatic identification of mixed bacterial species fingerprints in a MALDI-TOF mass-spectrum, *Bioinformatics*, 2014, **30**(9), 1280–1286.
66. Y. Yang, Y. Lin and L. Qiao, Direct MALDI-TOF MS Identification of Bacterial Mixtures, *Anal. Chem.*, 2018, **90**(17), 10400–10408.
67. N. AlMasoud, *et al.*, Discrimination of bacteria using whole organism fingerprinting: the utility of modern physicochemical techniques for bacterial typing, *Analyst*, 2021, **146**(3), 770–788.
68. N. M. Sindt, *et al.*, MALDI-TOF-MS with PLS Modeling Enables Strain Typing of the Bacterial Plant Pathogen Xanthomonas axonopodis, *J. Am. Soc. Mass Spectrom.*, 2018, **29**(2), 413–421.
69. E. A. Thevenot, *et al.*, Analysis of the Human Adult Urinary Metabolome Variations with Age, Body Mass Index, and Gender by Implementing a Comprehensive Workflow for Univariate and OPLS Statistical Analyses, *J. Proteome Res.*, 2015, **14**(8), 3322–3335.
70. P. S. Gromski, *et al.*, A tutorial review: Metabolomics and partial least squares-discriminant analysis – a marriage of convenience or a shotgun wedding, *Anal. Chim. Acta*, 2015, **879**, 10–23.
71. K. Bohme, *et al.*, SpectraBank: An open access tool for rapid microbial identification by MALDI-TOF MS fingerprinting, *Electrophoresis*, 2012, **33**(14), 2138–2142.
72. W. Beyer, *et al.*, A MALDI-TOF Mass Spectrometry Database for Identification and Classification of Highly Pathogenic Microorganisms from the Robert Koch-Institute (RKI).
73. A. Pinar-Mendez, *et al.*, Rapid and improved identification of drinking water bacteria using the Drinking Water Library, a dedicated MALDI-TOF MS database, *Water Res.*, 2021, **203**, 117543.
74. C. Weis, *et al.*, Direct antimicrobial resistance prediction from clinical MALDI-TOF mass spectra using machine learning, *Nat. Med.*, 2022, **28**(1), 164–174.
75. Y. Perez-Riverol, *et al.*, Making proteomics data accessible and reusable: Current state of proteomics databases and repositories, *Proteomics*, 2015, **15**(5–6), 930–949.
76. C. V. Weis, C. R. Jutzeler and K. Borgwardt, Machine learning for microbial identification and antimicrobial susceptibility testing on MALDI-TOF mass spectra: a systematic review, *Clin. Microbiol. Infect.*, 2020, **26**(10), 1310–1317.
77. K. De Bruyne, *et al.*, Bacterial species identification from MALDI-TOF mass spectra through data analysis and machine learning, *Syst. Appl. Microbiol.*, 2011, **34**(1), 20–29.
78. J. J. Bright, *et al.*, Rapid typing of bacteria using matrix-assisted laser desorption ionisation time-of-flight mass spectrometry and pattern recognition software, *J. Microbiol. Methods*, 2002, **48**(2–3), 127–138.
79. Y. Maeda, *et al.*, Colony Fingerprint-Based Discrimination of Staphylococcus species with Machine Learning Approaches, *Sensors*, 2018, **18**(9), 2789.

80. C. R. Chung, *et al.*, Incorporating Statistical Test and Machine Intelligence Into Strain Typing of Staphylococcus haemolyticus Based on Matrix-Assisted Laser Desorption Ionization-Time of Flight Mass Spectrometry, *Front. Microbiol.*, 2019, **10**, 2120.
81. H. Y. Wang, *et al.*, A new scheme for strain typing of methicillin-resistant *Staphylococcus aureus* on the basis of matrix-assisted laser desorption ionization time-of-flight mass spectrometry by using machine learning approach, *PLoS One*, 2018, **13**(3), e0194289.
82. L. N. Ikryannikova, *et al.*, Discrimination between Streptococcus pneumoniae and Streptococcus mitis based on sorting of their MALDI mass spectra, *Clin. Microbiol. Infect.*, 2013, **19**(11), 1066–1071.
83. B. Zielinski, *et al.*, Deep learning approach to bacterial colony classification, *PLoS One*, 2017, **12**(9), e0184554.
84. W. G. Shao, *et al.*, A peptide identification-free, genome sequence-independent shotgun proteomics workflow for strain-level bacterial differentiation, *Sci. Rep.*, 2015, **5**, 14337.
85. M. T. Gekenidis, *et al.*, Beyond the Matrix-Assisted Laser Desorption Ionization (MALDI) Biotyping Workflow: in Search of Microorganism-Specific Tryptic Peptides Enabling Discrimination of Subspecies, *Appl. Environ. Microbiol.*, 2014, **80**(14), 4234–4241.
86. R. Karlsson, *et al.*, Strain-Level Typing and Identification of Bacteria Using Mass Spectrometry-Based Proteomics, *J. Proteome Res.*, 2012, **11**(5), 2710–2720.
87. J. P. Dworzanski, *et al.*, Identification of bacteria using tandem mass spectrometry combined with a proteome database and statistical scoring, *Anal. Chem.*, 2004, **76**(8), 2355–2366.
88. J. P. Dworzanski, *et al.*, Discrimination and Phylogenomic Classification of Bacillus anthracis-cereus-thuringiensis Strains Based on LC–MS/MS Analysis of Whole Cell Protein Digests, *Anal. Chem.*, 2010, **82**(1), 145–155.
89. S. H. Chen, *et al.*, Genus, Species, and Subspecies Classification of Salmonella Isolates by Proteomics, *Appl. Sci.*, 2021, **11**(9), 4264.
90. P. Lasch, *et al.*, Identification of Microorganisms by Liquid Chromatography-Mass Spectrometry (LC–MS[1]) and in Silico Peptide Mass Libraries, *Mol. Cell. Proteomics*, 2020, **19**(12), 2125–2139.
91. K. Cervantes-Gracia, R. Chahwan and H. Husi, Integrative OMICS Data-Driven Procedure Using a Derivatized Meta-Analysis Approach, *Front. Genet.*, 2022, **13**, 828786.
92. Z. Wang, *et al.*, Multi-omic meta-analysis identifies functional signatures of airway microbiome in chronic obstructive pulmonary disease, *ISME J.*, 2020, **14**(11), 2748–2765.
93. A. T. Marees, *et al.*, A tutorial on conducting genome-wide association studies: Quality control and statistical analysis, *Int. J. Methods Psychiatr. Res.*, 2018, **27**(2), e1608.

CHAPTER 9

Rapid Detection and Evolution of SARS-CoV2 and Other Viruses with High Resolution Mass Spectrometry

KEVIN M. DOWNARD*[a]

[a] Infectious Disease Responses Laboratory, Prince of Wales Clinical Research Sciences, Sydney, NSW, Australia
*Email: kevin.downard@scientia.org.au

9.1 Prevalence and Diversity of Viruses

Viruses pervade our natural world and have plagued mankind throughout the centuries. Over 200 virus species are known to be able to infect humans,[1] with several new species being identified every year. The first human virus to be identified as such in 1901 was the yellow fever virus. More than two-thirds of human viruses can also infect non-human hosts. Typically, other mammals and birds provide routes to infection in humans. Scientists continue to catalogue viral diversity by sampling viruses in the population, and then characterizing these pathogens and their ability to replicate in the laboratory to assess those that pose the greatest threat.

Viruses that have had the greatest impact on human health during the past 100 years include the influenza virus, the smallpox virus, the measles virus, the hepatitis virus, the human immunodeficiency virus (HIV), the severe acute respiratory syndrome (SARS) associated coronavirus (CoV), the Middle East respiratory syndrome (MERS) virus, the Ebola virus and the Zika virus[2] (Figure 9.1).

New Developments in Mass Spectrometry No. 13
Detection and Analysis of Microorganisms by Mass Spectrometry
Edited by Liang Qiao and Jia Yi

Published by the Royal Society of Chemistry, www.rsc.org

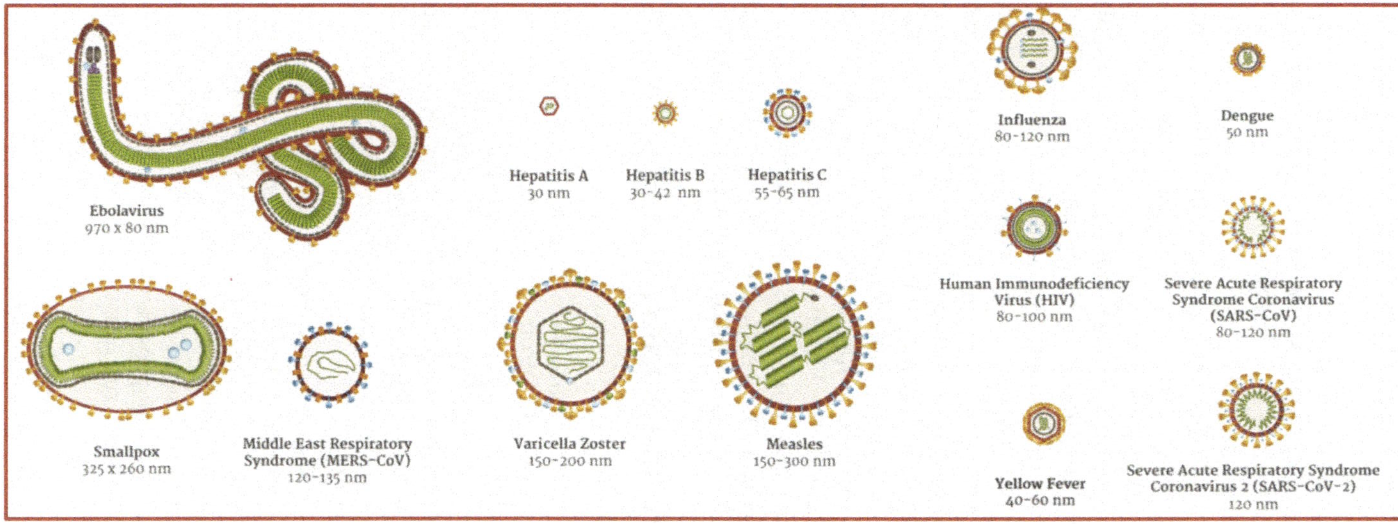

Figure 9.1 Schematic representation of the viruses responsible for the greatest impact on human health and their approximate virion diameter.

Pandemics associated with some of these viruses have resulted in tens of millions of deaths worldwide and far higher infection rates and disease levels within the human population.

Viruses show considerable genetic diversity, particularly those with RNA *versus* DNA genomes. RNA viruses are error prone and have mutations at high rates when copying their genetic material within a host cell. This generates a large amount of diversity within progeny RNA viruses that allows them to rapidly adapt to new environments, including new host species. This challenges efforts to effectively respond to viral outbreaks using vaccines and antiviral therapies in various formats.

Prior to the development of any treatment regime, there is a need to rapidly detect viruses in human specimens and combine such evidence with epidemiological data to manage and plan response strategies. Among the molecular based analytical detection methods,[3] mass spectrometry (MS) in various guises is emerging as a viable and sensitive alternative to conventional polymerase chain reaction (PCR)[4] and other assays for viral detection.[5–8] Combining its rapid speed of analysis, with the ability to accurately and reliably detect both genetic (DNA or RNA) and protein fragments, MS is also capable of approaching the sensitivity levels required for detecting viruses directly in clinical specimens.

This chapter illustrates protein MS-based approaches that have been employed to rapidly detect and characterize SARS-CoV2 and other respiratory viruses, which exploit the benefits of high resolution mass spectrometers.

9.2 Anatomy and Purification of a Virus Ahead of Analysis

A virus particle or virion contains a viral core consisting of either DNA or RNA in complex with capsid proteins. This nucleocapsid is protected by a shell built of proteins or a lipid bilayer into which proteins are embedded. This shell protects the viral genome from the hostile extracellular environment. Regardless of the virus type, the shell is always decorated with surface proteins responsible for attachment and entry into the host cell that initiates the replication cycle.[9] Virus particles range in diameter from 0.01 to as large as 1 μm (see Figure 9.1).

Detection of a virus is achieved at the molecular level by analysing segments of the viral genome or the associated proteins. The sizes and levels of these molecular entities vary from virus to virus; so depending on the type of analysis, it may be desirable to detect a particular genomic or protein component. The release of these components and their subsequent purification or semi-purification provides the first challenge to analysis. The purified or semi-purified virus must be recovered from other host cellular debris and media used to store or grow the virus.[10] The virion is then disrupted and the components are separated by repeated freeze-thawing, sample heating, irradiation, sonication, centrifugation methods and filtration. During such treatments care must be taken not to degrade the genetic or protein material of interest.[10,11] The steps taken prior to any subsequent analysis are critically important to the success of the latter and also typically consume more time. At a minimum, such treatments take several hours and considerably more time than that required for MS measurements themselves.

9.3 PCR Analysis *vs.* Protein MS with High Resolution

PCR-based approaches require pre-sample treatments and preparations that are no less time consuming.[4] DNA or RNA extraction and purification is followed by reverse transcription, as required, and the amplification of cDNA using a pair of upstream and downstream primers. The products are either detected using hybridization assays or sequenced employing chain termination chemistries. Total analysis takes at least 24–48 h. Furthermore, due to the continual evolution of a virus, the binding sites of primers and probes have to be continuously monitored. Despite their sensitivity, PCR methods suffer from a relatively low-throughput capacity due to the many biochemical steps involved,[12,13] false-positive and false-negative rates,[14] and issues concerning sampling and the potential for carry-over contamination,[15] as exemplified for the SARS-CoV2 virus. While real-time PCR still remains the "gold standard" for the molecular diagnosis of viruses, MS is beginning to play a greater role.

MS is particularly adept at detecting viral proteins or their peptide counterparts,[16,17] given their relative stability and ionization efficiency over genetic (DNA or RNA) fragments. MALDI-MS, in particular, offers advantages in terms of high sample throughput, short analysis times and relatively easy sample preparation[18,19] over ESI-based LC–MS. MALDI is also more tolerant to salts and contaminants often encountered in biological and clinical samples. MALDI MS is inherently faster than LC–MS-based strategies where the (often overlooked and under-reported) time to set up HPLC conditions and equilibrate and flush columns between runs adds to the analysis time and must be considered. The use of reverse phase columns is also accompanied by some unavoidable sample loss. MALDI-based instruments are also in widespread use throughout the world and particularly within clinical laboratories and settings for pathogen identification.[20,21]

Although MALDI has primarily been coupled to time-of-flight (TOF) instruments, given its compatibility with pulsed ion sources, its union with higher resolution analysers in the form of extended/multipass TOF[22] and ion traps such as the Orbitrap[23,24] and ion cyclotron resonance (ICR)[25] systems have all been achieved (Figure 9.2). All feature long ion trajectories that achieve, albeit differently, better separation of ions of similar masses. The ICR system provides the highest mass resolution ($m/\Delta m \geq 100\,000$) and thus confidence in mass assignment and molecular identification and routinely measures masses at or below 1 part-per-million (ppm) mass accuracy.

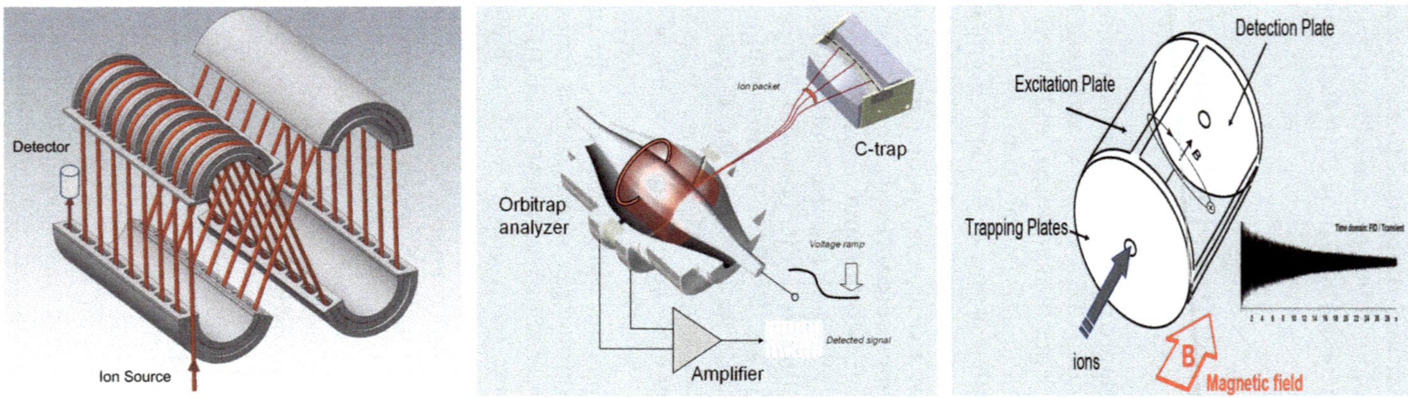

Figure 9.2 Side-by-side illustration of high resolution mass analysers: multipass TOF (left), Orbitrap with collection trap (middle) and ion cyclotron resonance (ICR) (right). In all analysers, ions have relatively long trajectories before detection to enhance ion separation.

9.4 Benefits of Detecting Signature Peptide Biomarkers by Mass Alone

For over a decade[26–28] the author's laboratory has identified and detected signature peptides following the digestion of viral proteins. Signature peptides are those of a conserved sequence within a viral protein across all known strains and whose mass is unique (with a low ppm error range) compared to all other peptides from other viral proteins and those of known contaminants.[16] Their identification and subsequent detection has enabled typing and subtyping[26,27] of respiratory viruses such as influenza and parainfluenza viruses and establishing their lineage[28] and origins.[16] Pandemic strains have also been differentiated from seasonal strains.[29]

These strategies have been more recently applied to identify unique signature peptide biomarkers that enable the SARS-CoV2 virus[30] and its various major variants of concern to be identified.[31,32] Such approaches offer the highest diagnostic reliability over studies that detect unidentified, or poorly identified, molecules unique to SARS-CoV2-positive clinical samples[33] that may or may not even be associated with the pathogen itself.[18] By contrast, the high mass resolution detection of specific viral peptide signatures in such specimens allows a confident diagnosis to be made.[16] Importantly, this is achieved without the need for additional time- and sample-consuming (by at least an order of magnitude) tandem sequencing experiments common to LC–MS strategies.

9.5 Peptide Biomarker Signatures for the Detection of the SARS-CoV2 Virus

The masses of peptide signatures for the spike (S) and nucleocapsid (N) proteins that can be used for the confident detection of the SARS-CoV2 virus in clinical specimens with high resolution MS (HR-MS) have been identified.[30] The masses of the former protein can also be used for the differentiation of the major variants of concern without the need for sequencing.[31,32] Identification of the S protein is particularly useful in this regard given it represents the main target of vaccination responses in the form of either RNA or protein subunit vaccines[34] as well as a focus of antiviral therapies.[35]

The basis of high resolution MALDI-MS analysis is to apply a two-stage strategy. The first is to identify the presence of the SARS-CoV2 virus in the specimen with a high degree of confidence, while the second optional analysis allows major variants of concern to be determined. From this standpoint, a single tryptic digest suffices for the initial stage where various protein segments across the S and N proteins can be detected by MALDI-MS following whole virus digestion.[30] These are listed in Table 9.1 and comprise 10 S and 8 N peptides. Three of the N peptides, marked in the right hand column, have also been detected in a separate study based on high resolution MALDI-MS using an alternative multipath TOF instrument.[36]

Table 9.1 Tryptic peptide mass biomarkers for spike (S) and nucleocapsid (N) proteins based on prior detection by high resolution MALDI-MS.

m/z (mono) theoretical	Viral protein code	Residues[a]	Sequence	Domain[b]	N segments detected in a separate study (Yoshinari *et al.*[36])
749.3536	N	178–185	GGSQASSR	RNA-binding domain	
831.4570	N	227–233	LNQLESK	Dimerization domain	
843.4247	S	1206–1211	YEQYIK	S2 subunit heptad repeat 2 (HR2) domain	
899.4945	S	409–417	QIAPGQTGK	S1 subunit C-terminal domain (CTD)	
1085.5374	S	311–319	GIYQTSNFR	Undefined	
1126.5640	N	267–276	AYNVTQAFGR	Dimerization domain	√
1224.6259	S	103–113	GWIFGTTLDSK	S1 subunit N-terminal domain (NTD)	
1267.6946	S	558–567 (1)	KFLPFQQFGR	Undefined	
1281.5892	S	836–847	QYGDCLGDIAAR	S2 subunit fusion peptide (FP)	
1455.7550	S	404–417 (1)	GDEVRQIAPGQTGK	S1 subunit CTD	
1670.8894	N	210–226	MAGNGGDAALALLLLDR	Undefined	
1788.8511	N	278–293	GPEQTQGNFGDQELIR	Dimerization domain	
1989.9738	S	387–403	LNDLCFTNVYADSFVIR	S1 subunit receptor-binding domain (RBD)	
2040.0007	S	329–346	FPNITNLCPFGEVFNATR	S1 subunit RBD	
2181.0207	N	69–88	GQGVPINTNSSPDDQIGYYR	RNA-binding domain	
2267.0807	N	108–127	WYFYYLGTGPEAGLPYGANK	NTD	√
2324.1894	N	41–61	RPQGLPNNTASWFTALTQHGK	RNA-binding domain	√
2742.2425	S	159–182	VYSSANNCTFEYVSQPFLMDLEGK	S1 subunit NTD	

[a] Residues marked with a (1) are associated with one missed cleavage site; all others contain no missed cleavage sites.
[b] As defined on the UniProt Knowledgebase (UniProtKB) at https://covid-19.uniprot.org/uniprotkb/.

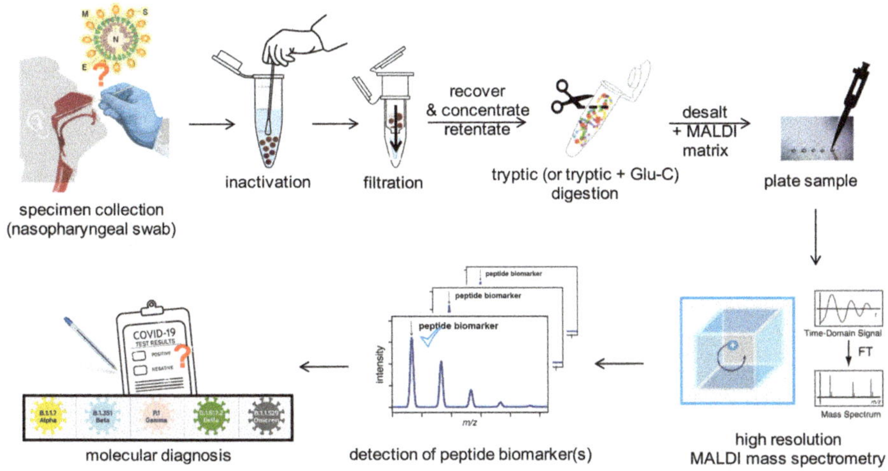

Figure 9.3 Schematic of the approach used to detect the SARS-CoV2 virus in clinical specimens using high resolution MALDI MS.

Samples are processed according to the procedures illustrated in Figure 9.3. Briefly, the collected clinical specimen is inactivated and filtered, the virus is precipitated and the retentate is digested and then prepared for MALDI-MS analysis to detect the peptide biomarker(s). The endoproteinase trypsin is favoured given its high specificity and the frequency of arginine and lysine residues in the S protein (some 8–10% of the total residues of the originating strains). Preparation of digests for MALDI mass analysis follows conventional protocols where pre-concentration and desalting steps are applied. The detection of any one or more of the peptide biomarkers allows the presence of SARS-CoV2 in a clinical specimen to be confirmed.

9.6 Detection of Peptide Biomarkers in a Set of SARS-CoV2 Clinical Specimens

A set of 16 specimens taken from nasopharyngeal swabs, comprising 16 confirmed to be positive and 4 to be negative by RT-PCR, were analysed (Table 9.2). The specimens contained a range of virus titres numbered in decreasing levels. One such sample in the mid-titre range (shown in bold font in Table 9.2) afforded the spectrum shown in Figure 9.4. The spectrum contains six S and N peptides above a signal-to-noise (S/N) ratio of 2 that are in boldface. These narrow range (5 u) segments, together with those for the other peptides of Table 9.1, were scanned to establish their presence (+) or absence (−). This is indicated in the column for each peptide in Table 9.2.

As the estimated titre levels drop from top to bottom, the number of peptides detected above a S/N ratio of 2 also falls, as is to be expected. At the highest virus titres ($1.5–4 \times 10^7$ copies per mL), 14 of the 18 peptides are

Table 9.2 Detection of tryptic peptide mass biomarkers for spike (S) and nucleocapsid (N) proteins in nasopharyngeal specimens.

Specimen number	RT-PCR SARS-CoV2 +/−	RT-PCR Copies per mL	Peptide biomarkers (*m*/*z* mono. as integer) and their detection (+/−)[a] 749	831	843	899	1085	1126	1224	1267	1281	1455	1670	1788	1989	2040	2181	2267	2324	2742
1	+	4.0×10^7	+	+	−	+	+	+	−	+	−	+	+	+	+	+	+	+	−	+
2	+	1.5×10^7	+	+	−	+	−	+	−	+	+	+	+	+	+	+	+	+	−	+
3	+	8.6×10^6	+	+	−	+	−	+	+	+	−	−	−	+	+	+	−	+	+	+
4	+	6.7×10^6	−	+	−	+	−	+	−	−	−	−	−	+	−	+	−	+	−	−
5	+	3.2×10^6	−	+	+	+	−	−	−	+	−	−	+	−	−	−	+	−	−	−
6	+	$\mathbf{7.5 \times 10^5}$	−	−	+	+	−	+	−	−	−	−	+	−	+	−	−	+	−	−
7	+	5.5×10^5	−	−	+	+	−	−	−	−	−	−	−	−	+	−	+	+	−	−
8	+	2.8×10^5	−	−	+	−	−	−	−	−	−	−	−	−	−	−	−	+	−	−
9	+	1.5×10^5	−	−	−	+	−	−	−	−	−	−	−	−	+	−	−	−	−	−
10	+	9.3×10^4	−	−	−	+	−	−	−	−	−	−	−	−	−	−	−	−	−	−
11	+	5.4×10^4	−	−	−	−	−	−	−	−	−	−	−	−	−	−	−	−	−	−
12	+	1.8×10^4	−	−	−	−	−	−	−	−	−	−	−	−	−	−	−	−	−	−
13	−	−	−	−	−	−	−	−	−	−	−	−	−	−	−	−	−	−	−	−
14	−	−	−	−	−	−	−	−	−	−	−	−	−	−	−	−	−	−	−	−
15	−	−	−	−	−	−	−	−	−	−	−	−	−	−	−	−	−	−	−	−
16	−	−	−	−	−	−	−	−	−	−	−	−	−	−	−	−	−	−	−	−

[a] Based on repeated detection of peptide ions in three separate spectra at predicted theoretical mass (Table 9.1) (within 2 ppm) and with a signal-to-noise (S/N) ratio >2.

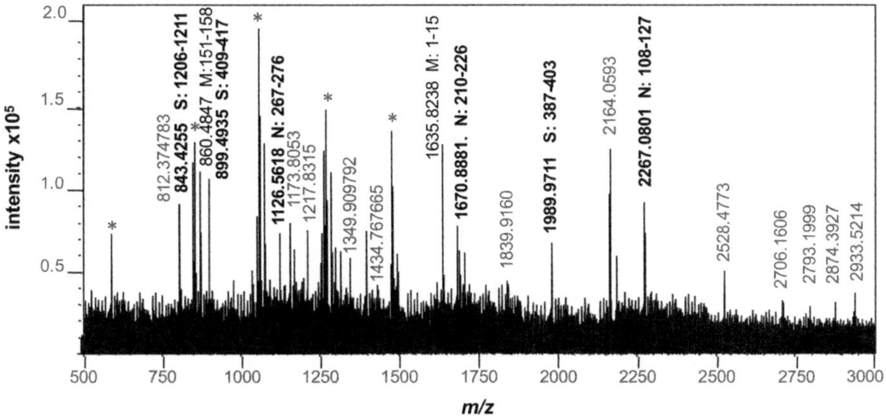

Figure 9.4 MALDI mass spectrum of the tryptic digest products of nasopharyngeal swab recovered the protein from whole SARS-CoV2 virus. Peptide ions corresponding to segments of the nucleocapsid (N) and spike (S) proteins are shown in boldface. Peaks marked with * correspond to MALDI matrix cluster ions. Reproduced from ref. 30 with permission from American Chemical Society, Copyright 2020.

detected. In the high 10^5 to 10^6 copies per mL titre range, this falls to 6–12 peptides. At 1.5×10^5 titres, only two peptides are detected and a single peptide of the set is detected in the 9.3×10^4 titre specimen. Below this level, the virus goes undetected. These values are consistent with our previous reports for the SARS-CoV2 virus[18,30–32] and those for the influenza virus,[28] which report detection limits in the 10^5 copy range.

While a lower order of magnitude (low 10^4) may be detectable on MALDI-TOF instruments, an inherent loss in mass accuracy (to some 50 ppm or less) results in a drop in confidence in the peptide assignment and another non-viral peptide or other contaminants may be falsely associated with the SAR-CoV2 virus. Consistent with all MS experiments, a balance must be struck between optimal mass resolution and detection sensitivity.

9.7 Identification of SARS-CoV2 Variants

To assign the nature of the variant, a series of trypsin–endoproteinase GluC peptides of the S protein are employed as biomarkers. Given that large segments (>6 kDa) are generated when the S protein is digested with each enzyme alone, a double digest ensures peptides are generated across a mass-to-charge (*m*/*z*) range typically of 400–4000 conducive to their detection on most MALDI-based instruments. It also generates peptides unique to a particular variant that mostly contain a single mutation[37] (or single deletion or insertion in the case of the omicron variant) with respect to the originating (Wuhan) strain sequence. These peptides are shown in Table 9.3. A larger number of peptide markers are associated with the strains of the omicron lineage (B.1.1.529) due to the higher number of mutations in the

Table 9.3 Tryptic + GluC peptide mass biomarkers for spike proteins that distinguish major variants of concern.

Lineage	Mutations	Tryptic + GluC segment[a]	Sequence (mutations shown underlined, except deletions)	Mass $[M+H]^+$ monoisotopic
B.1.17 (alpha)	HV69–70 del.	54–77 minus 69–70	LFLPFFSNVTWFHAISGTNGTK	2484.27
	Y144 del.	139–147 minus 144	PFLGVYHK	960.53
	P681H	664–682	IPIGAGICASYQTQTNSHR	2016.99
	T716I	703–725	NSVAYSNNSIAIPINFTISVTTE	2455.24
	D1118H	1112–1127	PQIITTHNTFVSGNCD	1746.81
B.1.351 (beta)	D80A	79–88	FANPVLPFND	1133.56
	D215G	215–224	GLPQGFSALE	1018.52
	LAL 242–244 del.	238–246 minus 242–244	FQTLHR	801.437
	R246I	238–253	FQTLLALHISYLTPGD	1788.95
	K417N	409–419	QIAPGQTGNIAD	1184.59
	E484K	472–484	IYQAGSTPCNGVK	1337.65
	A701V	686–702	SVASQSIIAYTMSLGVE	1755.88
B.1.617 (delta)	T95I	89–96	GVYFASIE	885.435
	E154K	151–154	SWMK	551.265
	E484Q	472–509	IYQAGSTPCNGVQGFNCYFPLQSYGFQPTNGVGYQPYR	4221.92
B.1.617.2 (delta plus)	T19R	1–19	MFVFLVLLPLVSSQCVNLR	2178.22
	R158G	154–169	SEFGVYSSANNCTFE	1654.67
	T478K	472–478	IYQAGSK	766.409
	D950N	948–964	LQNVVNQNAQALNTLVK	1867.04

P.1 (gamma)	T20N	1–21	MFVFLVLLPLVSSQCVNLTNR	2393.31
	P26S	22–34	TQLPSAYTNSFTR	1485.73
	D138Y	133–147	FQFCNYPFLGVYYHK	1925.9
	R190S	188–191	NLSE	462.22
	K417T	409–420	QIAPGQTGTIAD	1171.6
	E484K	472–484	IYQAGSTPCNGVK	1337.65
	N501Y	499–504	STYLVK	710.408
	H655Y	655–661	YVNNSYE	888.373
	T1027I	1019–1028	ASANLAAIK	858.504
	V1176F	1169–1181	ISGINASFVNIQK	1390.77
B.1.1.529 (omicron)	A67V, Δ69–70	54–77 minus 69–70	LFLPFFSNVTWFHVISGTNGTK	2512.3
	Δ211, L212I	207–214 minus 211	HTPIIVR	835.515
	G339D	329–339	FPNITNLCPFD	1280.6
	S371L, S373P	365–378	YSVLYNLAPFFTFK	1709.89
	G446S	445–454	VSGNYNYLYR	1248.6
	E484A, Q493R	479–493	PCNGVAGFNCYFPLR	1657.76
	G496S, Q498R	494–498	SYSFR	659.315
	N501Y, Y505H	499–509	PTYGVGHQPYR	1274.63
	T547K	538–547	CVNFNFNGLK	1155.56
	N679K	664–679	IPIGAGICASYQTQTK	1650.85
	N764K	749–764	CSNLLLQYGSFCTQLK	1817.89
	D796Y	796–808	YFGGFNFSQILPD	1504.71
	Q954H	951–964	VVNHNAQALNTLVK	1520.85
	N969K	965–969	QLSSK	562.32
	L981F	980–983	IFSR	522.304

[a] Residue numbering is based on the originating strain and may differ in some variants due to the presence of insertion and deletion sites.

S protein associated with this variant. The detection of any of these mass markers allows for the strain's lineage to be assigned in the same manner as has been reported for the type B influenza virus.[27] The more peptides detected for a particular lineage, the more confident the assignment.

To illustrate this, a MALDI spectrum of the double digest of specimen 1 was obtained, which is shown in Figure 9.5. About one half of the peaks (14 in number) that appear above a S/N ratio of 2 could be assigned to the S protein based on a comparison of their masses with the reference protein sequences. Three ions at *m/z* 562.3187, 835.5151 and 1155.5626 (shown in boldface in Figure 9.5) are within 2 ppm of the theoretical mass values of the peptide markers for the omicron (B1.1.529) variant (Table 9.3). This allows the strain to be identified as an omicron variant, over the other five major variants of concern, with a high degree of confidence. By conducting selected ion monitoring (SIM) across the masses for all peptide markers shown in Table 9.3, these variants can be distinguished where at least one or preferably more due to unique mutations that appear varying from strain to strain are detected. From this perspective, the precision of high resolution MS, achievable on FT-ICR, Orbitrap or extended or multipass TOF instruments, is important in order to assign strains to a particular variant with high confidence.

The specimens with the highest virus titres were used for this variant-based analysis, with a SIM experiment conducted for the double digest of the first eight (1–8) specimens. A summary of the data is shown in Table 9.4. Of the 8 specimens, the first four (1–4) of highest titres could be assigned to omicron (B.1.1.529) or delta (B.1.617.2) variants with high confidence through the detection of 3–5 peptide markers shown. One further specimen (no. 5) was predicted to be an omicron (B.1.1.529) variant, but since only a single peptide marker at *m/z* 1155 (nominal) was detected, it was assigned

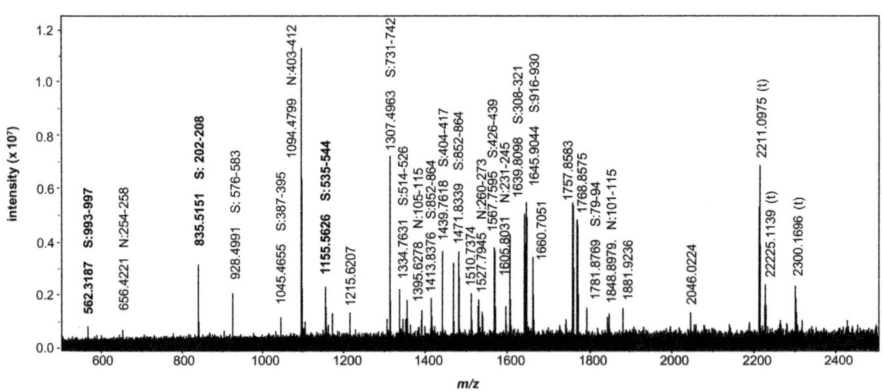

Figure 9.5 MALDI mass spectrum of the tryptic + GluC digest products of nasopharyngeal swab recovered the protein from whole SARS-CoV2 virus. Peptide ions corresponding to segments of the spike (S) proteins (with both 0 and 1 missed cleavages) of an omicron strain are shown in boldface. Peaks marked with (t) are predicted trypsin autolysis products.

Table 9.4 Tryptic + GluC variant-specific peptide mass biomarkers of spike proteins detected in nasopharyngeal specimens.

	RT-PCR		Variant-specific peptide biomarkers (m/z mono. as integer)[a]									
Specimen number	SARS-CoV2 +/−	Copies per mL	562	835	766	1155	1248	1650	1654	1817	1867	Variant determination
1	+	4.0×10^7	+	+	−	+	−	−	−	−	−	B.1.1.529 (omicron)
2	+	1.5×10^7	−	+	−	+	+	+	−	+	−	B.1.1.529 (omicron)
3	+	8.6×10^6	−	+	−	−	+	+	−	+	−	B.1.1.529 (omicron)
4	+	6.7×10^6	−	−	+	−	−	−	+	−	+	B.1.617.2 (delta plus)
5	+	3.2×10^6	−	−	−	+	−	−	−	−	−	B.1.1.529 (omicron)
6	+	7.5×10^5	−	−	−	−	−	−	−	−	−	Undetermined
7	+	5.5×10^5	−	−	−	−	−	−	−	−	−	Undetermined
8	+	2.8×10^5	−	−	−	−	−	−	−	−	−	Undetermined

[a] Based on repeated detection of peptide ions in three separate spectra. Only variant-specific peptide ions that were detected are shown; all others in Table 9.3 were not detected at the predicted theoretical mass (within 2 ppm) and with a signal-to-noise (S/N) ratio > 2.

with lower confidence. The remaining three were not assigned to a specific variant, although they were assigned to be SARS-CoV2 positive (see Table 9.2), as no variant-specific biomarker was detected. This is thought to be associated with the greater number of peptides which are produced from the viral proteins in a double digest (trypsin + endoproteinase GluC) of the whole virus. This can result in a greater ion suppression in the MALDI spectrum than is evident in the case of a single digest where fewer peptides are generated. Based on the results shown earlier, the nature of the variant for a strain can be confidently assigned through the analysis of a whole virus digest with high resolution MS at titres above the low 10^6 copies per mL level, corresponding to 10-fold more viruses than for SARS-CoV2 detection alone.

9.8 Typing and Subtyping Influenza Virus Strains

A similar proteotyping strategy has been applied to type and subtype influenza virus strains in whole virus and single viral protein digests. In this case, translated gene sequences for all proteins from a particular type, subtype and host of the influenza virus sourced from the NIAID Influenza Genome Sequence Database and GenBank were aligned. Consensus sequences were then established to identify proteolytic peptides with a high sequence conservation assessed based on the frequency of occurrence (P_o) at each residue of the peptide. The uniqueness of the theoretical monoisotopic mass of each signature peptide was then established using an algorithm (FluGest) specifically designed for this purpose. It was used to compare the masses of all proteolytic peptides across all viral influenza proteins of all known strains, together with those derived from potential protein biological contaminants such as albumins present in egg-grown viruses.

Signature peptides whose sequences are most conserved ($P_o \geq 0.95$) and whose masses differed by the largest value (>5 ppm) from other proteolytic peptides enable virus strains under analysis to be typed and subtyped with the greatest confidence,[16,26–28,38,39] where one or more signatures are detected in a high resolution mass spectrum. These are listed in Table 9.5 for hemagglutinin, neuraminidase, nucleoprotein and matrix M1 protein. Signatures for the two former proteins allow the common human subtypes and type B viral lineages to be determined.

A subsequent study demonstrated the ability to distinguish pandemic from seasonal strains of the virus following the 2009 H1N1 pandemic which resulted from a viral reassortment event. Signature peptides of N1 neuraminidase were obtained based on database sequences for all type A H1N1, human H5N1 and 2009 pandemic H1N1 strains. Three signature peptides at *m*/*z* values of 549.2991, 1380.6287 and 2543.1296 were unique to pandemic 2009 H1N1 strains that, if detected, enable this form to be unambiguously differentiated from all other H1N1 and H5N1 strains in circulation at that time.

Table 9.5 Signature peptides for influenza that enable the virus to be typed and subtyped with high resolution mass spectrometry.

Viral protein	Signature peptide mass[a]	Protein residues	Peptide sequence	Type	Subtype/ lineage
Hemagglutinin	780.4291	114–119	FEIFPK	A	H1
	1268.6117	103–113	EQLSSVSSFER	A	H1
	514.2660	257–260	GYFK	A	H3
	593.3075	266–270	SSIMR	A	H3
	937.5101	257–264	TGTIVYQR	B	Vic87
	939.4894	257–264	TGTITYQR	B	Yam88
	1035.5469	203–211	LYGDSKPQK	B	Yam88
	1280.6269	305–311	SKPYYTGEHAK	B	—
Neuraminidase	805.4567	78–84	GDVFVIR	A	N1
	592.2872	385–389	SCINR	A	N2
	948.5400	223–230	ILFIEEGK	A	N2
	1042.5389	390–397	CFYVELIR	A	N2
	1174.6215	77–87	LSAGGDIWVTR	A	N2
	1607.6941	142–163	QVCIAWSSSSCHDGK	A	N2
	1625.6802	320–333	GWAFDDGNDVWMGR	A	N2
Nucleoprotein	458.30850	153–156	ALVR	A	—
	609.26605	437–441	TSDMR	A	—
	622.30957	205–208	NFWR	A	—
	672.44025	262–267	SALILR	A	—
	676.30823	157–162	TGMDPR	A	—
	721.37347	222–227	MCNILK	A	—
	1067.51892	66–74	MVLSAFDER	A	—
	1323.62170	163–174	MCSLMQGSTLPR	A	—
Matrix protein M1	473.2831	73–76	GLQR	A	—
	555.2634	175–178	HENR	A	—
	849.4069	244–250	MGVQMQR	A	—
	921.5074	179–187	MVLASTTAK	A	—
	1125.6990	48–57	TRPILSPLTK	A	—
	1635.9104	58–72	GILGFVFTLTVPSER	A	—
	470.2470	161–164	AHSR	B	—
	547.2471	71–74	DQER	B	—
	660.3563	22–27	AELAEK	B	—
	675.3308	195–200	GEDVQK	B	—
	688.3624	221–227	NGEGIAK	B	—
	690.3781	214–220	SLGASQK	B	—
	701.3940	168–174	SSVPGVR	B	—
	738.3273	188–194	TMNGMGK	B	—
	771.4723	95–101	GLILAER	B	—
	820.4233	51–57	CLTDIQK	B	—
	822.3966	154–160	QASHSHR	B	—
	833.4437	228–234	DVMEVLK	B	—
	1050.5322	144–153	LGTLCALCEK	B	—
	1340.6007	176–187	EMQMVSAMNTAK	B	—
	1360.8021	58–70	ALIGASICFLKPK	B	—
	1465.72095	36–47	EFDLDSALEWIK	B	—
	1553.78796	78–92	FITEPLSGMGTTATK	B	—

[a] Only peptides with P_o values exceeding 0.95 are shown.

An additional alignment of translated hemagglutinin gene sequences of all characterized type A H5N1 strains, or subsets thereof, enabled signature peptides of these strains to be determined from the perspective of the period during which strains were isolated.[40] Due to the rapid cross-species transmission of H5N1 strains among migratory birds, poultry and humans throughout south-east Asia, no signatures specific to the host or region were found. Yet several period-specific signature peptides were identified that enable strains associated with the 1997 H5N1 pandemic to be rapidly differentiated from those in circulation across the subsequent decade.

9.9 Typing of Parainfluenza Virus Strains

The parainfluenza virus (PIV) is another virus that causes respiratory illnesses which present similar clinical symptoms to influenza. It must be distinguished from the more common influenza virus in order for effective treatments to be administered. If left untreated, parainfluenza can be life threatening and the virus is responsible for more hospitalizations in children under the age of five than the influenza virus.

Translated full length gene sequences for each of the six proteins encoded by the virus (hemagglutinin–neuraminidase (HN), fusion (F), nucleocapsid, large, phospho and matrix proteins) for each of the three most common serotypes (PIV1, PIV2 and PIV3) were separately aligned.[41] Tryptic peptide signatures were identified for each serotype based on their sequence conservation (with $P_o \geq 0.9$) and masses which differed by greater than 5 ppm from the mass of any other tryptic peptide derived from all characterized PIV and influenza viral proteins. Application of the method was demonstrated for separated viral proteins and two whole virus digests of egg-grown and cell-cultured PIV1 and PIV3 strains, respectively.[41]

9.10 Subtyping of the Hepatitis Virus

The hepatitis C virus (HCV) is associated with severe inflammation of the liver and long-term medical complications, including cirrhosis and liver cancer. The virus is characterized by six major genotypes (1–6) that are geographically dispersed. Each is further subdivided into subtypes, denoted 1a, 1b, 1c, *etc.* Correct genotype determination is clinically important as it allows for the customization of the correct dose and duration of available antiviral drug therapies.

The viral RNA genome encodes a single polyprotein that is cleaved by both host and viral proteases to yield ten individual viral proteins. Among them, the envelope glycoproteins E1 and E2 are particularly important given that they facilitate virus entry into the cell and initiate infection. Full length sequences for E1 and E2 envelope proteins from confirmed subtypes (1a, 1b, 2b, 3a) were sourced from the European HCV (euHCV) and UniProt databases and tryptic peptide segments with the highest sequence conservation of unique mass were determined based on a comparison to all tryptic peptides generated *in silico* from a HCV viral protein database of almost 100 000 sequences comprising known sequences for unprocessed polyprotein, core protein, envelope glycoprotein E1, envelope glycoprotein E2 and various HCV non-structural proteins. These were used to confidently assign subtypes from the high resolution mass spectral data of E1 or E2 protein and whole virus digests, demonstrating the broader applicability of a single-ion detection proteotyping approach beyond that described for the respiratory viruses earlier.

9.11 Use of High Resolution Mass Maps to Build Phylogenetic Trees

An additional benefit of high resolution mass map data is their use for the building of protein phylogenies. It has been shown that high resolution mass maps acquired on a FT-ICR instrument considerably improve protein identification in mass mapping or fingerprinting strategies over MALDI-TOF-based measurements.[42] Increases in the mass accuracy across the peptides of a map improve the confidence in the assigned top scored protein by better matching those sharing a common sequence based solely on a comparison of their masses.

The masses of peptides (m_1, m_2, …, m_n) produced by the proteolysis of a protein reflect the sequence of that protein. Homologous proteins yield peptides with identical masses in sections that share a common sequence. If the sequence in any segment of the protein is changed, its mass will change within that peptide segment and in the protein as a whole. Furthermore, the vast majority (94%) of single point mutations can be distinguished by mass alone when peptide masses are compared with high mass accuracy.[43] The sign (+/−) of the mass difference corresponding to a single point mutation between two peptide masses from different mass sets indicates the mutation direction.[43]

Such data enable phylogenetic trees ("mass trees") to be built from sets of peptide masses from the digestion of a series of proteins without the need for the sequences themselves. The approach uses a purpose built algorithm employing a distance score and has been termed phylonumerics (Figure 9.6).[43] Homologous proteins, which share the greatest number of common masses within their digests, are grouped on adjacent branches and clades, while those of different sequences with different peptide masses are positioned more remotely. Unlike conventional sequence-based phylogenetics, no sequence determination or alignment is required.

The mass tree concept has been conceived and implemented[44] and subsequently applied to chart and study viral protein evolution.[43] The MassTree algorithm (v4) calculates and displays single point mutations along the branches of the trees in a single tree building step.[45] These mutations have been found to be reliable in some 80% of cases in typical datasets. The majority of differences when compared with sequence alignment data were

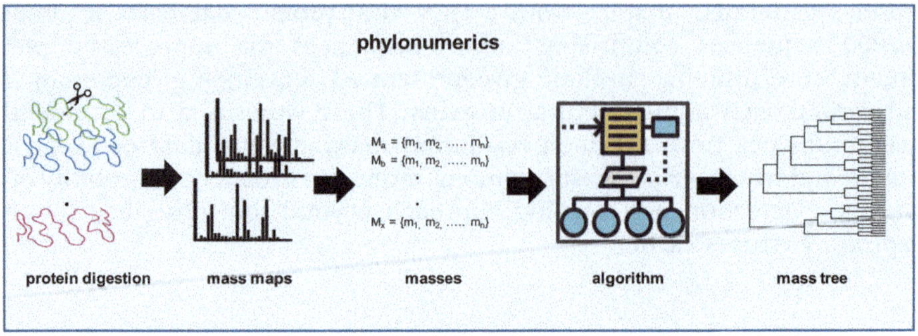

Figure 9.6 Schematic of the mass tree-based phylonumerics approach. Mass map or fingerprint data from a series of protein digests are input directly into the MassTree algorithm.

associated with mutations associated with the introduction or removal of a proteolytic cleavage residue.[45] Such mutations alter the mass map profile by generating two peptides from one or *vice versa*. Mass trees have been found to highly congruent to sequence-based trees based on a side-by-side comparison and through the use of several tree comparison algorithms.[43,45]

9.12 Mass Trees to Study Virus Evolution – Applications to the Influenza Virus and SARS-CoV2

Large mass trees have been built and interrogated to study mechanisms associated with epistasis and antiviral resistance in relation to the surface antigens (hemagglutinin and neuraminidase) of the influenza virus.[46–48] Putative epistatic mutations, determined based on their proximity along interconnected branches of the mass tree, were identified in H3 hemagglutinin from a tree built from 888 strains. Leading precursor or parent mutations were found to predominate within reported antigenic sites, while subsequent ancestral mutations resided exclusively in different antigenic regions, providing the virus with a possible immune escape mechanism, or at other remote sites that drive beneficial protein structural and functional change. Furthermore, the majority of frequent consecutive mutations exhibited less structural, and thus functional, change consistent with a "small steps" evolutionary model[46] reported for RNA viruses. Such a model favours the evolutionary survival of the virus over mutations associated with more substantive change that may cause or risk its own extinction.

The mechanisms associated with the emergence of antiviral resistance in human type A H1N1 viruses were also examined in the context of mutations that appeared in strains prior to, and post, the emergence of the H275Y mutation in N1 neuraminidase.[47] Mass maps were generated for N1 neuraminidase from a total of 6406 strains comprising 1092 seasonal and 5314 pandemic 2009 strains. Of these, a total of 593 were identified with the Y275 mutation and thus designated as oseltamivir resistant. A mass tree was built (partially shown in Figure 9.7), with the algorithm also outputting the distance between mutations along interconnected branches.

By far the most common ancestral mutation to these resistant strains (at node 94) was the S299A mutation (Figure 9.7). It preceded the S247N mutation that occurs in almost 20% of the A299 strains, followed by the N369S, F74S, then V13I, N28D, H275Y and I374V mutations across some 16.1%, 9.7% and 6.5% of the strains, respectively.[47] The S247N mutation has been reported to reduce the oseltamivir sensitivity and greatly increase the resistance in the context of the H275Y mutation and both S299 and S247 were shown to be proximal to the Y275 residue in a high resolution X-ray crystal structure of N1 neuraminidase.

A similar study for N2 neuraminidase, derived from 8561 human type A H3N2 strains containing 3216 with one or more antiviral resistance mutations to oseltamivir, found the majority of frequent ancestral and descendant mutations that precede and follow the manifestation of antiviral resistance mutations resided in the head region around the active site. Here, they drive changes in hydrophilicity that increase or reduce the accessibility of the site to the bulk solvent, primarily through the incorporation or loss of hydroxyl groups.[48]

A recent study was focused on resolving the major variants of the SARS-CoV2 virus from the perspective of the surface spike protein. Five proteins for each of the major variants (B.1.17 (alpha), B.1.351 (beta), P.1 (gamma), B.1.617.2 (delta plus) and B.1.1.529 (omicron)) were chosen from a range of geographic regions from the NCBI SARS-CoV2 Data Hub, with those of the omicron variant confirmed based on sequence alignments with the omicron reference sequence (GenBank UFO69279.1).

A sequence tree built from a ClustalW alignment is shown in Figure 9.8A. The tree clearly resolves the sequences for each of the 5 variants where the omicron strains lie on a distinct clade consistent with its separate evolution. A similar tree topology (Figure 9.8B) is evident in the mass tree built solely from the masses (theoretical between 400 and 6600) derived from the *in silico* digestion of each sequence with both trypsin and endoproteinase GluC. In this tree, the variants are resolved from one another based on the presence or absence of masses of particular peptide segments which are predicted by their sequence (Table 9.6).

For example, the peptide mass value of 2455.2351 is unique to B.1.1.7 (alpha) variants and corresponds to the spike protein residues 703–725 that contain the alpha-specific mutation T716I.[31] In this tree, the omicron strains are resolved onto two separate clades, though still diverged from all other variants, due to an extra R343K mutation present only in the lower three strains that results in an associated peptide mass difference (of −28.0061). These three strains were diverged, though to a lesser extent within the same clade, as shown at the base of the sequence tree (Figure 9.8B).

Such variant-specific masses can occasionally also be present in the data for single strains of other variants. For example, the sequence QWE68776.1

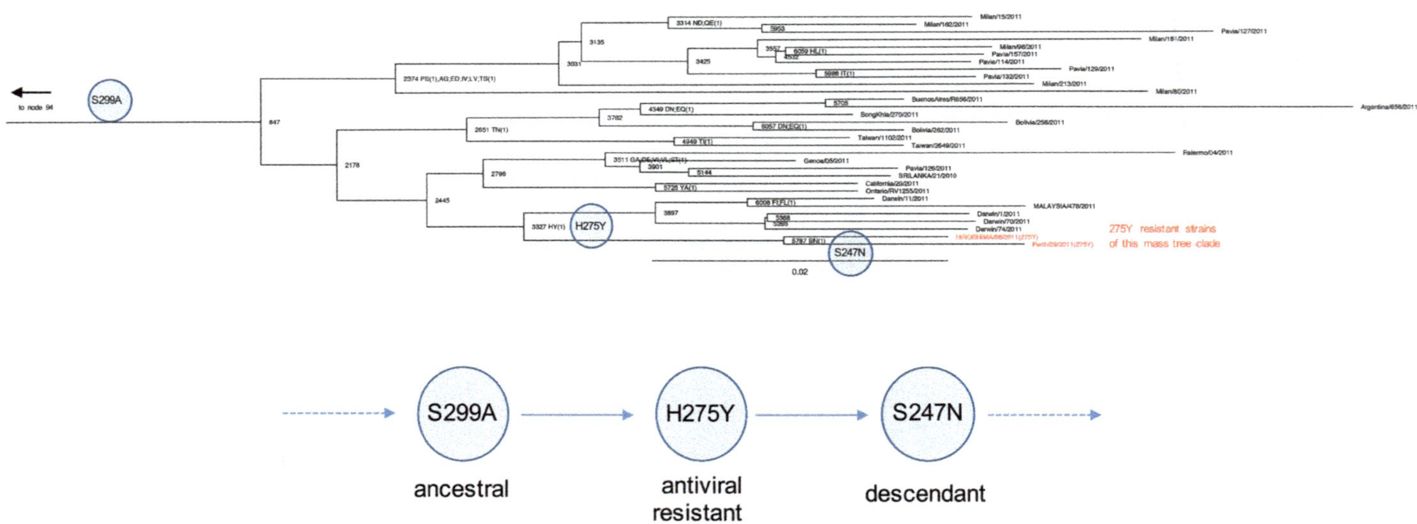

Figure 9.7 Partial mass tree of N1 neuraminidase of 6406 type A H1N1 strains from branch node 847 showing the mutation trajectory of two antiviral resistant strains (in red). Mutations annotated on this portion of the tree and their frequency (in brackets) at each branch are shown.

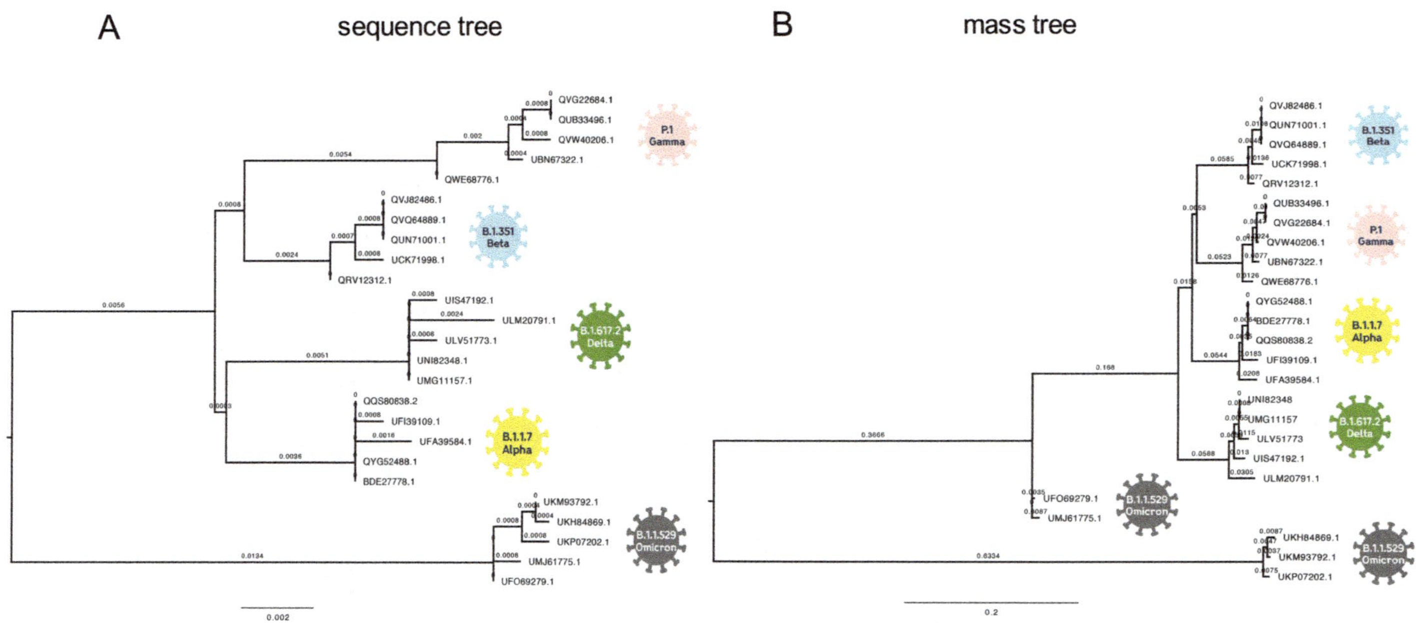

Figure 9.8 Sequence (A) and mass (B) trees for 25 spike proteins comprising five for each of the major variants (B.1.17 (alpha) yellow, B.1.351 (beta) blue, P.1 (gamma) pink, B.1.617.2 (delta plus) green and B.1.1.529 (omicron) orange).

Table 9.6 Masses of datasets unique to variants and their associated spike protein segments and peptide sequences.

Variant	Unique mass signatures	Spike protein segment[a]	Peptide sequence (underlined residues are variant unique mutations)	Presence (+) or absence (−) in a single strain by accession
B.1.1.7 (alpha)	960.53	139–147 minus 144	PFLGVYHK	+QWE68776.1 (P.1)
	1746.81	1112–1127	PQIITTHNTFVSGNCD	−UFA39584.1
	2016.99	664–682	IPIGAGICASYQTQTNSHR	−UFI39109.1
	2455.24	703–725	NSVAYSNNSIAIPINFTISVTTE	
	2484.27	54–77 minus 69–70	LFLPFFSNVTWFHAISGTNGTK	
B.1.351 (beta)	801.437	238–246 minus 242–244	FQTLHR	
	1018.52	215–224	GLPQGFSALE	
	1133.56	79–88	FANPVLPFND	
	1755.88	686–702	SVASQSIIAYTMSLGVE	+QUB33496.1, +QVG22684.1 (P.1)
B.1.617.2 (delta plus)	1867.04	948–964	LQNVVNQNAQALNTLVK	
	2178.22	1–19	MFVFLVLLPLVSSQCVNLR	
P.1 (gamma)	858.504	1019–1028	ASANLAAIK	
	1171.6	409–420	QIAPGQTGTIAD	
	1390.77	1169–1181	ISGINASFVNIQK	
	1485.73	22–34	TQLPSAYTNSFTR	
B.1.1.529 (omicron)	835.515	207–214 minus 211	HTPIIVR	+ULM20791.1 (B.1.617.2)
	1155.56	538–547	CVNFNFNGLK	
	1248.6	445–454	VSGNYNYLYR	
	1280.6	329–339	FPNITNLCPFD	
	1504.71	796–808	YFGGFNFSQILPD	
	1520.85	951–964	VVNHNAQALNTLVK	
	1650.85	664–679	IPIGAGICASYQTQTK	
	1709.89	365–378	YSVLYNLAPFFTFK	
	1817.89	749–764	CSNLLLQYGSFCTQLK	
	2512.3	54–77 minus 69–70	LFLPFFSNVTWFHVISGTNGTK	

[a] Residue numbering is based on the originating strain sequence (QHD43416.1) that may differ in some variants due to the presence of deletion and insertion sites.

of the P.1 variant contains the peptide segment (PFLGVYHK) of m/z 960.5302 (at residues 137–144 in this case) since it coincidentally contains the Y144 deletion common to the B.1.1.7 (alpha) variants (Table 9.6). In two other cases, variant-specific peptides are not present in their respective strains. This is the case of S protein for the sequence UFA39584.1 which contains a TI mutation at the second threonine residue (within the peptide sequence PQIITIHNTFVSGNCD), which alters its mass from 1746.8116 to 1758.8480.

In brief, given the high rate of mutations in the virus and strain-specific mutations in the S protein alter both the S protein's sequence and the mass of peptide segments generated from it, some variant-specific peptide masses may go undetected in some strains. Nevertheless, given that multiple unique mass signatures exist for each of the variants of concern, as is evident from Table 9.6, the detection of any number of them can be used to assign a variant with increasing confidence as more peptide masses match these values. This coupled to the ability to build trees without sequence data, where experimental mass data are generated, and without the need for any sequence alignment, favours the mass tree approach.

9.13 Conclusions and Outlook

Despite the importance and widespread application of PCR-based methods in clinical settings to detect and characterize respiratory and other viruses that impact human health, the growing role of MS is evident. Protein and peptide MS-based approaches offer benefits and provide complementary strategies to rapidly detect and characterize viruses among the range of MS strategies recently comprehensively reviewed for the SARS-CoV2 virus.[31] This chapter demonstrates the particular benefits of high resolution MS to enable confident viral peptide identification to be made by mass alone. This affords a considerable time saving over tandem MS (MS/MS) strategies as well as sensitivity benefits that allow viruses to be detected at lower levels of at least an order of magnitude. With the wider availability of such instruments (in several forms per Figure 9.2), and their increasingly easier and automated operation, the approaches described will find wider application in the near future.

References

1. M. Woolhouse, F. Scott, Z. Hudson, R. Howey and M. Chase-Topping, Human viruses: discovery and emergence, *Philos. Trans. R. Soc. London, Ser. B*, 2012, **367**, 2864–2871.
2. M. K. Parvez and S. Parveen, Evolution and Emergence of Pathogenic Viruses: Past, Present, and Future, *Intervirology*, 2017, **60**, 1–7.
3. B. Foxman, *Molecular Tools and Infectious Disease Epidemiology*, Elsevier, Burlington, MA, U.S.A., 2012.
4. I. M. Mackay, Real-time PCR in the microbiology laboratory, *Clin. Microbiol. Infect.*, 2004, **10**, 190–212.
5. K. M. Downard, B. Morrissey and A. B. Schwahn, Mass spectrometry analysis of the influenza virus, *Mass Spectrom. Rev.*, 2009, **28**, 35–49.

6. L. V. Kordyukova and M. V. Serebryakova, Mass Spectrometric Approaches to Study Enveloped Viruses: New Possibilities for Structural Biology and Prophylactic Medicine, *Biochemistry*, 2012, 77, 830–842.
7. L. M. Ganova-Raeva and Y. E. Khudyakov, Application of mass spectrometry to molecular diagnostics of viral infections, *Expert Rev. Mol. Diagn.*, 2013, **13**, 377–388.
8. A. Milewska, J. Ner-Kluza, A. Dabrowska, A. Bodzon-Kulakowska, K. Pyrc and P. Suder, Mass Spectrometry in Virological Sciences, *Mass Spectrom. Rev.*, 2020, **39**, 499–522.
9. H. R. Gelderblom, in *Structure and Classification of Viruses, in Medical Microbiology*, ed. S. Baron, University of Texas Medical Branch Galveston, Texas U.S.A., 4th edn, 1996.
10. S. L. Payne, Methods to Study Viruses, in *Viruses: From Understanding to Investigation*, Elsevier, Amsterdam, Netherlands, 1st edn, 2017, ch. 4, pp. 37–52.
11. E. K. Loveday, K. S. Hain, I. Kochetkova, J. F. Hedges, A. Robison, D. T. Snyder, S. K. Brumfield, M. J. Young, M. A. Jutila, C. B. Chang and M. P. Taylor, Effect of Inactivation Methods on SARS-CoV-2 Virion Protein and Structure, *Viruses*, 2021, **13**, 562.
12. W. Feng, A. M. Newbigging, C. Le, B. Pang, H. Peng, Y. Cao, J. Wu, G. Abbas, J. Song, D.-B. Wang, M. Cui, J. Tao, D. L. Tyrrell, X.-E. Zhang, H. Zhang and X. E. Le, Molecular Diagnosis of COVID-19: Challenges and Research Needs, *Anal. Chem.*, 2019, **92**, 10196–10209.
13. S. J. R. da Silva, C. T. A. da Silva, K. M. Guarines, R. P. G. Mendes, K. Pardee, A. Kohl and L. Pena, Clinical and Laboratory Diagnosis of SARS-CoV-2, the Virus Causing COVID-19, *ACS Infect. Dis.*, 2020, **6**, 2319–2336.
14. S. Woloshin, N. Patel and A. S. Kesselheim, False Negative Tests for SARS-CoV-2 Infection – Challenges and Implications, *N. Engl. J. Med.*, 2020, **383**, e38.
15. W. Ahmeda, S. L. Simpson, P. M. Bertscha, K. Bibby and A. Bivins, *et al.*, Minimizing errors in RT-PCR detection and quantification of SARS-CoV-2 RNA for wastewater surveillance, *Sci. Total Environ.*, 2022, **805**, 149877.
16. K. M. Downard, Proteotyping for the Rapid Identification of Pandemic Influenza Virus and other Biopathogens, *Chem. Soc. Rev.*, 2013, **42**, 8584–8595.
17. J. Zheng, R. J. Sugrue and K. Tanga, Mass spectrometry based proteomic studies on viruses and hosts – A review, *Anal. Chim. Acta*, 2011, **702**, 149–159.
18. J. H. Griffin and K. M. Downard, Mass spectrometry analytical responses to the SARS-CoV2 coronavirus in review, *Trends Anal. Chem.*, 2021, **142**, 116328.
19. M. Preianò, S. Correnti, C. Pelaia, R. Savino and R. Terracciano, MALDI MS-Based Investigations for SARS-CoV-2 Detection, *BioChem.*, 2021, **1**, 250–278.
20. A. Croxatto, G. Prod'hom and G. Greub, Applications of MALDI-TOF mass spectrometry in clinical diagnostic microbiology, *FEMS Microbiol. Rev.*, 2012, **36**, 380–407.

21. J. Kriegsmann, M. Kriegsmann and R. Casadonte, MALDI TOF imaging mass spectrometry in clinical pathology: a valuable tool for cancer diagnostics, *Int. J. Oncol.*, 2015, **46**, 893–906.
22. T. Satoh, T. Sato and J. Tamura, Development of a high-performance MALDI-TOF mass spectrometer utilizing a spiral ion trajectory, *J. Am. Soc. Mass Spectrom.*, 2007, **18**, 1318–1323.
23. R. A. Zubarev and A. Makarov, Orbitrap mass spectrometry, *Anal. Chem.*, 2013, **85**, 5288–5296.
24. B. Chen, C. B. Lietz and L. Li, Coupling matrix-assisted ionization with high resolution mass spectrometry and electron transfer dissociation to characterize intact proteins and post-translational modifications, *Anal. Bioanal. Chem.*, 2018, **410**, 1007–1017.
25. J. M. Spraggins, D. G. Rizzo, J. L. Moore, K. L. Rose, N. D. Hammer, E. P. Skaar and R. M. Caprioli, MALDI FTICR IMS of Intact Proteins: Using Mass Accuracy to Link Protein Images with Proteomics Data, *J. Am. Soc. Mass Spectrom.*, 2015, **26**, 974–985.
26. A. B. Schwahn, J. W. H. Wong and K. M. Downard, Signature peptides of influenza nucleoprotein for the typing and subtyping of the virus by high resolution mass spectrometry, *Analyst*, 2009, **134**, 2253–2261.
27. A. B. Schwahn and K. M. Downard, Proteotyping to establish the lineage of type A H1N1 and type B human influenza virus, *J. Virol. Methods*, 2011, **171**, 117–122.
28. N. D. Fernandes and K. M. Downard, Incorporation of a Proteotyping Approach using Mass Spectrometry for the Surveillance of the Influenza Virus in Cell Culture, *J. Clin. Microbiol.*, 2014, **52**, 725–735.
29. A. B. Schwahn, J. W. H. Wong and K. M. Downard, Rapid Differentiation of Seasonal and Pandemic H1N1 Influenza through Proteotyping of Viral Neuraminidase with Mass Spectrometry, *Anal. Chem.*, 2010, **82**, 4584–4590.
30. N. L. Dollman, J. H. Griffin and K. M. Downard, Detection, Mapping, and Proteotyping of SARS-CoV-2 Coronavirus with High Resolution Mass Spectrometry, *ACS Infect. Dis.*, 2020, **6**, 3269–3276.
31. C. Mann, J. H. Griffin and K. M. Downard, Detection and Evolution of SARS CoV-2 Coronavirus Variants of Concern with Mass Spectrometry, *Anal. Bioanal. Chem.*, 2021, **413**, 7241–7249.
32. C. Mann, J. S. Hoyle and K. M. Downard, Detection of SARS CoV-2 coronavirus omicron variant with mass spectrometry, *Analyst*, 2022, **147**, 1181–1190.
33. M. F. Rocca, J. C. Zintgraff, M. E. Dattero, L. S. Santos and M. Ledesma, *et al.*, A combined approach of MALDI-TOF mass spectrometry and multivariate analysis as a potential tool for the detection of SARS-CoV-2 virus in nasopharyngeal swabs, *J. Virol. Methods*, 2020, **286**, 113991.
34. F. X. Heinz and K. Stiasny, Distinguishing features of current COVID-19 vaccines: knowns and unknowns of antigen presentation and modes of action, *NPJ Vaccines*, 2021, **6**, 104.

35. J. van de Leemput and Z. Han, Understanding Individual SARS-CoV-2 Proteins for Targeted Drug Development against COVID-19, *Mol. Cell. Biol.*, 2021, **41**, e0018521.
36. T. Yoshinari, K. Hayashi, S. Hirose, K. Ohya and T. Ohnishi, *et al.*, Matrix-Assisted Laser Desorption and Ionization Time-of-Flight Mass Spectrometry Analysis for the Direct Detection of SARS-CoV-2 in Nasopharyngeal Swabs, *Anal. Chem.*, 2022, **94**, 4218–4226.
37. C. Mann and K. M. Downard, Evolution of SARS-CoV2 Coronavirus Surface Protein Investigated with Mass Spectrometry Based Phylogenetics, *Anal. Lett.*, 2022, **55**, 305–317.
38. A. B. Schwahn, J. W. H. Wong and K. M. Downard, Subtyping of the Influenza Virus by High Resolution Mass Spectrometry, *Anal. Chem.*, 2009, **81**, 3500–3506.
39. A. B. Schwahn, J. W. H. Wong and K. M. Downard, Typing of human and animal strains of influenza virus with conserved signature peptides of matrix M1 protein by high resolution mass spectrometry, *J. Virol. Methods*, 2010, **165**, 178–185.
40. J.-W. Ha and K. M. Downard, Evolution of H5N1 influenza virus through proteotyping of hemagglutinin with high resolution mass spectrometry, *Analyst*, 2011, **136**, 3259–3267.
41. A. P. Nguyen and K. M. Downard, Proteotyping of the Parainfluenza Virus with High Resolution Mass Spectrometry, *Anal. Chem.*, 2013, **85**, 1097–1105.
42. D. M. Horn, E. C. Peters, H. Klock, A. Meyers and A. Brock, Improved protein identification using automated high mass measurement accuracy MALDI FT-ICR MS peptide mass fingerprinting, *Int. J. Mass Spectrom.*, 2004, **238**, 189–196.
43. K. M. Downard, Sequence-free phylogenetics with mass spectrometry, *Mass Spectrom. Rev.*, 2022, **41**, 3–14.
44. A. T. L. Lun, K. Swaminathan, J. W. H. Wong and K. M. Downard, Mass Trees – A New Phylogenetic Approach and Algorithm to Chart Evolutionary History with Mass Spectrometry, *Anal. Chem.*, 2013, **85**, 5475–5482.
45. E. H. Akand and K. M. Downard, Mutational Analysis Employing a Phylogenetic Mass Tree Approach in a Study of the Evolution of the Influenza Virus, *Mol. Phylogenet. Evol.*, 2017, **112**, 209–217.
46. E. H. Akand and K. M. Downard, Identification of Epistatic Mutations and Insights into the Evolution of the Influenza Virus Using a Mass-Based Protein Phylogenetic Approach, *Mol. Phylogenet. Evol.*, 2018, **121**, 132–138.
47. E. H. Akand and K. M. Downard, Ancestral and Compensatory Mutations that Promote Antiviral Resistance in Influenza N1 Neuraminidase Revealed by a Phylonumerics Approach, *J. Mol. Evol.*, 2018, **86**, 546–553.
48. E. H. Akand and K. M. Downard, Mechanisms of Antiviral Resistance in Influenza Neuraminidase Revealed by a Mass Spectrometry Based Phylonumerics Approach, *Mol. Phylogen. Evol.*, 2019, **135**, 286–296.

CHAPTER 10

Mass Spectrometry-based Proteomics of Eukaryotes

JENNIFER GEDDES-McALISTER,*[a,b] TIA RIZAKOS[a] AND BENJAMIN MUSELIUS[a]

[a] Department of Molecular and Cellular Biology, University of Guelph, Guelph, Ontario, N1G 2W1, Canada; [b] Canadian Proteomics and Artificial Intelligence Consortium, Canada
*Email: jgeddesm@uoguelph.ca

10.1 Introduction

Over the past few decades, mass spectrometry (MS)-based proteomics has advanced rapidly in technological improvements and accessibility of bioinformatics tools and platforms. Its applications to diverse biological systems are also evident and provide new insight into mechanisms regulating many components of the world around us. For eukaryotes, including all animals, plants, and fungi, MS-based proteomics can highlight biological changes over time in response to environmental stimuli, such as climate change,[1] investigate the relationship between a host and a pathogen during disease,[2,3] and explore opportunities for novel drug discovery and development.[4,5] Many eukaryote-centric MS analyses focus on clinical applications of the technology. For instance, clinical proteomics enables improved diagnostic and prognostic methods that identify disease-causing agents, detect biomarkers associated with infection, track disease progression over space and time, and monitor treatment efficacy.[6–9] However, applications beyond the clinic are rapidly growing and increasing our understanding and awareness of diverse biological systems on a global scale. We highlight the progression of MS-based proteomics over the past two decades in

New Developments in Mass Spectrometry No. 13
Detection and Analysis of Microorganisms by Mass Spectrometry
Edited by Liang Qiao and Jia Yi

Published by the Royal Society of Chemistry, www.rsc.org

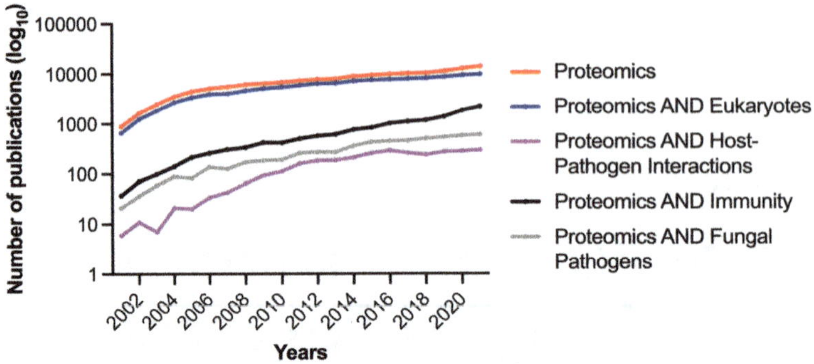

Figure 10.1 Mass spectrometry-based proteomics and eukaryotes research publications from 2002 to 2020. Search parameters set to: 'Proteomics' (red), 'Proteomics AND Eukaryotes' (blue), 'Proteomics AND Host–Pathogen Interactions' (purple), 'Proteomics AND Immunity' (black), and 'Proteomics AND Fungal Pathogens' (grey) in the Title and/or Abstract. Source: National Center for Biotechnology Information, database: PubMed. Figure generated in GraphPad Prism version 9.0.

alignment with investigation of eukaryotes and host–pathogen interactions (Figure 10.1).

In this chapter, we provide an overview of MS-based proteomics, including technologies and applications relevant to eukaryotic infections, as well as the role of bioinformatics in advancing innovation. We focus on applications of proteomic profiling to investigate the relationships and interactions between a host and a pathogen. We explore recent scientific discoveries pushing our knowledge of fungal pathogens forward and further defining how the relationship with host immunity drives disease progression. Furthermore, we present advantages and limitations of different methodologies as the design and development of MS experiments require careful consideration of multiple components, including genome annotation availability, instrumentation accessibility, bioinformatics expertise, cost, and prioritization of biological questions to pursue physiologically-relevant avenues of discovery.[10]

10.1.1 Overview of MS-based Proteomics

The field of MS-based proteomics is broad, encompassing biochemistry, computer science, mathematics, analytical chemistry, biology, and engineering with applicability across a plethora of additional scientific disciplines. Proteomics can provide global information on protein abundance and production across a biological system, explore contents released into an extracellular environment, define modifications regulating protein activity, and identify interactions between proteins or within complexes that drive signaling and communication.[11] A common initial distinction made in proteomics experiments is the difference in approach to define the composition of biological systems, *i.e.*, top-down *versus* bottom-up. Top-down

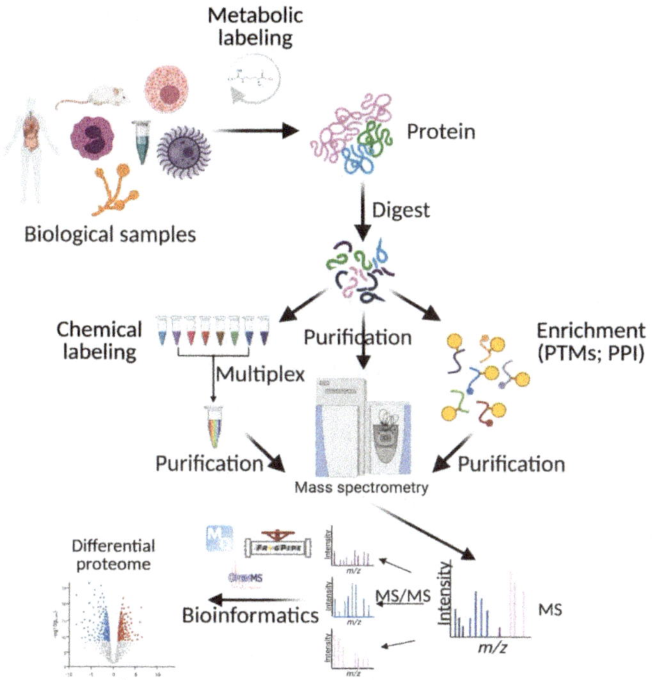

Figure 10.2 Summary of mass spectrometry-based proteomics platforms and techniques. Metabolic and chemical labeling, and enrichment are optional. PTMs = post-translational modifications; PPI = protein–protein interactions. Figure generated with Biorender.com.

proteomics focuses on detection and analysis of intact proteins, proteins within complexes, and proteoforms on the mass spectrometer.[12,13] Conversely, bottom-up proteomics, which is classically defined as 'discovery-based' proteomics, focuses on digestion of proteins into peptides followed by measurement on the mass spectrometer, and is the focus of this chapter.[14] Another highly recognizable method is targeted proteomics, which is often applied for biomarker discovery as it provides high specificity towards the detection and quantification of peptides within complex mixtures[15] (see Figure 10.2).

10.1.1.1 Data Acquisition

An important consideration for experimental design is the selected mode of data acquisition. For example, data-dependent acquisition (DDA) has been the predominant method of choice for quantitative proteomics to date, using a survey scan to select the top *n* most abundant peptides for sequential fragmentation.[14] DDA matches measured peptides to a protein sequence library (*e.g.*, FASTA file) for detection of proteins within a given sample, and, depending on the genome size, can yield thousands of protein identifications.

However, it is difficult to achieve reproducibility and precision with DDA during high peptide co-elution events as only the most abundant peptides within a given time are measured in the MS1 scan on the mass spectrometer.[16] Additionally, the quantification power for DDA is limited based on minimal (one or two) peptide sampling as a tradeoff for surveying as many peptides as possible. Conversely, data-independent acquisition (DIA) considers peptides over the entire mass-to-charge (*m*/*z*) range and fragments all peptides within a cycling *m*/*z* window.[17,18] As DIA samples all peptides within the given *m*/*z* window, this approach provides measurement information about all peptides identified within the sample.[19] DIA presents an opportunity to incorporate the breadth of DDA measurements with the precision of targeted proteomics. Generally, a DDA approach is more accessible to researchers and supports flexible experimental design with well-established data analysis pipelines in place; however, DDA may limit peptide identification rates compared to DIA. Conversely, for DIA, errors in peptide identification from complex spectra and mainstream applications of data analysis pipelines present ongoing challenges that need to be addressed.[16]

10.1.1.2 *Quantification*

Measuring changes in protein abundance across conditions is a hallmark of many proteomics experiments.[20] Quantification strategies are generally classified into three categories: metabolic, chemical, and label-free, with an emphasis on relative quantification approaches, although some strategies support absolute measurements.[21] Metabolic approaches include labeling proteins at the cellular or organismal level during growth with isotopically stable and non-radioactive amino acids to monitor protein synthesis and turnover. A routinely used metabolic labeling strategy is SILAC (stable isotope labeling with amino acids in cell culture), which incorporates light, medium, and heavy amino acids (*e.g.*, arginine) into the culture medium followed by sample pooling and label detection *via* shifts in amino acid mass on the mass spectrometer.[22] A drawback of this approach is the requirement for cell culturing to promote label incorporation and a limited range of multiplexing (*i.e.*, only three mass versions available). Chemical labeling represents an alternative approach to quantification by using derivatized peptides. Key examples include iTRAQ (isobaric tags for relative and absolute quantification),[23] TMT (tandem mass tags),[24] and EASI-tags (easily abstractable sulfoxide-based isobaric tags)[25] to add a chemical tag onto a protein (pre-digestion) or peptide (post-digestion) of interest to enable multiplexing of samples. A limitation of chemical labeling is the requirement for increased sample handling, supporting the potential to introduce technical variation into the experiment. Alternatively, label-free quantification (LFQ) approaches provide freedom for experimental design based on the measurement of peptide signal intensities following sample measurement, as well as lower costs as additional labeling reagents are not required.[26] However, LFQ strategies require reliable determination of protein concentration prior to measurement on the mass spectrometer to ensure consistency across sample loadings and are subject to missing values and the use of imputation during data analysis to generate a complete data set.

For metabolic and chemical labeling approaches, the ability to multiplex samples and measure on the instrument in smaller batches may reduce technical variability, as well as the amount of instrument time required, lowering experimental costs; however, upfront costs to purchase reagents can be limiting.[10] Conversely, LFQ does not require upfront costs but does increase the amount of measuring time needed on the mass spectrometer, increasing sample-to-sample variation and operational costs. Additionally, using labeling techniques demands the use of a high-resolution, high-sensitivity mass spectrometer to reliably and accurately measure small changes in mass shifts and differentiate between modified and unmodified peptides. The same is true for LFQ; however, lowered sample complexity provides flexibility on the selection of instrumentation. Notably, metabolic and chemical labeling techniques have evolved into 'plug-and-play' methodologies; however, critical considerations for researchers include the upfront costs of labeling reagents and the need for an internal control for normalization of technical variability across samples.

10.1.1.3 Post-translational Modifications

Post-translational modifications (PTMs) can modulate the production, activity, and function of a protein to influence diverse biological properties.[27] PTMs are encompassed within an expanding field of study, *i.e.*, proteoforms, which influences many aspects of disease-based research and is gaining traction and recognition through the Human Proteoform Project.[28,29] PTMs include the addition of a functional group to a protein *via* a covalent or noncovalent bond, as well as the degradation of tagged proteins using proteases. Phosphorylation is a commonly studied PTM that influences cell signaling cascades and communication networks within a biological system, but other PTMs, including acetylation, glycosylation, succinylation, and ubiquitination, also play diverse and critical roles in cellular regulation.[30–32] Some important limitations of assessing PTMs from complex MS datasets are the low stoichiometry of modified to unmodified peptides, variability in modification mass changes (*e.g.*, glycosylation), and poor reproducibility across enrichment techniques (*e.g.*, immortalized metal affinity chromatography). MS instrumentation capable of MS^n measurements and advanced bioinformatics tools capable of normalizing modification effects to total protein abundance, as well as open mass searches for all modifications, support the application and utility of PTM investigations.[33,34]

10.1.1.4 Protein–Protein Interactions

Protein–protein interactions (PPI) provide information about signaling networks, protein complexes, and localization, which are central to the structure and integrity of a cell.[35] Traditional techniques to identify interacting proteins relied on the yeast two-hybrid method (*i.e.*, testing physical interactions between two proteins *via* transcription factor activation) and immunoprecipitation (*i.e.*, antigen–antibody conjugation on an immobilized support). Alternative strategies include protein correlation profiling, which assumes shared compartmentalization of proteins based on common elution profiles[36–38] and, more recently, the development of proximity-based labeling strategies, such as BioID (proximity-dependent biotin identification). Proximity approaches have increased in popularity and are commonly used to identify the interactome of a specific (bait) protein to provide spatial and compartmentalization information about the selected protein within a cell.[39–41] BioID uses a non-specific biotin ligase fused to the protein of interest that, upon induction, will biotinylate interacting and proximal proteins; the reaction optimally occurs over 15–18 h.[42] Iterative advances to BioID include BioID2 (smaller biotin ligase),[43] TurboID (engineered biotin ligase that reacts in 10 min),[44] miniTurboID (smaller engineered ligase with a rapid reaction time),[44] and split-TurboID (contact-dependent labeling).[45] Other methodologies include APEX (ascorbic acid peroxidase), which fuses a protein of interest to a peroxidase in the presence of H_2O_2 for production of biotin phenoxyl radicals that label surrounding proteins.[46] These proximity-based approaches are powerful and robust, but the throughput is limited by specific labeling of the selected bait protein. On a more global scale, thermal proteome profiling (TPP),[47] cellular thermal shift assay (CETSA),[48] and proteome integral solubility alteration (PISA)[49] define PPI and interaction networks of proteome-wide drug and metabolite interactions by mapping thermal stability. When the techniques listed earlier are combined with MS, these affinity-based approaches support the identification of proteins of interest and interacting proteins with spatial and temporal resolution.

10.1.2 Bioinformatics Tools for Proteomics

The robust and reproducible analysis of proteomics datasets requires advanced bioinformatics tools for deconvoluting mass spectra, identifying and quantifying peptides within a sample, and mapping peptides to protein sequences. Some key components of bioinformatics platforms are accessibility for the user, transparency in processing methods, and resources for data sharing. For bottom-up proteomics, several publicly-available platforms exist, including FragPipe, which is built on MSFragger[50] to offer ultra-fast processing of MS raw data files and MaxQuant,[51] which integrates the Andromeda[52] search engine for peptide identification. These platforms offer graphical user interfaces, methods for protein identification, quantification, and modification, and interpretation of interaction datasets, and robust statistical tools for in-depth analysis. Alternative platforms optimized for targeted proteomics data processing include Skyline,[53] which is tailored for multiple reaction monitoring datasets and provides spectral library building and searching options, and OpenSWATH[54] and OpenMS[55] platforms for analysis of DIA datasets. Application of these bioinformatics tools is supported by well-annotated proteomes housed within publicly-available databases (*e.g.*, UniProt, https://www.uniprot.org), functional analysis platforms (*e.g.*, STRING[56]), and classification tools (*e.g.*, Gene Ontology), as well as data sharing strategies across the scientific community (*e.g.*, ProteomeXchange PRIDE).

10.2 Diverse Applications of MS-based Proteomics to Human Fungal Pathogens

Fungal pathogens represent an established and emerging threat to global health with over 300 million people inflicted annually, causing 1.6 million deaths.[57,58] The treatment of fungal pathogens is an ongoing challenge given the limited selection of clinically effective, non-toxic antifungal drugs, the emergence of new pathogens, and the evolution of drug resistant strains.[59–61] Strategies to overcome these challenges include the identification of novel druggable targets and new understanding into the biological mechanisms driving fungal pathogenesis and drug resistance. MS-based proteomics provides a platform to achieve these goals by detecting changes in protein abundance under diverse conditions to better understand pathogens, improve diagnostic platforms, and identify new proteins as putative inhibitory targets.[4] A powerful experimental approach is the application of proteomics to reveal global proteome or spatial changes under altered environmental conditions and integration of these results with phenotypic and virulence profiling, as well as other omics techniques (*e.g.*, transcriptomics, genomics, metabolomics) for a comprehensive view of biological regulation and mechanistic discoveries.[62] For example, the application of bottom-up proteomics combined with multiple reaction monitoring for a targeted analysis defined the first biomarkers of cryptococcal infection caused by the human fungal pathogen *Cryptococcus neoformans* in a murine model.[63] Another example is monitoring of bronchoalveolar lavage collected from infected human and murine specimens for biomarkers of the human fungal pathogen *Aspergillus fumigatus* infection by quantitative proteomics.[64] Profiling revealed AspF2,[65] a known immunoregulator produced by *A. fumigatus*, as a promising biomarker for disease. These studies provide a foundation for the discovery of new diagnostic methods, as well as strategies to monitor the progression of disease and treatment efficacy.

MS-based proteomics can also make important contributions to our overall understanding of fungal pathogenesis. For instance, quantitative proteomic profiling of *C. neoformans* under the regulation of the cAMP/Protein Kinase A signaling cascade,[66,67] which has important roles in modulating fungal virulence, revealed a novel drug repurposing strategy to combat the pathogen.[68] Specifically, suppression of *pka1* influenced the production of proteins within the ubiquitin proteosome pathway and disruption of this activity with bortezomib (an FDA-approved cancer therapeutic) led to reduction in fungal capsule formation, a key virulence factor mediating fungal survival within the natural environment and the host. Proteomics is also informative for discovery of antifungal targets and proteome-wide modulations in response to treatment. Such is the case for exposure of *Candida auris* (an emerging, deadly fungal pathogen with defined multidrug resistance[69]) to the echinocandin caspofungin, which led to changes in cell morphology, transcription, and increased production of cell wall-associated processes, and suppression of stress response pathways.[70] Proteomic analysis by LFQ revealed increased production of proteins associated with fungal biofilm formation, translation, glycosylphosphatidylinositol-anchored proteins, and mannose-related proteins, with the latter connected to enhanced immunogenicity properties. These findings support a new avenue for discovery of the relationship between mannan production, immunogenicity, and antifungal response for *C. auris*.

Recently, the combination of proteome and secretome profiling of *C. neoformans* identified an uncharacterized protein, Wos2 (homologous to a co-chaperone in other fungal pathogens[71]), to be regulated by the availability of zinc within culture medium.[72] Integration of phenotypic characterization with the proteomic findings supported a role for Wos2 in fungal growth, thermotolerance, and capsule production, emphasizing the importance of metal acquisition in fungal pathogenesis and proposing a novel target for inhibitory disruption. Another study using MS-based proteomics to explore the impact of copper toxicity in *C. neoformans* reported an increase in protein degradation and decrease in protein synthesis in the copper sensitive strain, potentially associated with the production of copper-induced reactive oxygen species.[73] Moreover, application of a proteasome inhibitor (MG132) demonstrated a protective function towards the pathogen under copper-induced cellular stress, providing new insight into copper-associated regulatory mechanisms and a putative new antifungal target.

Interactions between a pathogen and a host play critical roles in determining the activation of an immune response and the outcome of infection. For *A. fumigatus*, defining the cell surface proteome (*i.e.*, cell wall and surface) upon conidia formation revealed consistent production of CcpA (conidial cell wall protein A).[74] Phenotypic characterization of CcpA demonstrated a role in fungal virulence and altered recognition of the host immune system, supporting a role for the protein as a major structural component of the fungal cell surface that prevents recognition by the innate immune system of the host. Another proteome study of *A. fumigatus* conidia explored macrophage infection with melanized or melanin-deficient fungal strains to observe cell wall-associated melanin drivers of phagolysosome maturation.[75] This study used a magnetic separation technique combined with MS to enrich for phagolysosomes post-infection and observed differential productivity of several host proteins, including NADH dehydrogenase, Rab5, Lamp1, Vamp8, and Cathepsin Z, under the regulated melanin conditions. These proteins are critical drivers of phagolysosome maturation and function through regulation of acidification, endocytic trafficking, maturation signaling, and protease production, providing new insight into the mechanistic regulation of host immune response by fungal melanin.

For several fungal pathogens, extracellular vesicles (EVs) produced serve as 'virulence factor delivery bags'[76] for successful invasion and adaptation to the hostile environment of the host. Comparative proteomics of EVs produced by *C. auris* and *Candida albicans* (commensal and pathogenic fungi) revealed vesicle-specific proteome profiles compared to cellular composition, as well as substantial changes in protein profiles within the vesicles of each pathogen.[77] For *C. auris*, proteins involved in starch and sugar metabolism and protein processing were more abundant, whereas proteins associated with the tricarboxylic acid cycle showed increased production by *C. albicans*. These findings corroborate observed phenotypic differences in fungal killing (higher for *C. albicans*) *vs.* survival (higher for *C. auris*) between the pathogens upon exposure to host cells. Another example is proteomic profiling of the interaction between interferon-γ (IFN-γ)-primed or non-primed alveolar macrophages and the fungal pathogen *Paracoccidioides brasiliensis*.[78] As anticipated, primed macrophages exerted fungicidal activity towards the pathogen with minimal response by the pathogen. Conversely, within unprimed macrophages, the pathogen was able to survive through activation of the pentose-phosphate pathway, the methylcitrate cycle, cell wall synthesis, and mitochondrial activity, as well as induction of heat shock proteins, antioxidant response, and virulence factor accumulation, demonstrating a favorable environment for *P. brasiliensis* survival and replication.

With improved resolution from the mass spectrometer over the past few decades, along with advances in computational power and bioinformatics tools, proteomic profiling of co-infection models (*e.g.*, host infected with multiple pathogens) has enhanced our ability to monitor infection from a multitude of perspectives and propose new strategies to combat the pathogen(s). For instance, proteomic profiling *via* DDA and LFQ investigated the co-infection of macrophages with the fungal pathogen *C. neoformans* and the bacterial pathogen *Klebsiella pneumoniae* during immediate (acute) and established (chronic) states of infection.[79] The findings describe global reprogramming of host cells regardless of their infection state under singular and multipathogen invasion, including an initial defense response to activate the macrophages and protect the host from invasion. The host immune response returned to homeostasis (as indicated by few significant changes in protein production) upon adaptation of the host to the fungal pathogen (able to survive intracellularly) during an established infection. However, upon introduction of the bacterium for co-infection, there were significant changes in microbial virulence factor production (*e.g.*, extracellular enzymes, melanin) to modulate the host defense response, revealing putative targets for anti-virulence therapeutic approaches.

Integrating multi-omics platforms for comprehensive profiling of fungal pathogenesis has demonstrated power for improved biological understanding. For instance, total proteome and acetylome (antibody enrichment of PTMs following acetylation) profiling of host lung and brain tissue during cryptococcal infection, combined with transcriptome profiling, revealed correlation across 127 gene products at the protein and RNA levels.[80] The findings showed reciprocal trends in production between the organs and diversity of regulation among biological processes, including protein–DNA complex, nucleosome, DNA packaging complex, nuclear chromatin, and nuclear nucleosome pathways, supporting independent acetylation/deacetylation regulatory mechanisms specific to each tissue. Another study explored mechanisms of multidrug resistance between *C. auris* and *C. albicans* strains through integration of proteomic, metabolic, and lipidomic profiling to reveal roles for carbon metabolism and regulation of cell wall integrity driving resistance.[81] Specifically, for the proteome of *C. auris*, proteins involved in respiration showed higher abundance, providing a connection to fluconazole resistance,[82] as well as kinases (*e.g.*, Tpk2, Mck1, Pkc1) with defined roles in regulation of cell wall integrity, which may further influence antifungal resistance and underscore discrepancies in susceptibility between the species. Taken together, these diverse applications of proteomics delve into the complexities of fungal pathogenesis and highlight the importance of interactions with the host and across molecular levels to promote new discoveries into improved diagnostic practices and therapeutic opportunities.

10.3 Proteomic Profiling of the Host and Modulation of Immune Responses

The immune system plays critical roles in protecting the host from disease. MS-based proteomics presents an opportunity to define communication patterns across cells for increased knowledge of signaling cascades, to describe the interplay between a host and a pathogen, and to identify opportunities to enhance the immune response and clear infections.[83] One example is the application of MS-based proteomics to 28 primary human hematopoietic cell populations to explore communication patterns across activated immune cells.[84] The study reports previously undefined immune cell specializations associated with ligand–receptor expression that connect distinct immune cell functions. In addition, proteomics reported protein secretion profiles and their respective communication signals between previously unconnected immune cell types. Proteomic profiling of host–microbe interactions also provides insight into cross-sectional communication throughout the host. For instance, global phosphoproteome reprogramming of host signaling cascades upon fungal infection with *C. neoformans* revealed differential phosphorylation of macrophages following fungal phagocytosis.[85] Altered signaling events influenced the host autophagy initiation complex for promotion of intracellular fungal trafficking and replication. These findings support the role of autophagy in regulating host cell susceptibility to fungal pathogens and suggest new approaches to activate host signaling networks for an efficient defense response to pathogen invasion.

Given the resolving power of MS-based proteomics techniques, tissue-, fluid-, or organ-specific host responses to pathogen invasion can be well described. For example, LFQ proteome analysis of ocular fluids from specific-pathogen-free mice treated with antibiotics *versus* germ-free mice subjected to singular or multimicrobial colonization, influenced the ocular surface and, in particular, immunoglobulin levels.[86] These data supported the presence of a gut–eye axis and the role of commensal microbes within the gut in generating B-cell memory to promote health at diverse sites throughout the host. Further studies applying proteomic profiling at the ocular surface upon bacterial infection revealed acute neutrophil recruitment to the site of infection and described novel biomarkers for ocular infection.[87,88] Conversely, proteomic profiles of tears from keratitis patients infected with the fungal pathogen *Aspergillus flavus*, compared to samples collected from healthy individuals, revealed significant differences in the core proteome for each host.[89] Of these differences, glycosylation events modulated host response to infection, as well as the production of complement, neutrophil extracellular traps (NETs), and anti-inflammatory proteins. These findings were anticipated given the infectious state of the host and the need to mount an effective antifungal response and promote wound healing. Taken together, these studies demonstrate the power of proteomics to discover putative non-invasive biomarkers of infection for disease diagnosis and monitoring.

Cell-type-specific responses to disease also provide crucial information about the mechanistic roles of different immune cells and opportunities to activate such cells for enhanced protection. For example, MS-based proteomic profiling of neutrophils (leukocytes; important for proper immune function)[90,91] derived from murine models (*i.e.*, resistant or susceptible to bacterial infection) revealed distinct immune cell proteomes correlated with susceptibility to infection.[92] Specifically, strain-specific and infection-induced changes in the neutrophil proteomes were observed and the elevation of enzymes (*e.g.*, α-mannosidase) suggested a role during infection. Integration of proteomics and *in vitro* data showed that inhibition of α-mannosidase limited the antimicrobial properties of neutrophils, whereas exogenous application of the enzyme reduced microbial biofilm production and the burden of disease. Overall, these data demonstrate the power of proteomics to identify drivers of disease and propose novel therapeutic options.

As described earlier, neutrophils have important roles in the defense against pathogens, and upon infection, these host cells will produce EVs with antimicrobial properties. For instance, during *A. fumigatus* infection of differentiated PLB-985 neutrophil-like cells (human diploid myeloid leukemia cell line), the neutrophils will phagocytose fungal conidia and acidify conidia-containing phagolysosomes, as well as release NETs and produce EVs with antifungal properties.[93] To gain better insight into the mechanisms of antifungal activity deployed within the EVs,[94] isolation and proteomic profiling revealed production of well-characterized EV markers, such as CD63 and CD81 tetraspanins and TSG101 (tumor susceptibility gene 101), and a decrease in cytoplasmic proteins (*e.g.*, CANX (calnexin)). A core proteome was defined between infection-associated and spontaneously-derived neutrophil EVs with known antimicrobial agents, including azurocidin, cathepsin G, and defensin, specific to *A. fumigatus* infection. Another study exploring the production of EVs by neutrophils in response to *A. fumigatus* infection used MS-based proteomics combined with TMT labeling for sample quantification and multiplexing.[94] Here, EVs collected from 20 donors were classified as afEVs (antifungal EVs, EVs induced by wild-type *A. fumigatus*), pksPEVs (EVs induced by the pksP (polyketide synthase) mutant *A. fumigatus*), and sEVs (spontaneously released EVs), with proteome profiling revealing enhanced antimicrobial protein and peptide production from the infected EVs. In addition, the afEVs demonstrated increased metabolic enzyme production (specific to the presence of conidial pigmentation (melanin)) to promote a tailored antifungal response of the host cells. To investigate the interactions between NETs and fungal pathogens, proteome profiling of cell wall-associated proteins from *C. albicans* hyphae was performed, which revealed the activation of granular enzymes, including elastase, myeloperoxidase, and lactotransferrin, as well as the production of antimicrobial proteins, such as cathepsin G, proteinase-3, and cathelicidin-derived peptide LL-37, by neutrophils in defense to the hyphae.[95] Further, assessment of the interplay between glycosylphosphatidylinositol-anchored adhesin proteins of *C. albicans*, including enolase (Eno1), phosphoglycerate mutase (Gpm1), and triosephosphate isomerase 1 (Tpi1), showed that the interplay did not inactivate the antimicrobial enzymes of NETs. However, regulation between the biological systems indicated that fungal cells may successfully escape the host immune response.

Macrophages are another class of important immune cells that bridge the innate and adaptive immune systems through the role of antigen presentation with dynamic communication signals decipherable with MS-based proteomics.[83] During infection, macrophages are often the first line of defense against pathogenic invasion and support destruction of the pathogen as well as recruitment of additional immune cells; however, macrophage recruitment can also promote the production of pathogenic virulence factors. For instance, the mechanism of activation behind filamentation of *C. albicans* upon macrophage phagocytosis has not been fully elucidated but is a driver of fungal pathogenesis. Recently, bioactivity-guided fractionation of macrophage lysates combined with MS-based proteomics assessed the induction of *C. albicans* filamentation by macrophages.[96] The study detected over 200 host proteins from fractions associated with robust *C. albicans* filamentation induction with prothymosin-α, an immunomodulator, defined as the lead candidate. Combining proteomics data with *in vitro* characterization demonstrated that immunoneutralization of prothymosin-α blocked fungal filamentation. These data support the role of macrophages in igniting fungal virulence and suggest a valuable opportunity to block the host response and promote fungal clearance with improved efficiency. Taken together, exploration of host responses to infection by MS-based proteomics provides an unparalleled assessment of activities and functions determining the outcome of disease and reveals strategies to disarm the pathogen and empower the host for improved protective mechanisms.

10.4 Conclusion

MS-based proteomics is a powerful platform for diverse applications in infectious disease research. Proteomics is used to investigate differences between homeostatic and altered states from both host and pathogen perspectives, define protein biomarkers of disease for diagnostic and prognostic purposes, and discover and characterize novel therapeutic options to treat disease. The integration of information gleaned from global proteome profiling, detection of PTMs, and identification of PPIs provides new insight into the complexity of biological systems and the communication patterns driving disease. In this chapter, we focused on proteomics applications in eukaryotes, emphasizing fungal pathogens and host immune responses; however, applications of the technology far exceed these areas of research and provide valuable mechanistic insight into a plethora of natural sources found across the globe.

Abbreviations

APEX	Ascorbate peroxidase
BioID	Proximity-dependent biotin identification
CANX	Calnexin
CcpA	Conidial cell wall protein A
CETSA	Cellular thermal shift assay
DDA	Data-dependent acquisition
DIA	Data-independent acquisition
EASI-tags	Easily abstractable sulfoxide-based isobaric tags
Eno1	Enolase 1
EV	Extracellular vesicle
Gpm1	Phosphoglycerate mutase 1
iTRAQ	Isobaric tags for relative and absolute quantification
LFQ	Label-free quantification
m/*z*	Mass-to-charge
MS	Mass spectrometry
NETs	Neutrophil extracellular traps
PISA	Proteome integral solubility alteration

PPI	Protein–protein interactions
PTM	Post-translational modification
SILAC	Stable isotope labeling with amino acids in cell culture
TMT	Tandem mass tags
Tpi1	Triosephosphate isomerase 1
TPP	Thermal proteome profiling
Tsg101	Tumor susceptibility gene 101

Author Contributions

J. G.-M. conceptualized the chapter and prepared the figures. T. R., B. M., and J. G.-M. wrote the first draft of the chapter, edited, and wrote the final chapter version. All authors read and approved the submitted manuscript.

Funding

This project is supported, in part, by the University of Guelph, Canadian Institutes of Health Research (Project Grant), and Natural Sciences and Engineering Research Council of Canada (Discovery Grant) for J. G.-M.

Conflicts of Interest

The authors declare no conflicts of interest.

Acknowledgements

We thank members of the Geddes-McAlister lab for constructive feedback on the chapter preparation and support with design of Figure 10.2, and we extend appreciation to Dr Jan Rieckmann for writing the R script used to generate Figure 10.1.

References

1. M. Domnauer, F. Zheng, L. Li, Y. Zhang, C. E. Chang, J. R. Unruh, J. Conkright-Fincham, S. McCroskey, L. Florens, Y. Zhang, C. Seidel, B. Fong, B. Schilling, R. Sharma, A. Ramanathan, K. Si and C. Zhou, *Mol. Cell*, 2021, **81**, 3294–3309.e12.
2. B. Ball, A. Bermas, D. Carruthers-Lay and J. Geddes-McAlister, *J. Fungi*, 2019, **5**, 52.
3. A. Sukumaran, E. Woroszchuk, T. Ross and J. Geddes-McAlister, *Can. J. Microbiol.*, 2021, **67**, 213–225.
4. F. Meissner, J. Geddes-McAlister, M. Mann and M. Bantscheff, *Nat. Rev. Drug Discovery*, 2022, **21**, 637–645.
5. J. Geddes-McAlister, N. Prudhomme, D. Dutierrez Gongora, D. Cossar and M. D. McLean, *Curr. Opin. Chem. Biol.*, 2022, **68**, 102133.
6. A. Macklin, S. Khan and T. Kislinger, *Clin. Proteomics*, 2020, **17**, 17.

7. Y. Zhu, R. Aebersold, M. Mann and T. Guo, *Cell*, 2021, **184**, 4840–4840.e1.
8. S. P. Mann, P. V. Treit, P. E. Geyer, G. S. Omenn and M. Mann, *Mol. Cell. Proteomics*, 2021, **20**, 100046.
9. C. A. Crutchfield, S. N. Thomas, L. J. Sokoll and D. W. Chan, *Clin. Proteomics*, 2016, **13**, 1–12.
10. A. A. Houfani and L. J. Foster, *Methods in Molecular Biology*, 2022, vol. 2456, pp. 1–14.
11. R. Aebersold and M. Mann, *Nature*, 2016, **537**, 347–355.
12. T. K. Toby, L. Fornelli and N. L. Kelleher, *Annu. Rev. Anal. Chem.*, 2016, **9**, 499–519.
13. J. C. Tran, L. Zamdborg, D. R. Ahlf, J. E. Lee, A. D. Catherman, K. R. Durbin, J. D. Tipton, A. Vellaichamy, J. F. Kellie, M. Li, C. Wu, S. M. M. Sweet, B. P. Early, N. Siuti, R. D. Leduc, P. D. Compton, P. M. Thomas and N. L. Kelleher, *Nature*, 2011, **480**, 254–258.
14. L. C. Gillet, A. Leitner and R. Aebersold, *Annu. Rev. Anal. Chem.*, 2014, 7, 427–454.
15. R. Aebersold, A. Bensimon, B. C. Collins, C. Ludwig and E. Sabido, *Proteomics*, 2016, **16**, 2065–2067.
16. A. Hu, W. S. Noble and A. Wolf-Yadlin, *F1000Research*, 2016, **5**, 419.
17. L. C. Gillet, P. Navarro, S. Tate, H. Röst, N. Selevsek, L. Reiter, R. Bonner and R. Aebersold, *Mol. Cell. Proteomics*, 2012, **11**, O111.016717.
18. A. Hu, W. S. Noble and A. Wolf-Yadlin, *F1000Research*, 2016, **5**, 419.
19. A. Bilbao, E. Varesio, J. Luban, C. Strambio-De-Castillia, G. Hopfgartner, M. Müller and F. Lisacek, *Proteomics*, 2015, **15**, 964–980.
20. S.-E. Ong and M. Mann, *Nat. Chem. Biol.*, 2005, **1**, 252–262.
21. O. T. Schubert, H. L. Röst, B. C. Collins, G. Rosenberger and R. Aebersold, *Nat. Protoc.*, 2017, **12**, 1289–1294.
22. S.-E. Ong, *Mol. Cell. Proteomics*, 2002, **1**, 376–386.
23. P. L. Ross, Y. N. Huang, J. N. Marchese, B. Williamson, K. Parker, S. Hattan, N. Khainovski, S. Pillai, S. Dey, S. Daniels, S. Purkayastha, P. Juhasz, S. Martin, M. Bartlet-Jones, F. He, A. Jacobson and D. J. Pappin, *Mol. Cell. Proteomics*, 2004, **3**, 1154–1169.
24. A. Thompson, J. Schäfer, K. Kuhn, S. Kienle, J. Schwarz, G. Schmidt, T. Neumann and C. Hamon, *Anal. Chem.*, 2003, **75**, 1895–1904.
25. S. Virreira Winter, F. Meier, C. Wichmann, J. Cox, M. Mann and F. Meissner, *Nat. Methods*, 2018, **15**, 527–530.
26. S. Nahnsen, C. Bielow, K. Reinert and O. Kohlbacher, *Mol. Cell. Proteomics*, 2013, **12**, 549–556.
27. M. Larance and A. I. Lamond, *Nat. Rev. Mol. Cell Biol.*, 2015, **16**, 269–280.
28. L. M. Smith and N. L. Kelleher, *Science*, 2018, **359**, 1106–1107.
29. L. M. Smith, J. N. Agar, J. Chamot-Rooke, P. O. Danis, Y. Ge, J. A. Loo, L. Paša-Tolić, Y. O. Tsybin and N. L. Kelleher, *Sci. Adv.*, 2021, 7, eabk0734.
30. C. Retanal, B. Ball and J. Geddes-McAlister, *J. Fungi*, 2021, 7, 124.
31. V. Csizmok and J. D. Forman-Kay, *Curr. Opin. Struct. Biol.*, 2018, **48**, 58–67.

32. B. Macek, K. Forchhammer, J. Hardouin, E. Weber-Ban, C. Grangeasse and I. Mijakovic, *Nat. Rev. Microbiol.*, 2019, **17**, 651–664.
33. M. M. Savitski, M. L. Nielsen and R. A. Zubarev, *Mol. Cell. Proteomics*, 2006, **5**, 935–948.
34. F. Yu, G. C. Teo, A. T. Kong, S. E. Haynes, D. M. Avtonomov, D. J. Geiszler and A. I. Nesvizhskii, *Nat. Commun.*, 2020, **11**, 4065.
35. K. Titeca, I. Lemmens, J. Tavernier and S. Eyckerman, *Mass Spectrom. Rev.*, 2019, **38**, 79–111.
36. L. J. Foster, C. L. de Hoog, Y. Zhang, Y. Zhang, X. Xie, V. K. Mootha and M. Mann, *Cell*, 2006, **125**, 187–199.
37. D. N. Itzhak, S. Tyanova, J. Cox and G. H. H. Borner, *eLife*, 2016, **5**, e16950.
38. A. R. Kristensen, J. Gsponer and L. J. Foster, *Nat. Methods*, 2012, **9**, 907–909.
39. K. J. Roux, D. I. Kim, B. Burke and D. G. May, *Curr. Protoc. Protein Sci.*, 2018, **91**, 19.23.1–19.23.15.
40. D. I. Kim, S. C. Jensen, K. A. Noble, B. Kc, K. H. Roux, K. Motamedchaboki and K. J. Roux, *Mol. Biol. Cell*, 2016, **27**, 1188.
41. R. M. Sears, D. G. May and K. J. Roux, *Methods in Molecular Biology*, 2019, vol. 2012, pp. 299–313.
42. K. J. Roux, D. I. Kim, M. Raida and B. Burke, *J. Cell Biol.*, 2012, **196**, 801–810.
43. D. I. Kim, S. C. Jensen, K. A. Noble, B. KC, K. H. Roux, K. Motamedchaboki and K. J. Roux, *Mol. Biol. Cell*, 2016, **27**, 1188–1196.
44. T. C. Branon, J. A. Bosch, A. D. Sanchez, N. D. Udeshi, T. Svinkina, S. A. Carr, J. L. Feldman, N. Perrimon and A. Y. Ting, *Nat. Biotechnol.*, 2018, **36**, 880–887.
45. K. F. Cho, T. C. Branon, S. Rajeev, T. Svinkina, N. D. Udeshi, T. Thoudam, C. Kwak, H. W. Rhee, I. K. Lee, S. A. Carr and A. Y. Ting, *Proc. Natl. Acad. Sci. U. S. A.*, 2020, **117**, 12143–12154.
46. S. S. Lam, J. D. Martell, K. J. Kamer, T. J. Deerinck, M. H. Ellisman, V. K. Mootha and A. Y. Ting, *Nat. Methods*, 2015, **12**, 51–54.
47. K. V. M. Huber, K. M. Olek, A. C. Müller, C. S. H. Tan, K. L. Bennett, J. Colinge and G. Superti-Furga, *Nat. Methods*, 2015, **12**, 1055–1057.
48. D. Martinez Molina and P. Nordlund, *Annu. Rev. Pharmacol. Toxicol.*, 2016, **56**, 141–161.
49. M. Gaetani, P. Sabatier, A. A. Saei, C. M. Beusch, Z. Yang, S. L. Lundström and R. A. Zubarev, *J. Proteome Res.*, 2019, **18**, 4027–4037.
50. A. T. Kong, F. V. Leprevost, D. M. Avtonomov, D. Mellacheruvu and A. I. Nesvizhskii, *Nat. Methods*, 2017, **14**, 513–520.
51. J. Cox and M. Mann, *Nat. Biotechnol.*, 2008, **26**, 1367–1372.
52. J. Cox, N. Neuhauser, A. Michalski, R. A. Scheltema, J. V. Olsen and M. Mann, *J. Proteome Res.*, 2011, **10**, 1794–1805.
53. B. MacLean, D. M. Tomazela, N. Shulman, M. Chambers, G. L. Finney, B. Frewen, R. Kern, D. L. Tabb, D. C. Liebler and M. J. MacCoss, *Bioinformatics*, 2010, **26**, 966–968.

54. H. L. Röst, G. Rosenberger, P. Navarro, L. Gillet, S. M. Miladinoviä, O. T. Schubert, W. Wolski, B. C. Collins, J. Malmström, L. Malmström and R. Aebersold, *Nat. Biotechnol.*, 2014, **32**, 219–223.
55. H. L. Röst, T. Sachsenberg, S. Aiche, C. Bielow, H. Weisser, F. Aicheler, S. Andreotti, H. C. Ehrlich, P. Gutenbrunner, E. Kenar, X. Liang, S. Nahnsen, L. Nilse, J. Pfeuffer, G. Rosenberger, M. Rurik, U. Schmitt, J. Veit, M. Walzer, D. Wojnar, W. E. Wolski, O. Schilling, J. S. Choudhary, L. Malmström, R. Aebersold, K. Reinert and O. Kohlbacher, *Nat. Methods*, 2016, **13**, 741–748.
56. D. Szklarczyk, A. L. Gable, D. Lyon, A. Junge, S. Wyder, J. Huerta-Cepas, M. Simonovic, N. T. Doncheva, J. H. Morris, P. Bork, L. J. Jensen and C. Von Mering, *Nucleic Acids Res.*, 2019, **47**, D607–D613.
57. F. Bongomin, S. Gago, R. Oladele and D. Denning, *J. Fungi*, 2017, **3**, 57.
58. S. F. Dufresne, D. C. Cole, D. W. Denning and D. C. Sheppard, *Eur. J. Clin. Microbiol. Infect. Dis.*, 2017, **36**, 987–992.
59. J. R. Perfect, *Nat. Rev. Drug Discovery*, 2017, **16**, 603–616.
60. N. P. Wiederhold, *Infect. Drug Resist.*, 2017, **10**, 249–259.
61. J. Geddes-McAlister and R. S. Shapiro, *Ann. N. Y. Acad. Sci.*, 2019, **1435**, 57–78.
62. B. Ball, M. Langille and J. Geddes-McAlister, *mBio*, 2020, **11**, e01020-20.
63. J. M. H. Geddes, D. Croll, M. Caza, N. Stoynov, L. J. Foster and J. W. Kronstad, *BMC Microbiol.*, 2015, **15**, 1–26.
64. S. Machata, M. M. Müller, R. Lehmann, P. Sieber, G. Panagiotou, A. Carvalho, C. Cunha, K. Lagrou, J. Maertens, H. Slevogt and I. D. Jacobsen, *Virulence*, 2020, **11**, 1337–1351.
65. A. Fekkar, V. Balloy, C. Pionneau, C. Marinach-Patrice, M. Chignard and D. Mazier, *J. Infect. Dis.*, 2012, **205**, 1163–1172.
66. J. W. Kronstad, G. Hu, J. Choi, J. W. Kronstad, G. Hu and J. Choi, *Mycobiology*, 2011, **39**, 143–150.
67. M. Caza and J. W. Kronstad, *Front. Cell. Infect. Microbiol.*, 2019, **9**, 1–15.
68. J. M. H. Geddes, M. Caza, D. Croll, N. Stoynov, L. J. Foster and J. W. Kronstad, *mBio*, 2016, **7**, 1–15.
69. K. Forsberg, K. Woodworth, M. Walters, E. L. Berkow, B. Jackson, T. Chiller and S. Vallabhaneni, *Med. Mycol.*, 2019, **57**, 1–12.
70. D. Zamith-Miranda, R. F. Amatuzzi, I. F. Munhoz da Rocha, S. T. Martins, A. C. R. Lucena, A. Z. Vieira, G. Trentin, F. Almeida, M. L. Rodrigues, E. S. Nakayasu, J. D. Nosanchuk and L. R. Alves, *Comput. Struct. Biotechnol. J.*, 2021, **19**, 5264–5277.
71. M. J. Muñoz, E. R. Bejarano, R. R. Daga and J. Jimenez, *Genetics*, 1999, **153**, 1561–1572.
72. B. Ball, E. Woroszchuk, A. Sukumaran, H. West, A. Afaq, D. Carruthers-Lay, B. Muselius, L. Gee, M. Langille, S. Pladwig, S. Kazi, A. Hendriks and J. Geddes-McAlister, *BMC Microbiol.*, 2021, **21**, 341.
73. T. Sun, Y. Li, Y. Li, H. Li, Y. Gong, J. Wu, Y. Ning, C. Ding and Y. Xu, *Front. Cell. Infect. Microbiol.*, 2017, **60**, 1317–1330.

74. V. Voltersen, M. G. Blango, S. Herrmann, F. Schmidt, T. Heinekamp, M. Strassburger, T. Krüger, P. Bacher, J. Lother, E. Weiss, K. Hünniger, H. Liu, P. Hortschansky, A. Scheffold, J. Löffler, S. Krappmann, S. Nietzsche, O. Kurzai, H. Einsele, O. Kniemeyer, S. G. Filler, U. Reichard and A. A. Brakhage, *mBio*, 2018, **9**, e01557-18.
75. H. Schmidt, S. Vlaic, T. Krüger, F. Schmidt, J. Balkenhol, T. Dandekar, R. Guthke, O. Kniemeyer, T. Heinekamp and A. A. Brakhage, *Mol. Cell. Proteomics*, 2018, **17**, 1084–1096.
76. M. L. Rodrigues, E. S. Nakayasu, D. L. Oliveira, L. Nimrichter, J. D. Nosanchuk, I. C. Almeida and A. Casadevall, *Eukaryotic Cell*, 2008, 7, 58–67.
77. D. Zamith-Miranda, H. M. Heyman, S. P. Couvillion, R. J. B. Cordero, M. L. Rodrigues, L. Nimrichter, A. Casadevall, R. F. Amatuzzi, L. R. Alves, E. S. Nakayasu and J. D. Nosanchuk, *mSystems*, 2021, **6**, e00822-21.
78. E. G. A. Chaves, J. A. Parente-Rocha, L. C. Baeza, D. S. Araújo, C. L. Borges, M. A. P. de Oliveira and C. M. de A. Soares, *Front. Microbiol.*, 2019, **10**, 1–14.
79. A. Sukumaran, B. Ball, J. R. Krieger and J. Geddes-McAlister, *mBio*, 2022, **13**, e0168722.
80. H. Li, Y. Li, T. Sun, W. Du, Z. Zhang, D. Li and C. Ding, *Front. Microbiol.*, 2020, **11**, 1–14.
81. D. Zamith-Miranda, H. M. Heyman, L. G. Cleare, S. P. Couvillion, G. C. Clair, E. L. Bredeweg, A. Gacser, L. Nimrichter, E. S. Nakayasu and J. D. Nosanchuk, *mSystems*, 2019, **4**, e00257-19.
82. H. Guo, S. M. Xie, S. X. Li, Y. J. Song, X. Y. Zhong and H. Zhang, *J. Mycol. Med.*, 2017, **27**, 339–344.
83. A. Sukumaran, J. Coish, J. Yeung, B. Muselius, M. Gadjeva, A. MacNeil and J. Geddes-McAlister, *J. Leukocyte Biol.*, 2019, **106**, 1221–1232.
84. J. C. Rieckmann, R. Geiger, D. Hornburg, T. Wolf, K. Kveler, D. Jarrossay, F. Sallusto, S. S. Shen-Orr, A. Lanzavecchia, M. Mann and F. Meissner, *Nat. Immunol.*, 2017, **18**, 583–593.
85. A. Pandey, S. L. Ding, Q. M. Qin, R. Gupta, G. Gomez, F. Lin, X. Feng, L. Fachini da Costa, S. P. Chaki, M. Katepalli, E. D. Case, E. J. van Schaik, T. Sidiq, O. Khalaf, A. Arenas, K. S. Kobayashi, J. E. Samuel, G. M. Rivera, R. C. Alaniz, S. H. Sze, X. Qian, W. J. Brown, A. Rice-Ficht, W. K. Russell, T. A. Ficht and P. de Figueiredo, *Cell Host Microbe*, 2017, **21**, 637–649.e6.
86. A. Kugadas, Q. Wright, J. Geddes-McAlister and M. Gadjeva, *Invest. Ophthalmol. Visual Sci.*, 2017, **58**, 4593–4600.
87. J. Yeung, M. Gadjeva and J. Geddes-McAlister, *Proteomics*, 2019, **20**, 1900290.
88. J. Yeung, J. Lamb, J. R. Krieger, M. Gadjeva and J. Geddes-McAlister, *Curr. Protoc. Mouse Biol.*, 2020, **1**, e83.
89. J. Kandhavelu, N. L. Demonte, V. P. Namperumalsamy, L. Prajna, C. Thangavel, J. M. Jayapal and D. Kuppamuthu, *J. Proteomics*, 2017, **152**, 13–21.
90. K. Ley, H. M. Hoffman, P. Kubes, M. A. Cassatella, A. Zychlinsky, C. C. Hedrick and S. D. Catz, *Sci. Immunol.*, 2018, **3**, eaat4579.

91. T. N. Mayadas, X. Cullere and C. A. Lowell, *Annu. Rev. Pathol.: Mech. Dis.*, 2014, **9**, 181–218.
92. A. Kugadas, J. Geddes-Mcalister, A. Thanabalasuriar, E. Guy, A. DiGiandomenico, D. Sykes, M. Mansour, R. Mirchev and M. Gadjeva, *J. Leukocyte Biol.*, 2019, **105**, 1099–1110.
93. M. Rafiq, F. Rivieccio, A.-K. Zimmermann, C. Visser, A. Bruch, T. Krüger, K. González Rojas, O. Kniemeyer, M. G. Blango and A. A. Brakhage, *mSphere*, 2022, 7, e0094021.
94. I. A. Shopova, I. Belyaev, P. Dasari, S. Jahreis, M. C. Stroe, Z. Cseresnyés, A.-K. Zimmermann, A. Medyukhina, C.-M. Svensson, T. Krüger, V. Szeifert, S. Nietzsche, T. Conrad, M. G. Blango, O. Kniemeyer, M. von Lilienfeld-Toal, P. F. Zipfel, E. Ligeti, M. T. Figge and A. A. Brakhage, *mBio*, 2020, **11**, e00596-20.
95. J. Karkowska-Kuleta, M. Smolarz, K. Seweryn-Ozog, D. Satala, M. Zawrotniak, E. Wronowska, O. Bochenska, A. Kozik, A. H. Nobbs, M. Gogol and M. Rapala-Kozik, *Cells*, 2021, **10**, 2736.
96. N. T. Case, K. Duah, B. Larsen, C. J. Wong, A.-C. Gingras, T. R. O'Meara, N. Robbins, A. O. Veri, L. Whitesell and L. E. Cowen, *Cell Rep.*, 2021, **36**, 109584.

CHAPTER 11

Mass Spectrometry Imaging in Microbiology

YUCHEN ZOU AND BIN LI*

State Key Laboratory of Natural Medicines and School of Traditional Chinese Pharmacy, China Pharmaceutical University, Nanjing, 210009, China
*Email: binli@cpu.edu.cn

11.1 Introduction

Microbial metabolism is a series of complex procedures involving dynamic molecular processes of many aspects, such as synthesis, decomposition, and transportation. These molecular reactions are the basic units that compose all kinds of life events of microorganisms, from intra-individual regulations to inter-species competitions. Transcriptomics and genomics studies provide scientists with valuable insights into the regulation and expectations of related activities.[1] However, it is important to note that metabolite molecules, which are regulated by the transcriptome and genome, directly participate in the fundamental chemical processes of metabolism. Their spatiality and temporality are the most evident indications of microbial metabolism. Chromatography coupled with mass spectrometry (MS) utilized in microbiological studies usually involves bulk sampling, which neutralizes the spatial information within the sample. Targeted approaches like fluorescence imaging and immunohistochemistry require prior knowledge of analytes of interest to pinpoint the spatial distribution of certain molecules. While these imaging techniques are handy at annotating macromolecules like proteins, it is seriously difficult for them to label every single small molecule among thousands.

New Developments in Mass Spectrometry No. 13
Detection and Analysis of Microorganisms by Mass Spectrometry
Edited by Liang Qiao and Jia Yi

Published by the Royal Society of Chemistry, www.rsc.org

The emergence of mass spectrometry imaging (MSI) for the first time allowed for *in situ* analysis of molecules directly from samples. Contrary to targeted approaches, MSI can detect hundreds to thousands of molecules in their native locations without target-specific reagents or probes. In MSI, an artificially pixelated region of the sample is ionized pixel-by-pixel, and the molecular profile for each pixel is collected and analyzed one at a time. Eventually, the chemical profile of the whole sample is formed with each mass spectrum from every single pixel. Since its first appearance in the 1960s,[2] MSI has contributed to countless breakthroughs in a variety of disciplines, such as medicine,[3,4] biology,[5] food science,[6] environmental science,[7] materials science,[8] and pharmacy.[9,10]

MSI debuted in microbiological studies in the first decade of this century, where secondary ion mass spectrometry (SIMS) was utilized to study element assimilation by microbes.[11] More applications have been established in the past two decades, mainly focusing on the dynamic molecular alterations of intra-microbe metabolism or microbial community interaction, and spatio-temporal localization of molecules. In this chapter, an overview of MSI methodology and applications in microbiology will be provided, aiming to demonstrate the unique feasibility of this technology in microbial disciplines.

11.2 The MSI Techniques

Many common MS techniques utilized in microbiological studies are capable of image analysis, such as matrix-assisted laser desorption/ionization (MALDI), SIMS, and desorption electrospray ionization (DESI). Imaging or not, these techniques work based on similar principles, but their performance may vary due to the diversity of composition of microbes and the different detecting capabilities of each ionization technique. Currently, MALDI, DESI, and SIMS are the three most common techniques for MSI and are well-developed to be commercially available.[12] Several cutting-edge ionization techniques have also been reported to be utilized in microbial studies.[13,14]

MALDI is the most common MSI technique, for its capability of mapping multiple important molecule families: lipids,[15] peptides,[16] proteins,[17] polysaccharides,[18] and other small molecules.[19,20] Unlike how MALDI MS analyzes extracted samples spotted on the target, in an imaging experiment, the laser irradiates regions of interest that are predefined as pixels. Datasets are then obtained, which allow individual *m*/*z* images to be generated by splicing all pixels according to their spatial coordinates and corresponding signal intensities. Other MSI methods share similar principles, differing mainly in how the analytes are desorbed and ionized. In MALDI MSI, the size of the pixel, which determines the spatial resolution of the imaging results, is limited by the diameter of the laser spot and the distance between neighboring raster spots. The introduction of advanced laser sources and optics and the emergence of more sophisticated designs of instrumentation have increased the resolution prevailing in MSI communities from more than 100 μm to less than 5 μm in the last decade.[21,22] Reports of applications with ever-advancing resolution seem to mark the future ample scope of single-cell-level or even sub-cell-level imaging.[23] Considering how quickly advancements have been made in these fields recently, it is reasonable to expect the emergence of single-cell MALDI MSI within the following quarter decade.

SIMS imaging could date back to the 1960s.[2,24] It ionizes the sample by bombarding the sample surface with a charged ion beam to generate secondary ions for detection,[25] hence the name. SIMS is well-known for its high spatial resolution, which even allows imaging at the sub-micron level. Static SIMS has been applied to image many biological systems but with a restricted mass range, typically below *m*/*z* 1500.[26] It is one of the major limitations of SIMS brought by the high energy of primary ion beams, as the structure of big molecules usually cannot survive such powerful impacts. Therefore, one of the major modifications of SIMS aims to improve the detectable mass range by using cluster and polyatomic ion sources, such as metal and gas cluster ion beams, which reduce the fractionation of ions and damage on the surface by spreading the energy among multiple molecules in the projectile.[27] One particular type of SIMS instrumentation that is equipped with an ion microprobe and a magnetic sector mass spectrometer and uses Cs^+ or O^- as the primary ion beam is commonly referred to as nano-SIMS, named for its achievable spatial resolution of less than 50 nm.[28] SIMS is also a highly expandable and modular technique and reported to combine with other imaging modalities like confocal Raman microscopy (CRM)[29] and fluorescence *in situ* hybridization (FISH),[30] and MALDI MSI.[31] Adoption of mass analyzers like Orbitrap and Fourier transform ion cyclotron resonance (FT-ICR) analyzers further stretches the mass-resolving power and mass accuracy of SIMS imaging.[32–34]

DESI is probably the most iconic method of ambient MSI, famous for its minimal requirements for sample pretreatment and environmental conditions. Currently, the spatial resolution of DESI MSI typically ranges from 50 to 200 μm.[35] As a solvent-dependent ionization technique, a critical aspect of a successful DESI experiment is the production of a focused, coaxially aligned spray. More specifically, the speed and size of the droplets in the spray govern the effectiveness of the spray, which is configured by users through the nebulizing gas flow rate and the distance between the sprayer and sample surface.[36] A faster gas flow rate leads to smaller and faster droplets, which is usually preferable as it could increase the desolvation efficiency. But the smaller size of the droplets also increases the risk of solvent evaporating before reaching the surface and thus the potential signal loss. Therefore, the relationship between solvent and gas flow rates strongly determines the final imaging resolution. In 2010, a variant of DESI, *i.e.*, nano-DESI, was reported and soon well-accepted by the community, where two capillaries were used to form a liquid bridge on the sample surface for analyte extraction, instead of gas jets and inlets in traditional DESI.

Analytes were extracted into solvent and removed from the surface by a self-aspirating nanospray.[37] Nano-DESI can achieve much better resolution, even under 12 μm. Nano-DESI is a highly feasible method in microbial studies,[35] as its capability of sampling on rough topography by precisely controlling the distance between the liquid bridge and sample surface makes it a much more feasible tool for wet and fresh samples.[38] However, when imaging microbial samples with nano-DESI, complete recovery of the solvent utilized can be hindered due to the differences in hydrophobicity for some of the microbial colonies, leading to low overall signal intensities.[39] The transfer capillary may clog as it moves along the sample surface due to the surface relief of a microbial colony.[38]

The development of MSI has brought a significant number of techniques besides the three mentioned earlier. Some of them are still under development and specialize in certain disciplines. Laser ablation electrospray ionization (LAESI)[40] is another MSI approach utilized relatively commonly by researchers in microbiological studies. It is also an ambient MS technique, requiring little to no sample preparation. Unlike SIMS and MALDI which require dehydration of the sample to be vacuum compatible, LAESI needs the water in the sample to function properly, which makes it well-suited for the investigation of fresh biological tissues and cells. Initially, the sample continuously absorbs excess energy from the IR radiation that vaporizes molecules from the sample. The superfluous energy that goes beyond the consumption of vaporization then excites the hydroxyl vibrations of the water in the sample.[41] The water eventually cannot withstand such overheating and would relax to a stable state by releasing excessive energy with explosions,[42] which then results in the release of a plume consisting mostly of neutral matter in the form of nanoparticles, droplets, and large particulates from the sample. Subsequently, this reverberated plume would be intercepted by the spray from the ESI source, where the droplets in the spray then fuse the plume and ionize the analytes in it, like how ESI works.[40] One of the downsides of LAESI as well as other IR-powered ionization is the collateral damage to the sample, which can be reflected by the craters visible on the sample surface. This destruction limits the subsequent analysis of the sample such as histological staining after the LAESI experiment.[43,44] Currently, LAESI as a standalone platform is commercially available and can be compatible with many commercial mass spectrometers. Rapid evaporative ionization mass spectrometry (REIMS) and the liquid microjunction surface sampling probe (LMJ-SSP) technique were two other ambient real-time surface analysis approaches that have been widely applied to the analysis of microbial colonies. REIMS works by rapidly heating the sample and transferring the aerosol created on the ablation into an MS system for detection. In clinical practice, REIMS is also known as an intelligent knife (iKnife),[45] for its primary purpose is to identify tissues during surgical interventions. REIMS imaging has a poor resolution (500–750 μm) but has demonstrated high flexibility and feasibility in distinguishing bacterial strains and healthy/cancerous regions.[46] The LMJ-SSP extracts analytes from the sample surface by using coaxial capillaries to form a liquid microjunction at the extremity of the probe. It usually operates at high-μm to low-mm spatial resolutions,[47] as exemplified in a recent paper regarding imaging microbial biofilms using the LMJ-SSP,[48] but novel modifications on probes have achieved resolutions of 35 μm.[49]

11.3 Sample Preparation for Microbial MSI

Since the implementation of preparation has a huge impact on the ionization process, sample preparation has always been one of the core parts of microbial MSI experimental design.[50] Compared with animal or plant tissue samples, the diverse status and composition of microbial samples, which usually consist of microorganisms with culture media, nutrient substances, excreted metabolites, and other secondary products, complicate the sample preparation protocol for most MSI techniques. More importantly, cryosection of most microbial samples is currently impracticable. Application of the MALDI matrix would bring the first problem, due to the inhomogeneous coverage of the matrix on fresh culture caused by lateral diffusion and absorption of agar,[51] which is the most common medium for microbial cultivation. Yang *et al.*[52] concluded the preparation process with three phases: grow microbes on thin agar, saturate the sample with matrix *via* a sieve, and dehydrate while checking for flaking and sample height. 'Thin' agar means a height of less than 2 mm (usually 1–1.5 mm), to achieve the surface conductivity required for a high-vacuum MALDI TOF instrument. However, this thickness of the sample still may lead to the decay of molecular signals, because of the insulation of agar. This impediment can be overcome by using agar with an ultimate dry thickness of less than 50 μm or potentially by sputter-coating the sample with a few nanometers of a conductive material prior to imaging.[1] The order of matrix application and dehydration is flexible, but dehydrated agar could not dissolve the solid matrix filtered from sieves. Although most modern MALDI instrumentation now can work under atmospheric pressure or partial vacuum, dehydration is still required in most cases, which could be another potentially problematic aspect and thus require careful attention and mild drying conditions (typically 37–40 °C for 4–6 hours or overnight)[53] to prevent undesired incidents like sample deformation, degradation of labile compounds, and delocalization of analytes. The selection of matrices should be made based on pre-experiments, despite the existing utilization of several universal matrices such as 2,5-dihydroxybenzoic acid (DHB), α-cyano-4-hydroxycinnamic acid (CHCA), and 9-aminoacridine (9-AA). It is because certain vital analytes may have better ionization efficiency under a specific matrix.[54] The instruments for matrix application have evolved rapidly over the last decade, from steel sieves and hand-held airbrushes to a programmable robotic matrix application platform. It is quick and easy to implement direct application of a sieve-filtered matrix onto wet agar-based culture but it may result in an uneven coating and matrix aggregation that may produce artifacts in the ion images obtained, which could be partially resolved by applying a suspension of mixed matrices.[55] Use of airbrushes is the simplest method to apply matrix solution, but it is time-consuming and occasionally produces uneven crystallization due to the water content. Although use of airbrushes generates more favorable coatings than that of sieves, both approaches that require manual operation are limited by the lack of reproducibility, whereas automatic matrix sprayers are invincibly superior, providing a reliable, homogeneous, and precise coating of the matrix. The quality of the coating can be significantly affected by several key parameters in spraying-based approaches. These parameters include the pressure of the carrier gas (typically N_2), the flow rate of the solvent, the distance between the sprayer and the sample stage, and the composition of the matrix solution.[51] It is worth noting that spraying matrix on a wet sample may cause analyte delocalization. Although it was reported that this delocalization was not significant enough to make any difference,[51] with the development of the advancing optical instrumentation adopted in current MSI techniques, such tiny migration of analyte may lead to deviated imaging results. Although the extensive commercialization of matrix sprayers has made them included in most MSI configurations available in the market, primal approaches like sieves and airbrushes still have their places in labs, for their convenience and competence for less rigorous experiments and other activities such as teaching and showcasing (Figure 11.1A). Nevertheless, as for non-agar-based microbial samples, which are collected from the natural environment or cultivated on other materials, a specific optimized protocol (sometimes a simpler one) may be required for more satisfactory results.[56,57]

SIMS instrumentation requires a strict vacuum environment; therefore, samples have to be stable under vacuum. The traditional route for microbial samples was to isolate cells before sampling for analysis. For instance, Tyler *et al.*[58] grew biofilms in a Tubular Biofilm Flowcell (TBF) system, where tubing was first inoculated with microbes. Following the growth period, the cube was cut into 0.5 cm long sections, frozen in liquid propane, and then stored under liquid nitrogen until they were transferred to the SIMS instrument. In this case, a modified TOF-SIMS instrument was adopted, outfitted with a uniquely designed in-vacuum cryosectioning chamber that was directly attached to the analysis chamber, and consequently temperature control and airtightness could be ensured during the whole transferring process. Dehydration was also allowed or inhibited through the regulation of temperature. But in most cases, a frozen sample needs to be thawed before sending it into an MS system. For instance, before air dehydration or heat fixation, Vaidyanathan *et al.*[59] first thawed frozen mycelia with a warm water bath, cleaned them with distilled deionized water, and resuspended them in water. Single bacterial cells can also be imaged by SIMS. Tian *et al.*[60] deposited washed single-cell suspension obtained from centrifuging on a silicon wafer by directly spin-coating. The samples were then plunged into liquid ethane and swiftly transferred to liquid nitrogen. Under the liquid nitrogen, the samples were loaded onto a precooled sample holder at 153 K in a vacuum followed by freeze-drying for 4 h. It used to be difficult to image

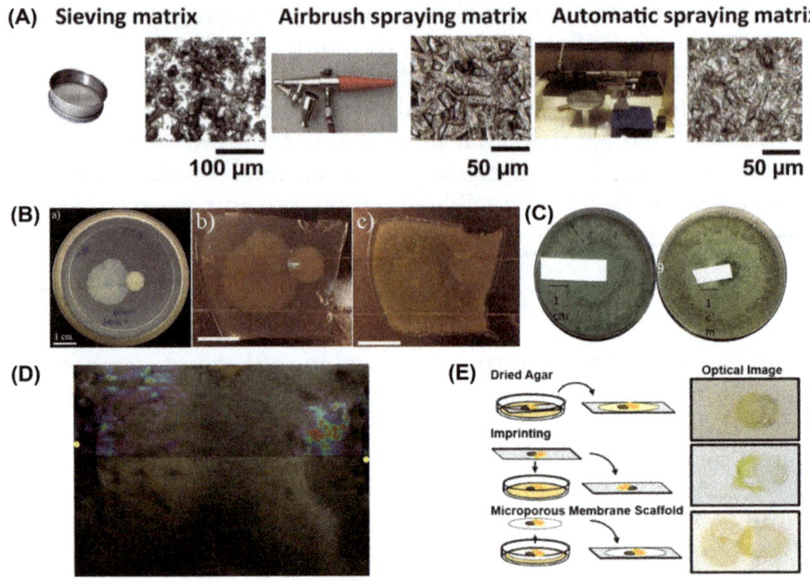

Figure 11.1 (A) Three common matrix application methods and the respective resulting crystallization of the matrix. Reproduced from ref. 51 with permission from John Wiley & Sons, Copyright 2016. (B) Sample preparation of a biofilm for SIMS analysis including cultivation (a), transferring (b), and dehydration (c). Reproduced from ref. 64 with permission from American Chemical Society, Copyright 2018. (C) Imprinting *Trichoderma harzianum* culture with a section of PTFE (left) and a flat silicon wafer (right). Reproduced from ref. 70 with permission from the Royal Society of Chemistry. (D) A *Clohesyomyces aquaticus* strain (G100) grown on a cardboard insert. The upper half of the image was overlaid with an ion image of phomopsinone A obtained by DESI MSI. The yellow circles indicate the points of inoculation. Reproduced from ref. 72 with permission from the Royal Society of Chemistry. (E) MMS-DESI MSI. Reproduced from ref. 73 with permission from American Chemical Society, Copyright 2019.

the sample natively with culture media, because conductivity and certain surface flatness are also required for SIMS samples. Imprint transfer is an alternative to circumvent issues with charging and surface architecture, which works by imprinting agar-based culture onto a silicon wafer and subjecting it to analysis. The advantages of this method are attributed to silicon wafers being conductive and having a far flatter surface than the culture, which guarantees few differences for primary and secondary ions and leads to a higher mass resolution and an easier peak attribution for a common TOF analyzer.[61] Another common approach is to cultivate the sample directly on the conductive surface to avoid an intricate transferring process. Silicon wafers, indium-tin-oxide (ITO)-coated glass slides,[57] and silicon nitride membranes[62] have been reported to be feasible materials.

Despite all these restrictions, direct SIMS analysis on agar-based microbial culture was successfully performed and reported in 2018,[63] where the agar containing the microbes was excised from the Petri dish and placed on a custom aluminum SIMS plate covered with double-sided conductive copper tape. Dehydration was accomplished with a gentle stream of nitrogen, which inevitably caused the nonlinear and inhomogeneous deformation of the whole sample. The deformation, however, was found to be uniform and predictable, with the greatest contraction occurring near the agar edges (Figure 11.1B). Thus, the overall morphology of the culture was preserved. And nitrogen-assisted dehydration was reported to provide the best retention of the sample and the most consistent result.[64] But notably, the direct analysis still risked the uncertainty of diffusion of analytes (especially polar ones) through the sample as well as the inhomogeneous ionization of the sample.[65] In 2020, Zhang *et al.*[14] reported an analysis of a hydrated bacterial biofilm with a novel cryo-Orbitrap-SIMS technique. In particular, the grown biofilm was frozen with a high-pressure freezing device and the addition of ammonium formate as a cryoprotectant before being stored in liquid nitrogen. The SIMS instrumentation in this study was also equipped with a cryogenic sample transfer shuttle to maintain the temperature and airtightness during the transferring process, thus avoiding water condensation on the sample surface. Apart from the terrain mapping of the sample, another demonstration emphasized in this study was the enhanced intensities of signals in the frozen-hydrated biofilm, especially the intensities of polar analytes like amino acids, which yielded a ten-thousand-fold increase. Despite the tremendous cost of the cryo-related equipment and early development stage of Orbitrap SIMS, this method might shed light on the expansion of SIMS imaging to metabolites and pharmaceuticals.

The main challenge of sample preparation in a microbial DESI experiment is the high-pressure spray that could deform and even damage the soft agar-based culture and the requirement of a flat sample surface, but there have been several strategies proven to be practical in previous reports. Watrous *et al.*[66] adopted the imprinting method mentioned earlier using mixed cellulose ester filter membranes to make an imprint of bacterial culture from solid agar, which could be directly analyzed with DESI with no additional preparation procedure needed (sometimes dehydration of the imprint is needed depending on the sample). Polytetrafluoroethylene (PTFE) was also used to imprint microbial samples in culturing dish.[67] Porous Teflon,[68] TLC plates,[69,70] and C-18 beads could be viable (Figure 11.1C), as their uses for imprinting plant tissues or microbial samples were reported as well. One simpler approach is to cultivate microbes in a thin-layered agar plate, where the reduced volume agar could avoid the undesired growth of microorganisms beyond the culture that could turn the surface irregular. The agar-based culture was then dehydrated, resulting in a hard and uniform surface. This method, according to Angolini *et al.*,[71] was superior to the imprinting method, which led to the loss or inconsistency of spatial molecular information. Despite these flaws, it is worth noting that imprint transfer is a highly feasible strategy. With appropriate imprint materials, it applies to all kinds of samples and is compatible with most MSI techniques.[70] Therefore, this method could be particularly useful for atypical and complex samples when speed and cost are the main considerations. Another alternative method is to cultivate microbes on solid materials that could be easily separated after growth. Sica *et al.*[72] used cellulosic materials in the cultivation of fungi to provide a carbon source and a firm surface that was less malleable by the DESI spray. Among several common materials, cardboard (Figure 11.1D) turned out to be the best option compared to cheesecloth, filter paper, and wood, which rotted into pieces after cultivation, absorbed too much solvent from the spray, or produced an uneven surface, respectively. Later, Ellis *et al.*[73] described a microporous membrane scaffold (MMS) method for the sample preparation of microbial DESI (Figure 11.1E). Specifically, the microbes were spotted on a sterile nylon membrane, and the inoculated membrane was placed on top of the medium. After cultivation and prior to sampling, the membrane was removed from the medium, dried for 10 minutes, and adhered to a glass slide with double-sided Scotch tape. A membrane without inoculation of microbes was also incubated on another of the same medium as blanks. Compared to the imprinting method and direct analysis on dried agar-based culture, the MMS method had better sensitivity and repeatability. The most vital advantage of the MMS method was the minimal delocalization of analytes, which was particularly meaningful for the analysis of multiple cultures. As for nano-DESI, the preparation process could be easier, as one particular benefit of nano-DESI is that the analysis is not limited by sample conditions, and thus microbes could be cultivated in a more ideal environment to achieve the growth of the microbial sample as designed.[38] Usually, the microbial culture can be subjected to nano-DESI analysis directly without any sort of preparation by cutting pieces of agar from the culture and placing them on a sterile glass or polystyrene slide[74] or mounting the whole Petri dish containing the sample onto the sample stage using double-sided tape.[38] Besides, the analysis itself is nondestructive, so the time-course experiment is also practicable by simply placing back the sample in the incubator after swift acquisition of data at each time point.[75,76] Overall, a minimal preparation process of microbial samples is required for most ambient MSI techniques.

11.4 MSI on Microbial Monocultures

The size of microbial monocultures allows for a variety of instrumentation for MSI experiments, typically at the millimeter level. The aims of monoculture studies are diverse, such as identifying species, investigating production and spatial distribution of metabolites, and studying the influence of exogeneous substances. Some basic experiments have been reported to simply demonstrate the feasibility of the proposed methodology or to provide a chemical overview of the sample, carried out as a precursor to or in conjunction with experiments on co-cultures.[53,77]

Certain species of microbes are more commonly analyzed than others because these microbes are known as model species, and their genomes are well-characterized and able to make detectable or analyzable reactions under a wide range of customized experimental conditions. *Bacillus subtilis*, for example, is a Gram-positive bacterium omnipresent in soil and the gastrointestinal tract of animals, utilized by humans in many industries. In the lab, *B. subtilis* has the advantages of a fast growth rate, mild environmental and nutritional requirements, excellent fermentation efficiency, low toxicity, and high genetic amenability,[78] *Escherichia coli* is the model organism for Gram-negative species, due to the in-depth knowledge of *E. coli* obtained from studies over the past several decades. These properties make these species ideal model microorganisms for all kinds of biochemical exploration as well as methodological studies.[79] For example, Li *et al.*[51] compared three matrix application approaches for a microbial MALDI MSI experiment, where *B. subtilis* was used as a test microbe. Tian *et al.*[60] demonstrated imaging of antibiotics in single *E. coli* with TOF-C_{60}-SIMS, aiming to present its technical capability in molecule tracing at the subcell level.

In monoculture studies, the spatial aspect seems to be less noticeable, because microbe cells of the same species cultivated in the same dish and treated with the same conditions are usually considered to be homogeneous. But in fact, there is possible spatial heterogeneity in a metabolic process of a monoculture, which might not be noticed when analyzed with a homogeneous approach. In a study by Morales-Soto *et al.*,[63] the alkyl quinolone (AQ) signaling responses of *Pseudomonas aeruginosa* swarms upon exposure to two antibiotics, tobramycin and carbenicillin separately, were found to be spatially dependent. A multiplexing approach of CRM in combination with SIMS imaging was adopted to explore the phenotypic responses and spatial variability of AQs, including the *Pseudomonas* quinolone signal (PQS) and members of the alkyl hydroxyquinoline (AQNO) subclass. A drastic spatial heterogeneity of PQSs and other AQs was observed in *P. aeruginosa* swarm communities. Specifically, the researchers found that PQSs and AQNO signaling were independent and responsible for communicating in a different regional range of the swarm. Upon exposure to tobramycin, PQSs only reach a part of the colony near the closest cells, while AQNO response could be found in the whole community (Figure 11.2A). These results showed that multiple intercellular signals acting on different scales can be triggered by one common cue.

MSI could also be used to directly compare the relative contents of analytes among multiple monoculture samples. One of the examples involves the cannibalism of *B. subtilis*: Si *et al.*[80] employed the combination of MALDI MSI and fluorescence imaging to reveal the different spatial distributions of certain metabolites responsible for cannibalistic behavior among wild-type and two mutants, Δ*spo0A* and Δ*abrB* (Figure 11.2B). These differences suggested further differentiation in the cannibal/matrix-producing subpopulation in the *B. subtilis* colony. Condren *et al.*[81] studied the mechanism of metabolic alteration in *P. aeruginosa* caused by taurolithocholic

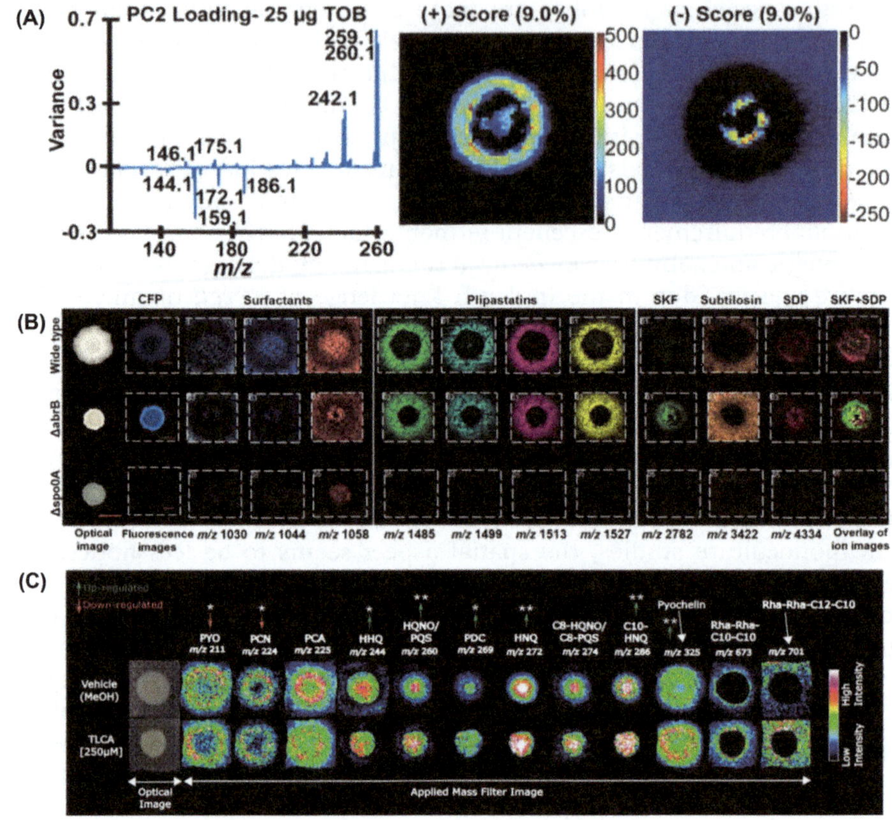

Figure 11.2 (A) Loadings and score images from principal component analysis of SIMS imaging show the distinct distribution patterns between PQS (positive loadings) and AQNO (negative loadings) upon exposures to 25 μg of tobramycin, suggesting differing chemical messages carried by these molecules. Reproduced from ref. 63, https://doi.org/10.1074/jbc.RA118.002605, under the terms of CC BY 4.0 license, https://creativecommons.org/licenses/by/4.0/. (B) Fluorescence imaging and MALDI MSI of *B. subtilis* biofilms of NCIB3610 and its mutants integrated with the PyqxM-CFP reporter confirm the important role of Spo0A as a transcriptional regulator that enables temporospatial cellular differentiation and AbrB as a negative regulator of the biofilm development. Reproduced from ref. 80 with permission from American Chemical Society, Copyright 2016. (C) MALDI MSI analysis of 12 altered metabolites in *P. aeruginosa* after exposure to TLCA. Reproduced from ref. 81 with permission from American Chemical Society, Copyright 2020.

acid (TLCA), a mammalian endogeneous enteric metabolite that demonstrated inhibition of biofilm formation against *P. aeruginosa* (Figure 11.2C). MALDI MSI results indicated an increase of pyochelin and rhamnolipids, which have been known for their association with increased virulence, in the

TLCA-treated group compared to the vehicle control group. These results suggested that biofilm inhibition (or dispersion) ultimately led to the bacteria becoming more virulent in a host model. The researchers further imaged the pyochelin production of the wild-type and Δ*phz* and Δ*pqsH* mutants, suggesting that the production of quinolones and phenazines might not be related to how TLCA induced the increase of pyochelin.

11.5 MSI on Complex Microbial Communities

Most creatures in the wilderness are surrounded by not only fellows of the same kind but also numerous organisms from across the four kingdoms, and so are microbes. A big portion of the metabolism of microbes was determined by their relationships with nearby organisms, such as predation, defense, competition, and symbiosis. With the capability of profiling molecules *in situ*, researchers gain deeper insights into the spatial aspects of activities occurring in microbial communities in their natural state or simulated in the lab.

One crucial step in natural microbial samples is to identify the species composition and community. Traditional genetic or biomarker-based approaches are both hindered by their limited spatial resolution for some complex multilayered samples, where MSI methods are advantageous. Wormer *et al.*[82] took advantage of MALDI FT-ICR MSI to spatially resolve the biological composition of microbial mats sampled from Octopus Spring in Yellowstone National Park. The MALDI MSI of chloropigments, quinones, and intact polar lipids (IPLs) was able to describe the spatial distribution of different groups of chlorophototrophs along the mat, which combined with the assignment of specific members of the community to the molecular profile allowed researchers to spatially locate the microbes and sequence the chlorophototrophs along the mat layers (Figure 11.3A). Fluorescence imaging could also be used to complementarily provide spatial information on microorganisms, while MSI focused on chemical profiling. Dekas *et al.*[83] used FISH to label microbes of different species in microbial consortia with different colors and nano-SIMS to trace the flow of nitrogen with isotopes, revealing nitrogen fixation of deep-sea anaerobic methane-oxidizing archaea and metabolite exchange between them and sulfate-reducing bacterial symbionts.

To explore the complicated interaction patterns between species of different relationships, the single variable principle should be rigidly followed to obtain a confident result. Thus, these kinds of studies are usually conducted on microbial co-cultures designed in the lab with environmental conditions under strict control, such as salt concentration, pH, and exposure to light. Moree *et al.*[84] published an exemplary work of capturing metabolic transformations between *P. aeruginosa* and *Aspergillus fumigatus* with MALDI MSI and discovered that phenazine metabolites produced by *P. aeruginosa* were converted by *A. fumigatus* into other chemical entities with alternative

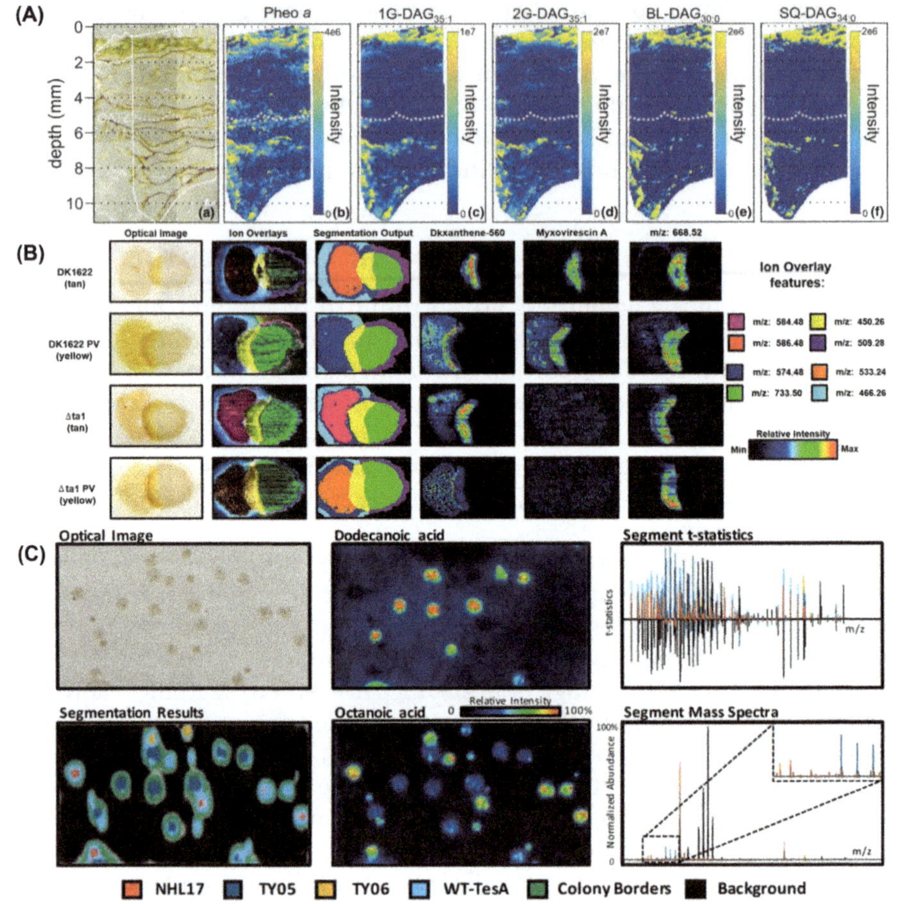

Figure 11.3 (A) From left to right: Images of the mat slice and 75 μm resolution MALDI MSI of Pheo *a*, which represents the distribution of *Synechococcus* spp., and co-locating IPLs, which is therefore attributed to the cyanobacterial layers. Reproduced from ref. 82, https://doi.org/10.1111/gbi.12411, under the terms of CC BY 4.0 license, https://creativecommons.org/licenses/by/4.0/. (B) Optical images, ion overlays, and unsupervised segmentation results of wild-type (WT) and *M. xanthus* Δta1 strains and their respective phase variants (PV) suggest the potential roles of these metabolites in predation. Reproduced from ref. 73 with permission from American Chemical Society, Copyright 2019. (C) Optical image, ion images, and unsupervised segmentation results of a mixed culture comprised of the control strain TY06 and FFA-overproducing strains TY05, NHL17, and WT-TesA. The segments can distinguish the molecular phenotypes of engineered microorganisms based on measured metabolomes. Reproduced from ref. 87 with permission from PNAS, Copyright 2021.

properties, including enhanced toxicities and the ability to induce fungal siderophores. Traditional chromatography-based MS and MS/MS were used to determine the accurate mass of metabolites detected in MSI analysis for further construction of molecule networking. The combination method alike was also adopted by Michelsen *et al.*[85] to study the differential polymicrobial interactions between a human-adapted strain (DK2-P2M24-2003) and a lab-used strain (PAO1) of *P. aeruginosa* when cultivated with *Staphylococcus aureus*. Their findings suggested that the human-adapted strain of *P. aeruginosa* had divergent metabolic interactions with *S. aureus*, which enhanced a proto-cooperative relationship between the two bacteria. Ellis *et al.*[73] applied the novel MMS-DESI protocol mentioned earlier on a predation model of *Myxococcus xanthus* preying *E. coli* and used unsupervised spatial segmentation and principal component analysis to screen potential ions of interest related to the predation. The developed workflow was further employed on wild-type and mutated *M. xanthus* Δ*ta1* strains and their respective phase variants to explore the roles of DKxanthene-560, Myxovirescin A, and an unknown ion in the predation (Figure 11.3B).

MSI techniques can also be effectively used as a high-throughput approach, enabling applications in engineering and biosynthesis, such as rapid phenotyping of metabolic activities by detecting metabolites like peptides and lipids. Compared to MS-based high-throughput methods, MSI approaches aim for *in situ* detection with less sample preparation and additional labor like extraction, meaning the spatial information was preserved even though it might not be needed. Si *et al.*[86] reported a label-free method for high-throughput engineering of multistep biochemical reactions based on optically guided MALDI-TOF MS analysis of bacterial colonies. The follow-up procedures included utilizing the multivariant statistical analysis to process and visualize the resulting mass spectral datasets for rapid identification as well as overlaying optical images for further localization and recovery. Two practical scenarios were exemplified in this work, which surveyed the structural diversity of plantazolicin analogs that are created with a substrate library and sought to quantify changes in the relative abundance of the reaction products due to modified enzymatic specificities, demonstrating the simplicity and effectiveness in a wide range of multistep enzyme reactions and high-throughput screening of microbial libraries. Another example was presented by Ellis *et al.*,[87] who ulteriorly took advantage of the combination of MMS-DESI MSI and unsupervised spatial segmentation, and applied it to accelerating synthetic biological workflows. Specifically, MMS-DESI was used to directly analyze microbial colonies, followed by species identification of each colony in the mixed culture through untargeted acquisitions and unsupervised segmentation (Figure 11.3C). Of note, the effects of data mining were emphasized in this work, which included the ranking producers of a specific biosynthetic product, revealing any secondary or off-target production, and profiling molecular information about engineered host cells.

11.6 MSI on Single Microbial Cells

Single-cell imaging has been regarded as a promising area where MSI could thrive,[23,88] due to the rapid advancement of MSI instrumentation. Single-cell scales are necessary to elucidate changes in cell biochemistry that occur, for example, in the early stages of a disease, as the heterogeneity of minor physiologically distinct cells within a cell population could be overwritten by averaging out and thus might not be resolved by bulk omics measurements.[88] Compared to plant cells or animal cells, most microbial cells are smaller, which typically range from 0.2 μm to 5 μm for bacteria and 2 μm to 10 μm for fungi, approximately, and thus it is harder to analyze them. In addition, many unique chemical compositions could only be found in certain microbial species, such as chitin and silica, which make single-cell-level microbial MSI further challenging.[89] Nevertheless, single-cell MSI has been applied to interpret biochemical differences between individual microbe cells in the past decade.

Due to its highest achievable spatial resolution among all MSI techniques, nano-SIMS is commonly utilized in single-cell imaging studies. But because of the limitation of SIMS itself, the focus of these studies is mainly on the elemental aspect of metabolism rather than metabolite molecules. Zimmermann *et al.*[90] reported a workflow to measure the heterogeneous fixation of N_2 and CO_2 in a single cell of the green sulfur bacterium *Chlorobium phaeobacteroides* collected from the meromictic lake Lago di Cadagno. The cells were cultivated with ^{15}N and ^{13}C incubated lake water and concentrated, then nano-SIMS was used to measure the ^{15}N and ^{13}C incorporated after identification with catalyzed-reporter deposition fluorescence *in situ* hybridization (CARD-FISH) targeting 16S rRNA (Figure 11.4A). In 2021, a study by Philippi *et al.*[91] provided the first-ever evidence of *in situ* N_2 fixation of purple sulfur bacteria in the same lake, where nano-SIMS determined single-cell N_2 fixation of five species (Figure 11.4B). They found that the N_2 fixation of purple sulfur bacteria was exclusively and highly efficiently catalyzed by molybdenum nitrogenase, despite the low molybdenum condition of the lake. These results challenged the current general knowledge that cyanobacteria are the main contributor to N_2 fixation on the early earth planet. The two aforementioned examples involve combined analysis by fluorescence imaging and MSI by nano-SIMS, which is a common strategy in element-level research. Lake Cadagno is frequently used as an analog system for the euxinic continental ocean margins on the billion-year-ago earth due to the special structure of the lake.[92]

As for other elements, the determination process could be more complex. Schoffelen *et al.*[93] managed to quantify the uptake of dissolved inorganic phosphorus (DIP) in combination with cellular CO_2 and N_2 fixation in three cyanobacteria species with nano-SIMS (Figure 11.4C). Their approach involved measuring the ^{33}S content, which was the beta decay product of unstable ^{33}P incorporated into cellular biomass, and calculating the $^{33}S/^{32}S$ ratio to indicate the phosphorus uptake. Compared to phosphorus isotopes, the sulfur isotope was better ionized by the cesium ion beam used in SIMS and therefore improved statistics for determining the cellular phosphorus uptake rates. Sekine *et al.*[94] imaged the interactions of silver nanoparticles with green algae with dark-field microscopy, electron microscopy, and

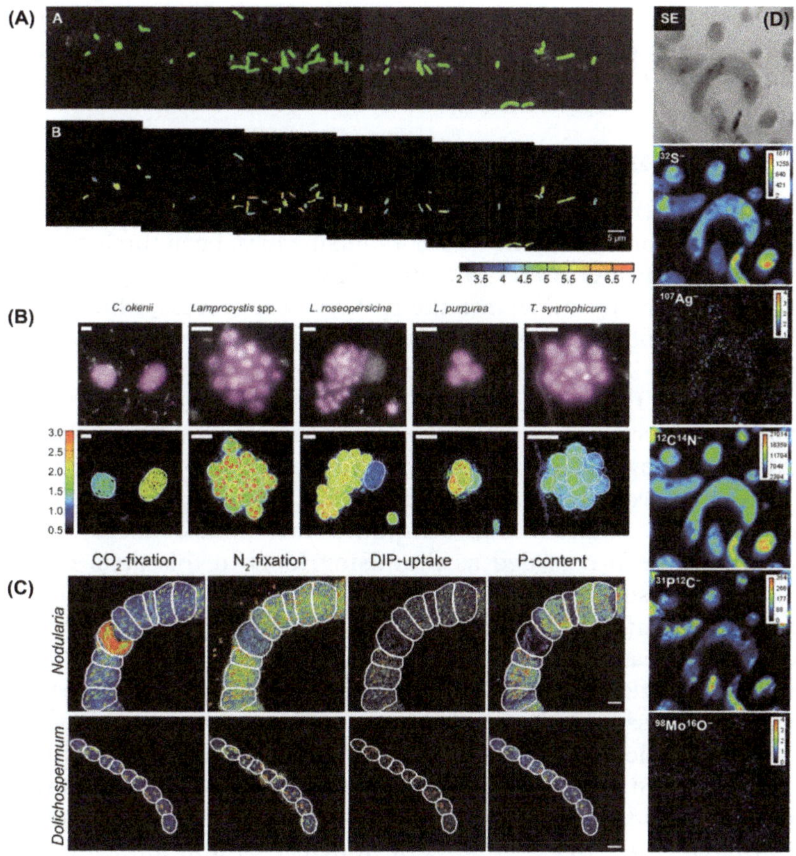

Figure 11.4 (A) Top/A: Flow sorted cells hybridized with a CARD-FISH probe targeted against *C. phaeobacteroides* (green), stained with Hoechst general DNA stain (gray); bottom/B: the corresponding ^{15}N atom fractions according to the color scale calculated from nitrogen ion counts obtained with nano-SIMS. Reproduced from ref. 90, https://doi.org/10.3389/fmicb.2015.00243, under the terms of CC BY 4.0 license, https://creativecommons.org/licenses/by/4.0/. (B) Representative epifluorescence (top row) and nano-SIMS ion images (bottom row) of the individual PSB populations. Colors in ion images indicated the fixation of N by $^{15}N/(^{15}N + ^{14}N)$. Reproduced from ref. 91, https://doi.org/10.1038/s41467-021-25000-z, under the terms of CC BY 4.0 license, https://creativecommons.org/licenses/by/4.0/. (C) CO_2 ($^{13}C/^{12}C$), N_2 ($^{15}N/^{14}N$) fixation, DIP ($^{33}S/^{32}S$) uptake, and cellular P distribution ($^{31}P/^{12}C$) measurement through nano-SIMS for *Nodularia* and *Dolichospermum*. Reproduced from ref. 93, https://doi.org/10.1038/s41598-018-35310-w, under the terms of CC BY 4.0 license, http://creativecommons.org/licenses/by/4.0/. (D) Nano-SIMS ion images and the secondary electron (SE) image of *Raphidocelis subcapitata* exposed to 2 $\mu g\,L^{-1}$ Ag^{+}. Reproduced from ref. 93, https://doi.org/10.1021/acsnano.7b04556, under the terms of CC BY 4.0 license, http://creativecommons.org/licenses/by/4.0/.

nano-SIMS (Figure 11.4D). The distribution of $^{12}C^{14}N^{-}$ was used to indicate the cell morphology; thus the accumulation of silver in the cells could be directly identified by co-locating $^{12}C^{14}N^{-}$ and Ag^{-}.

11.7 MSI on Microbial Infection

Parasitism is another vital subdiscipline in microbiology that has attracted a lot of attention, for pathogenic microorganisms have been threatening many aspects of human life, from agriculture and industry to the health of humans.[95] MSI offers a window into the pathological process, which allows for rapid detection of infections and may inspire the development of more countermeasures against infectious pathogens.

In agricultural fields, microbial infection has always been a major issue that harms production and the safety of the outcomes. With the emergence of chromatography and MS techniques, the identification of infection could be done utterly accurately, turning the pursuit of innovation into efficiency and convenience. Ambient MSI techniques have both of these advantages and thus have been regarded as promising tools in this area. de Moraes Pontes *et al.*[96] used DESI MSI to detect the alteration of metabolites in Huanglongbing-infected orange tree leaves. Among multiple molecules found to be increased in the infected tissues, global natural product social molecular networking (GNPS) analysis further suggested two metabolites that were most associated with the disease. These two metabolites (abieta-8,11,13-trien-18-oic acid and 4-acetyl-1-methylcyclohexene) could be detected by DESI MSI as biomarkers for rapid diagnosis, which was much more efficient than traditional methods like polymerase chain reaction (PCR). Da Silva Lima *et al.*[97] used LAESI MSI to detect patulin diffusion in mold-infected apples and strawberries (Figure 11.5A), demonstrating high sample preservation of the method permitted by little sample preparation. Costa *et al.*[98] applied DESI MSI to monitor the secondary metabolites produced on the orange surface during the *Penicillium digitatum* infection to gain insights into the mechanisms of interaction between the bacterium and the citrus. Through the combination of DESI MSI and molecular networking, the production of tryptoquivalines and fumiquinazolines by *P. digitatum* and the time-course of the accumulation of tryptoquialanines on the surface of the inoculated fruit were reported for the first time.

The complex interactions between microbes and their hosts are shaped by dynamic metabolic crosstalk and chemical communication. Mapping the metabolic landscape of dynamic processes with bulk analysis might overlook rare subpopulations and mask metabolic alterations that originate within them. Insecticidal bioassays suggested an important insecticidal action of tryptoquialanine A during orange decay. Schleyer *et al.*[99] combined the classical plaque assay method with MALDI MSI, to study the metabolic basis of host–virus interactions between the bloom-forming alga *Emiliania huxleyi* and its specific virus (Figure 11.5B). This innovative method termed 'in-plaque MSI' allowed for mapping the metabolic landscape induced during

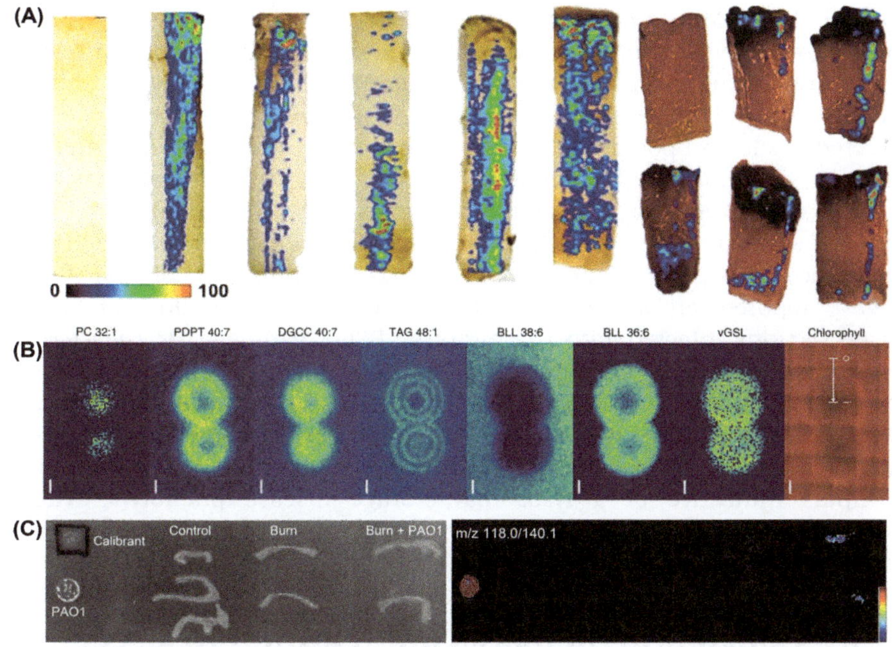

Figure 11.5 (A) Overlaid optical and LAESI MSI images of patulin in respective control (no signal) and molded apple (left) and strawberry (right) slices show the distribution of patulin even in areas without visible mold. Reproduced from ref. 97 with permission from Elsevier, Copyright 2022. (B) MALDI MSI analysis of virus-infected *E. huxleyi* culture shows unique distribution patterns of several lipid biomarkers of the infection. Reproduced from ref. 99 with permission from Springer Nature, Copyright 2019. (C) Optical (left) and MALDI MSI (right) images of the sample slide containing tissue sections in different groups reveal that the signal of ion 118.0/140.1 *m*/*z* is exclusively found in infected burnt tissues. Reproduced from ref. 105 with permission from Elsevier, Copyright 2017.

infection at a 100 μm resolution and unfolding the infection process in a continuous manner. As a result, this work revealed a systematic metabolic shift during infection towards lipids containing the odd-chain fatty acid pentadecanoic acid. As for animal infection, Perry *et al.*[100] visualized the host–pathogen iron competition by mapping siderophores within mouse tissues infected with *S. aureus*. Heterogeneous distributions of *S. aureus* siderophores across infectious foci were observed, challenging the paradigm that the vertebrate host is a uniformly iron-depleted environment to invading microbes.

MALDI MS has been the most versatile method for clinical microbiological analysis, including pathogenic microbe identification, virulence detection, and diagnosis of infectious disease.[101–104] MALDI MSI, on the other hand, as a tool for *in situ* biomarker investigation has been used mostly outside the clinic. But it may be about to change soon. Hamerly *et al.*[105]

applied MALDI MSI to the *in situ* detection of *P. aeruginosa* infecting a burn wound using a well-studied thermal-injury mouse model, which is the first report of detecting a pathogen *in situ* using MALDI MSI directly from tissue sections. Multiple molecular features were found to be present exclusively in infected tissue and positive *P. aeruginosa* controls and absent in uninfected tissues, suggesting the potential use of MSI in clinical scenarios (Figure 11.5C). As for analyzing pathogens that require high-level biocontainment, the incompatibility of the necessary inactivation or sterilization procedures poses as another barrier to the utilization of MSI. Wang *et al.*[106] reported an on-slide heat sterilization method to allow for MSI of tissue sections infected with the high-risk BSL-3 pathogen, *Mycobacterium tuberculosis*, outside of biocontainment. LC–MS was used to validate the integrity of metabolites and drugs in the mimetic model. Then the *M. tuberculosis*-infected tissue sterilized with the developed method was analyzed by MALDI MSI, enabling the first detection and spatial distribution of lipids indicative of a lysosomal storage disease phenotype within tuberculosis granuloma macrophages, in addition to the differential distribution of metabolites central to the fatty acid oxidation pathway.

11.8 Concluding Remarks and Future Prospects

The rapid development of MSI has extended the workable areas of this technique beyond people's realization. As for microbiology, MSI is expected to have a vital role in solving the conundrums of the past decades, such as cellular heterogeneity, secondary metabolism, drug resistance, and community-to-community signaling. The improved capability of MSI could lead to new perspectives and new means to these problems, which will greatly benefit fundamental biology discovery as well as the well-being of humanity. Despite many having been overcome or worked around through the advancement of technology and methodology, some major shortcomings of microbial MSI still exist and hinder its wider application. Inconsistent ionization and unpredicted sample status still challenge the applicability and reliability of MSI quantification analysis. Currently, SIMS is the most applicable ionization source for quantification,[107,108] while others are still reported as methodological progress, where traditional quantification methods or repetitions are used for validation.[109] The advantage of wide ion coverage could bring a rich harvest of chemical information, which not only profiles the sample more completely but also makes it more difficult to annotate the peaks and identify the ions. On the instrumentation side, high-resolution mass analyzers like Orbitrap and FT-ICR increase the mass resolution, making ions with the same nominal *m*/*z* more distinguishable,[32] while the addition of a second imaging modality could provide chemical interpretation from a different perspective. From the literature of the past few years, a continuation of the upward trend in the utilization of multimodal analysis could be easily observed.[110] The integration of different imaging modalities has also reached a new level to further increase chemical coverage and specificity. For example, image infusion was implemented to precisely register images from two modalities with drastically different resolutions, aiming to obtain or predict the chemical profile of single cells rather than pixels.[111,112] The MSI techniques are also incorporated within, usually called hybrid MSI, to complement the performance of a single ionization source.[113,114] On the data side, public authenticated databases have been providing a valid and updated reference for the identification of ions and affiliated compounds. But compared to all creations, what is known for now is merely a drop in the ocean.[115] Nevertheless, altruistic effort from the community for the development of packaged computational tools dedicated to biology has brought powerful brute force to save researchers from intensive labor.[116] The recent introduction of artificial intelligence (AI) has offered a unique opportunity for addressing the challenges of interpreting MSI data with ever-increasing size.[117,118] With more high-quality ground truth data created by researchers, AI models will be a critical tool for the annotation of the unknown. In the foreseeable future, MSI will only continue to provide more splendid pictures of microbial worlds, with non-stop progress of this technique and devotion from generations of fellow researchers.

References

1. S. J. B. Dunham, J. F. Ellis, B. Li and J. V. Sweedler, Mass spectrometry imaging of complex microbial communities, *Acc. Chem. Res.*, 2017, **50**(1), 96–104.
2. R. Castaing and G. Slodzian, Microanalysis by secondary ionic emission, *J. Microsc.*, 1962, **1**, 395–410.
3. R. M. Caprioli, Imaging mass spectrometry: Molecular microscopy for the new age of biology and medicine, *Proteomics*, 2016, **16**(11–12), 1607–1612.
4. R. D. Addie, B. Balluff, J. V. M. G. Bovée, H. Morreau and L. A. McDonnell, Current State and Future Challenges of Mass Spectrometry Imaging for Clinical Research, *Anal. Chem.*, 2015, **87**(13), 6426–6433.
5. B. Spengler, Mass Spectrometry Imaging of Biomolecular Information, *Anal. Chem.*, 2015, **87**(1), 64–82.
6. B. Li, S. J. B. Dunham, Y. Dong, S. Yoon, M. Zeng and J. V. Sweedler, Analytical capabilities of mass spectrometry imaging and its potential applications in food science, *Trends Food Sci. Technol.*, 2016, **47**, 50–63.
7. K. A. Maloof, A. N. Reinders and K. R. Tucker, Applications of mass spectrometry imaging in the environmental sciences, *Curr. Opin. Environ. Sci. Health*, 2020, **18**, 54–62.
8. D. S. McPhail, Applications of Secondary Ion Mass Spectrometry (SIMS) in materials science, *J. Mater. Sci.*, 2006, **41**(3), 873–903.
9. A. Nilsson, R. J. A. Goodwin, M. Shariatgorji, T. Vallianatou, P. J. H. Webborn and P. E. Andrén, Mass spectrometry imaging in drug development, *Anal. Chem.*, 2015, **87**(3), 1437–1455.

10. S. Schulz, M. Becker, M. R. Groseclose, S. Schadt and C. Hopf, Advanced MALDI mass spectrometry imaging in pharmaceutical research and drug development, *Curr. Opin. Biotechnol.*, 2019, **55**, 51–59.
11. J. B. Cliff, D. J. Gaspar, P. J. Bottomley and D. D. Myrold, Exploration of inorganic C and N assimilation by soil microbes with time-of-flight secondary ion mass spectrometry, *Appl. Environ. Microbiol.*, 2002, **68**(8), 4067–4073.
12. J. C. Vickerman, Molecular imaging and depth profiling by mass spectrometry—SIMS, MALDI or DESI?, *Analyst*, 2011, **136**(11), 2199–2217.
13. E. U. Brockmann, D. Steil, A. Bauwens, J. Soltwisch and K. Dreisewerd, Advanced Methods for MALDI-MS Imaging of the Chemical Communication in Microbial Communities, *Anal. Chem.*, 2019, **91**(23), 15081–15089.
14. J. Zhang, J. Brown, D. J. Scurr, A. Bullen, K. Maclellan-Gibson and P. Williams, *et al.*, Cryo-OrbiSIMS for 3D Molecular Imaging of a Bacterial Biofilm in Its Native State, *Anal. Chem.*, 2020, **92**(13), 9008–9015.
15. D. Gode and D. A. Volmer, Lipid imaging by mass spectrometry – a review, *Analyst*, 2013, **138**(5), 1289–1315.
16. J. Hanrieder, H. Zetterberg and K. Blennow, MALDI imaging mass spectrometry: neurochemical imaging of proteins and peptides, *Neuroproteomics*, Springer, 2019, pp. 179–197.
17. D. J. Ryan, J. M. Spraggins and R. M. Caprioli, Protein identification strategies in MALDI imaging mass spectrometry: a brief review, *Curr. Opin. Chem. Biol.*, 2019, **48**, 64–72.
18. L. Qin, Y. Zhang, Y. Liu, H. He, M. Han and Y. Li, *et al.*, Recent advances in matrix-assisted laser desorption/ionisation mass spectrometry imaging (MALDI-MSI) for in situ analysis of endogenous molecules in plants, *Phytochem. Anal.*, 2018, **29**(4), 351–364.
19. M. Shariatgorji, A. Nilsson, E. Fridjonsdottir, T. Vallianatou, P. Källback and L. Katan, *et al.*, Comprehensive mapping of neurotransmitter networks by MALDI–MS imaging, *Nat. Methods*, 2019, **16**(10), 1021–1028.
20. H. Liu, Y. Pan, C. Xiong, J. Han, X. Wang and J. Chen, *et al.*, Matrix-assisted laser desorption/ionization mass spectrometry imaging (MALDI MSI) for in situ analysis of endogenous small molecules in biological samples, *TrAC, Trends Anal Chem.*, 2022, 116809.
21. M. Kompauer, S. Heiles and B. Spengler, Atmospheric pressure MALDI mass spectrometry imaging of tissues and cells at 1.4-μm lateral resolution, *Nat. Methods*, 2017, **14**(1), 90–96.
22. M. Niehaus, J. Soltwisch, M. E. Belov and K. Dreisewerd, Transmission-mode MALDI-2 mass spectrometry imaging of cells and tissues at subcellular resolution, *Nat. Methods*, 2019, **16**(9), 925–931.
23. I. S. Gilmore, S. Heiles and C. L. Pieterse, Metabolic Imaging at the Single-Cell Scale: Recent Advances in Mass Spectrometry Imaging, *Annu. Rev. Anal. Chem.*, 2019, **12**, 201–224.

24. H. Liebl, Ion microprobe mass analyzer, *J. Appl. Phys.*, 1967, **38**(13), 5277–5283.
25. C. R. Anderton and L. J. Gamble, Secondary Ion Mass Spectrometry Imaging of Tissues, Cells, and Microbial Systems, *Microsc. Today*, 2016, **24**(2), 24–31.
26. J. D. Watrous and P. C. Dorrestein, Imaging mass spectrometry in microbiology, *Nat. Rev. Microbiol.*, 2011, **9**(9), 683–694.
27. J. C. Vickerman and N. Winograd, SIMS-A precursor and partner to contemporary mass spectrometry, *Int. J. Mass Spectrom.*, 2015, **377**(1), 568–579.
28. D. Gao, X. Huang and Y. Tao, A critical review of NanoSIMS in analysis of microbial metabolic activities at single-cell level, *Crit. Rev. Biotechnol.*, 2016, **36**(5), 884–890.
29. D. R. Ahlf, R. N. Masyuko, A. B. Hummon and P. W. Bohn, Correlated mass spectrometry imaging and confocal Raman microscopy for studies of three-dimensional cell culture sections, *Analyst*, 2014, **139**(18), 4578–4585.
30. C. Tominski, T. Lösekann-Behrens, A. Ruecker, N. Hagemann, S. Kleindienst and C. W. Mueller, *et al.*, Insights into carbon metabolism provided by fluorescence in situ hybridization-secondary ion mass spectrometry imaging of an autotrophic, nitrate-reducing, Fe(II)-oxidizing enrichment culture, *Appl. Environ. Microbiol.*, 2018, **84**(9), e02166-17.
31. E. J. Lanni, S. J. B. Dunham, P. Nemes, S. S. Rubakhin and J. V. Sweedler, Biomolecular imaging with a C_{60}-SIMS/MALDI dual ion source hybrid mass spectrometer: Instrumentation, matrix enhancement, and single cell analysis, *J. Am. Soc. Mass Spectrom.*, 2014, **25**(11), 1897–1907.
32. M. K. Passarelli, A. Pirkl, R. Moellers, D. Grinfeld, F. Kollmer and R. Havelund, *et al.*, The 3D OrbiSIMS – Label-free metabolic imaging with subcellular lateral resolution and high mass-resolving power, *Nat. Methods*, 2017, **14**(12), 1175–1183.
33. D. F. Smith, A. Kiss, F. E. Leach, E. W. Robinson, L. Paša-Tolić and R. M. A. Heeren, High mass accuracy and high mass resolving power FT-ICR secondary ion mass spectrometry for biological tissue imaging, *Anal. Bioanal. Chem.*, 2013, **405**(18), 6069–6076.
34. K. Schaepe, D. R. Bhandari, J. Werner, A. Henss, A. Pirkl and M. Kleine-Boymann, *et al.*, Imaging of Lipids in Native Human Bone Sections Using TOF–Secondary Ion Mass Spectrometry, Atmospheric Pressure Scanning Microprobe Matrix-Assisted Laser Desorption/Ionization Orbitrap Mass Spectrometry, and Orbitrap–Secondary Ion Mass Spectrometry, *Anal. Chem.*, 2018, **90**(15), 8856–8864.
35. D. Parrot, S. Papazian, D. Foil and D. Tasdemir, Imaging the Unimaginable: Desorption Electrospray Ionization - Imaging Mass Spectrometry (DESI-IMS) in Natural Product Research, *Planta Med.*, 2018, **84**(9–10), 584–593.

36. E. Claude, E. A. Jones and S. D. Pringle, DESI Mass Spectrometry Imaging (MSI), in *Imaging Mass Spectrometry: Methods and Protocols*, ed. L. M. Cole, Springer New York, New York, NY, 2017, pp. 65–75.
37. P. J. Roach, J. Laskin and A. Laskin, Nanospray desorption electrospray ionization: An ambient method for liquid-extraction surface sampling in mass spectrometry, *Analyst*, 2010, **135**(9), 2233–2236.
38. J. Watrous, P. Roach, B. Heath, T. Alexandrov, J. Laskin and P. C. Dorrestein, Metabolic profiling directly from the petri dish using nanospray desorption electrospray ionization imaging mass spectrometry, *Anal. Chem.*, 2013, **85**(21), 10385–10391.
39. J. Fang and P. C. Dorrestein, Emerging mass spectrometry techniques for the direct analysis of microbial colonies, *Curr. Opin. Microbiol.*, 2014, **19**(1), 120–129.
40. P. Nemes and A. Vertes, Laser Ablation Electrospray Ionization for Atmospheric Pressure, in Vivo, and Imaging Mass Spectrometry, *Anal. Chem.*, 2007, **79**(21), 8098–8106.
41. D. W. Etalo, C. Díez-Simón, R. C. H. de Vos and R. D. Hall, Laser Ablation Electrospray Ionization-Mass Spectrometry Imaging (LAESI-MS) for Spatially Resolved Plant Metabolomics, in *Plant Metabolomics: Methods and Protocols*, ed. C. António, Springer New York, New York, NY, 2018, pp. 253–267.
42. A. Vogel and V. Venugopalan, Mechanisms of Pulsed Laser Ablation of Biological Tissues, *Chem. Rev.*, 2003, **103**(2), 577–644.
43. B. Shrestha, C. M. Walsh, G. R. Boyce and P. Nemes, Microprobe MS Imaging of Live Tissues, Cells, and Bacterial Colonies Using LAESI, in *Advances in MALDI and Laser-Induced Soft Ionization Mass Spectrometry*, ed. R. Cramer, Springer International Publishing, Cham, 2016, pp. 149–167.
44. M. W. F. Nielen and T. A. Van Beek, Macroscopic and microscopic spatially-resolved analysis of food contaminants and constituents using laser-ablation electrospray ionization mass spectrometry imaging, *Anal. Bioanal. Chem.*, 2014, **406**(27), 6805–6815.
45. J. Balog, T. Szaniszlo, K.-C. Schaefer, J. Denes, A. Lopata and L. Godorhazy, *et al.*, Identification of Biological Tissues by Rapid Evaporative Ionization Mass Spectrometry, *Anal. Chem.*, 2010, **82**(17), 7343–7350.
46. O. Golf, N. Strittmatter, T. Karancsi, S. D. Pringle, A. V. M. Speller and A. Mroz, *et al.*, Rapid evaporative ionization mass spectrometry imaging platform for direct mapping from bulk tissue and bacterial growth media, *Anal. Chem.*, 2015, **87**(5), 2527–2534.
47. D. Simon and R. Oleschuk, The liquid micro junction-surface sampling probe (LMJ-SSP); A versatile ambient mass spectrometry interface, *Analyst*, 2021, **146**(21), 6365–6378.
48. C. L. Walton, M. Khalid, A. N. Bible, V. Kertesz, S. T. Retterer and J. Morrell-Falvey, *et al.*, In Situ Detection of Amino Acids from Bacterial Biofilms and Plant Root Exudates by Liquid Microjunction Surface-Sampling Probe Mass Spectrometry, *J. Am. Soc. Mass Spectrom.*, 2022, **33**(9), 1615–1625.

49. M. Lorenz, O. S. Ovchinnikova, V. Kertesz and G. J. Van Berkel, Controlled-Resonant Surface Tapping-Mode Scanning Probe Electrospray Ionization Mass Spectrometry Imaging, *Anal. Chem.*, 2014, **86**(6), 3146–3152.
50. R. J. A. Goodwin, Sample preparation for mass spectrometry imaging: Small mistakes can lead to big consequences, *J. Proteomics*, 2012, **75**(16), 4893–4911.
51. B. Li, T. J. Comi, T. Si, J. B. Dunham, J. V. Sweedler and S. J. B. Dunham, *et al.*, A one-step matrix application method for MALDI mass spectrometry imaging of bacterial colony biofilms, *J. Mass Spectrom.*, 2016, **51**(11), 1030–1035.
52. J. Y. Yang, V. V. Phelan, R. Simkovsky, J. D. Watrous, R. M. Trial and T. C. Fleming, *et al.*, Primer on Agar-Based Microbial Imaging Mass Spectrometry, *J. Bacteriol.*, 2012, **194**(22), 6023–6028.
53. J. E. Porter and L. M. Sanchez, MALDI Mass Spectrometry Imaging for Microbiology, *MALDI Mass Spectrometry Imaging*, 2021, pp. 291–320.
54. Y. Chen, D. Hu, L. Zhao, W. Tang and B. Li, Unraveling metabolic alterations in transgenic mouse model of Alzheimer's disease using MALDI MS imaging with 4-aminocinnoline-3-carboxamide matrix, *Anal. Chim. Acta*, 2021, 339337.
55. S. Vergeiner, L. Schafferer, H. Haas and T. Müller, Improved MALDI-TOF microbial mass spectrometry imaging by application of a dispersed solid matrix, *J. Am. Soc. Mass Spectrom.*, 2014, **25**(8), 1498–1501.
56. K. E. Zink, D. A. Tarnowski, M. J. Mandel and L. M. Sanchez, Optimization of a minimal sample preparation protocol for imaging mass spectrometry of unsectioned juvenile invertebrates, *J. Mass Spectrom.*, 2020, **55**(4), e4458.
57. E. Gemperline, H. A. Horn, K. Delaney, C. R. Currie and L. Li, Imaging with Mass Spectrometry of Bacteria on the Exoskeleton of Fungus-Growing Ants, *ACS Chem. Biol.*, 2017, **12**(8), 1980–1985.
58. B. J. Tyler, S. Rangaranjan, J. Möller, A. Beumer and H. F. Arlinghaus, TOF-SIMS imaging of chlorhexidine-digluconate transport in frozen hydrated biofilms of the fungus Candida albicans, *Appl. Surf. Sci.*, 2006, **252**(19), 6712–6715.
59. S. Vaidyanathan, J. S. Fletcher, R. Goodacre, N. P. Lockyer, J. Micklefield and J. C. Vickerman, Subsurface biomolecular imaging of Streptomyces coelicolor using secondary ion mass spectrometry, *Anal. Chem.*, 2008, **80**(6), 1942–1951.
60. H. Tian, D. A. Six, T. Krucker, J. A. Leeds and N. Winograd, Subcellular Chemical Imaging of Antibiotics in Single Bacteria Using C60-Secondary Ion Mass Spectrometry, *Anal. Chem.*, 2017, **89**(9), 5050–5057.
61. D. Debois, K. Hamze, V. Guérineau, J. P. Le Caër, I. B. Holland and P. Lopes, *et al.*, In situ localisation and quantification of surfactins in a Bacillus subtilis swarming community by imaging mass spectrometry, *Proteomics*, 2008, **8**(18), 3682–3691.

62. S. J. B. Dunham, T. J. Comi, K. Ko, B. Li, N. F. Baig and N. Morales-Soto, *et al.*, Metal-assisted polyatomic SIMS and laser desorption/ionization for enhanced small molecule imaging of bacterial biofilms, *Biointerphases*, 2016, **11**(2), 02A325.
63. N. Morales-Soto, S. J. B. Dunham, N. F. Baig, J. F. Ellis, C. S. Madukoma and P. W. Bohn, *et al.*, Spatially dependent alkyl quinolone signaling responses to antibiotics in Pseudomonas aeruginosa swarms, *J. Biol. Chem.*, 2018, **293**(24), 9544–9552.
64. S. J. B. Dunham, J. F. Ellis, N. F. Baig, N. Morales-Soto, T. Cao and J. D. Shrout, *et al.*, Quantitative SIMS Imaging of Agar-Based Microbial Communities, *Anal. Chem.*, 2018, **90**(9), 5654–5663.
65. A. M. Abdel-Mawgoud, F. Lépine and E. Déziel, Rhamnolipids: diversity of structures, microbial origins and roles, *Appl. Microbiol. Biotechnol.*, 2010, **86**(5), 1323–1336.
66. J. Watrous, N. Hendricks, M. Meehan and P. C. Dorrestein, Capturing bacterial metabolic exchange using thin film desorption electrospray ionization-imaging mass spectrometry, *Anal. Chem.*, 2010, **82**(5), 1598–1600.
67. M. Figueroa, A. K. Jarmusch, H. A. Raja, T. El-Elimat, J. S. Kavanaugh and A. R. Horswill, *et al.*, Polyhydroxyanthraquinones as quorum sensing inhibitors from the guttates of Penicillium restrictum and their analysis by desorption electrospray ionization mass spectrometry, *J. Nat. Prod.*, 2014, **77**(6), 1351–1358.
68. B. Li, C. Knudsen, N. K. Hansen, K. Jørgensen, R. Kannangara and S. Bak, *et al.*, Visualizing metabolite distribution and enzymatic conversion in plant tissues by desorption electrospray ionization mass spectrometry imaging, *Plant J.*, 2013, **74**(6), 1059–1071.
69. E. C. Cabral, M. F. Mirabelli, C. J. Perez and D. R. Ifa, Blotting assisted by heating and solvent extraction for DESI-MS imaging, *J. Am. Soc. Mass Spectrom.*, 2013, **24**(6), 956–965.
70. A. Tata, C. J. Perez, M. O. Ore, D. Lostun, A. Passas and S. Morin, *et al.*, Evaluation of imprint DESI-MS substrates for the analysis of fungal metabolites, *RSC Adv.*, 2015, **5**(92), 75458–75464.
71. C. F. F. Angolini, P. H. Vendramini, F. D. S. Araújo, W. L. Araújo, R. Augusti and M. N. Eberlin, *et al.*, Direct Protocol for Ambient Mass Spectrometry Imaging on Agar Culture, *Anal. Chem.*, 2015, **87**(13), 6925–6930.
72. V. P. Sica, H. A. Raja, T. El-Elimat and N. H. Oberlies, Mass spectrometry imaging of secondary metabolites directly on fungal cultures, *RSC Adv.*, 2014, **4**(108), 63221–63227.
73. B. M. Ellis, C. N. Fischer, L. B. Martin, B. O. Bachmann and J. A. McLean, Spatiochemically Profiling Microbial Interactions with Membrane Scaffolded Desorption Electrospray Ionization-Ion Mobility-Imaging Mass Spectrometry and Unsupervised Segmentation, *Anal. Chem.*, 2019, **91**(21), 13703–13711.
74. B. S. Heath, M. J. Marshall and J. Laskin, The Characterization of Living Bacterial Colonies Using Nanospray Desorption Electrospray Ionization Mass Spectrometry, *Methods in Molecular Biology*, 2014, pp. 199–208.

75. I. Lanekoff, O. Geydebrekht, G. E. Pinchuk, A. E. Konopka and J. Laskin, Spatially resolved analysis of glycolipids and metabolites in living Synechococcus sp. PCC 7002 using nanospray desorption electrospray ionization, *Analyst*, 2013, **138**(7), 1971.
76. S. N. Nguyen, A. V. Liyu, R. K. Chu, C. R. Anderton and J. Laskin, Constant-distance mode nanospray desorption electrospray ionization mass spectrometry imaging of biological samples with complex topography, *Anal. Chem.*, 2017, **89**(2), 1131–1137.
77. F. Araújo, R. Vieira, E. Molano, H. Máximo, R. Dalio and P. Vendramini, *et al.*, Desorption electrospray ionization mass spectrometry imaging reveals chemical defense of Burkholderia seminalis against cacao pathogens, *RSC Adv.*, 2017, 7(48), 29953–29958.
78. J. Errington and L. T. van der Aart, Microbe Profile: Bacillus subtilis: Model organism for cellular development, and industrial workhorse, *Microbiology*, 2020, **166**(5), 425.
79. Z. D. Blount, The natural history of model organisms: The unexhausted potential of E. coli, *eLife*, 2015, **4**, e05826.
80. T. Si, B. Li, K. Zhang, Y. Xu, H. Zhao and J. V. Sweedler, Characterization of bacillus subtilis colony biofilms via mass spectrometry and fluorescence imaging, *J. Proteome Res.*, 2016, **15**(6), 1955–1962. https://pubs.acs.org/doi/10.1021/acs.jproteome.6b00127. Further permissions related to the material excerpted should be directed to the ACS.
81. A. R. Condren, L. J. Kahl, G. Boelter, G. Kritikos, M. Banzhaf and L. E. P. Dietrich, *et al.*, Biofilm Inhibitor Taurolithocholic Acid Alters Colony Morphology, Specialized Metabolism, and Virulence of Pseudomonas aeruginosa, *ACS Infect. Dis.*, 2020, **6**(4), 603–612.
82. L. Wörmer, N. Gajendra, F. Schubotz, E. D. Matys, T. W. Evans and R. E. Summons, *et al.*, A micrometer-scale snapshot on phototroph spatial distributions: mass spectrometry imaging of microbial mats in Octopus Spring, Yellowstone National Park, *Geobiology*, 2020, **18**(6), 742–759.
83. A. E. Dekas, R. S. Poretsky and V. J. Orphan, Deep-Sea archaea fix and share nitrogen in methane-consuming microbial consortia, *Science*, 2009, **326**(5951), 422–426.
84. W. J. Moree, V. V. Phelan, C.-H. Wu, N. Bandeira, D. S. Cornett and B. M. Duggan, *et al.*, Interkingdom metabolic transformations captured by microbial imaging mass spectrometry, *Proc. Natl. Acad. Sci. U. S. A.*, 2012, **109**(34), 13811–13816.
85. C. F. Michelsen, S. M. H. Khademi, H. K. Johansen, H. Ingmer, P. C. Dorrestein and L. Jelsbak, Evolution of metabolic divergence in Pseudomonas aeruginosa during long-term infection facilitates a proto-cooperative interspecies interaction, *ISME J.*, 2016, **10**(6), 1330–1336.
86. T. Si, B. Li, T. J. Comi, Y. Wu, P. Hu and Y. Wu, *et al.*, Profiling of Microbial Colonies for High-Throughput Engineering of Multistep Enzymatic Reactions via Optically Guided Matrix-Assisted Laser

Desorption/Ionization Mass Spectrometry, *JACS*, 2017, **139**(36), 12466–12473.
87. B. M. Ellis, P. K. Babele, J. C. May, C. H. Johnson, B. F. Pfleger and J. D. Young, *et al.*, Accelerating strain phenotyping with desorption electrospray ionization-imaging mass spectrometry and untargeted analysis of intact microbial colonies, *Proc. Natl. Acad. Sci. U. S. A.*, 2021, **118**(49), e2109633118.
88. D. Petras, A. K. Jarmusch and P. C. Dorrestein, From single cells to our planet—recent advances in using mass spectrometry for spatially resolved metabolomics, *Curr. Opin. Chem. Biol.*, 2017, **36**, 24–31.
89. Y. Dong, B. Li, S. Malitsky, I. Rogachev, A. Aharoni and F. Kaftan, *et al.*, Sample preparation for mass spectrometry imaging of plant tissues: A review, *Front. Plant Sci.*, 2016, **7**, 60.
90. M. Zimmermann, S. Escrig, T. Hübschmann, M. K. Kirf, A. Brand and R. F. Inglis, *et al.*, Phenotypic heterogeneity in metabolic traits among single cells of a rare bacterial species in its natural environment quantified with a combination of flow cell sorting and NanoSIMS, *Front. Microbiol.*, 2015, **6**, 1–11.
91. M. Philippi, K. Kitzinger, J. S. Berg, B. Tschitschko, A. T. Kidane and S. Littmann, *et al.*, Purple sulfur bacteria fix N_2 via molybdenum-nitrogenase in a low molybdenum Proterozoic ocean analogue, *Nat. Commun.*, 2021, **12**(1), 1–12.
92. Y. Xiong, R. Guilbaud, C. L. Peacock, R. P. Cox, D. E. Canfield and M. D. Krom, *et al.*, Phosphorus cycling in Lake Cadagno, Switzerland: a low sulfate euxinic ocean analogue, *Geochim. Cosmochim. Acta*, 2019, **251**, 116–135.
93. N. J. Schoffelen, W. Mohr, T. G. Ferdelman, S. Littmann, J. Duerschlag and M. V. Zubkov, *et al.*, Single-cell imaging of phosphorus uptake shows that key harmful algae rely on different phosphorus sources for growth, *Sci. Rep.*, 2018, **8**(1), 1–13.
94. R. Sekine, K. L. Moore, M. Matzke, P. Vallotton, H. Jiang and G. M. Hughes, *et al.*, Complementary Imaging of Silver Nanoparticle Interactions with Green Algae: Dark-Field Microscopy, Electron Microscopy, and Nanoscale Secondary Ion Mass Spectrometry, *ACS Nano*, 2017, **11**(11), 10894–10902.
95. P. Rajapaksha, A. Elbourne, S. Gangadoo, R. Brown, D. Cozzolino and J. Chapman, A review of methods for the detection of pathogenic microorganisms, *Analyst*, 2019, **144**(2), 396–411.
96. J. G. de Moraes Pontes, P. H. Vendramini, L. S. Fernandes, F. H. de Souza, E. J. Pilau and M. N. Eberlin, *et al.*, Mass spectrometry imaging as a potential technique for diagnostic of Huanglongbing disease using fast and simple sample preparation, *Sci. Rep.*, 2020, **10**(1), 1–12.
97. G. da Silva Lima, G. Franco dos Santos, R. R. F. Ramalho, D. V. A. de Aguiar, J. V. Roque and L. I. L. Maciel, *et al.*, Laser ablation electrospray ionization mass spectrometry imaging as a new tool for accessing

patulin diffusion in mold-infected fruits, *Food Chem.*, 2022, **373**(Part B), 131490.

98. J. H. Costa, J. M. Bazioli, E. de Vilhena Araújo, P. H. Vendramini, M. C. de Freitas Porto and M. N. Eberlin, *et al.*, Monitoring indole alkaloid production by Penicillium digitatum during infection process in citrus by Mass Spectrometry Imaging and molecular networking, *Fungal Biol.*, 2019, **123**(8), 594–600.
99. G. Schleyer, N. Shahaf, C. Ziv, Y. Dong, R. A. Meoded and E. J. N. Helfrich, *et al.*, In plaque-mass spectrometry imaging of a bloom-forming alga during viral infection reveals a metabolic shift towards odd-chain fatty acid lipids, *Nat. Microbiol.*, 2019, **4**(3), 527–538.
100. W. J. Perry, J. M. Spraggins, J. R. Sheldon, C. M. Grunenwald, D. E. Heinrichs and J. E. Cassat, *et al.*, Staphylococcus aureus exhibits heterogeneous siderophore production within the vertebrate host, *Proc. Natl. Acad. Sci. U. S. A.*, 2019, **116**(44), 21980–21982.
101. E. Torres-sangiao, C. Leal Rodriguez and C. García-riestra, Application and perspectives of maldi–tof mass spectrometry in clinical microbiology laboratories, *Microorganisms*, 2021, **9**(7), 1–19.
102. P. M. Vaysse, R. M. A. Heeren, T. Porta and B. Balluff, Mass spectrometry imaging for clinical research-latest developments, applications, and current limitations, *Analyst*, 2017, **142**(15), 2690–2712.
103. *The Use of Mass Spectrometry Technology (MALDI-TOF) in Clinical Microbiology,* ed. F. Cobo, Academic Press, 2018.
104. M. Kostrzewa, Application of the MALDI Biotyper to clinical microbiology: progress and potential, *Expert Rev. Proteomics*, 2018, **15**(3), 193–202.
105. T. Hamerly, J. A. Everett, N. Paris, S. T. Fisher, A. Karunamurthy and G. A. James, *et al.*, Detection of Pseudomonas aeruginosa biomarkers from thermally injured mice in situ using imaging mass spectrometry, *Anal. Biochem.*, 2017, **539**, 144–148.
106. N. Wang, J. P. Sarathy, M. Zimmerman, F. Kaya, H. Wang and V. Dartois, *et al.*, On-Slide Heat Sterilization Enables Mass Spectrometry Imaging of Tissue Infected with High-Threat Pathogens Outside of Biocontainment: A Study Directed at Mycobacterium tuberculosis, *J. Am. Soc. Mass Spectrom.*, 2021, **32**(11), 2664–2674.
107. K. Wittmaack, Aspects of quantitative secondary ion mass spectrometry, *Nucl. Instrum. Methods*, 1980, **168**(1–3), 343–356.
108. H. Werner, Quantitative secondary ion mass spectrometry: A review, *Surf. Interface Anal.*, 1980, **2**(2), 56–74.
109. D. Unsihuay, D. Mesa Sanchez and J. Laskin, Quantitative Mass Spectrometry Imaging of Biological Systems, *Annu. Rev. Phys. Chem.*, 2020, **72**, 307–329.
110. E. K. Neumann, K. V. Djambazova, R. M. Caprioli and J. M. Spraggins, Multimodal Imaging Mass Spectrometry: Next Generation Molecular Mapping in Biology and Medicine, *J. Am. Soc. Mass Spectrom.*, 2020, **31**(12), 2401–2415.

111. R. Van de Plas, J. Yang, J. Spraggins and R. M. Caprioli, Image fusion of mass spectrometry and microscopy: A multimodality paradigm for molecular tissue mapping, *Nat. Methods*, 2015, **12**(4), 366–372.
112. O. Ryabchykov, J. Popp and T. Bocklitz, Fusion of MALDI spectrometric imaging and Raman spectroscopic data for the analysis of biological samples, *Front. Chem.*, 2018, **6**, 1–10.
113. E. U. Brockmann, A. Potthoff, S. Tortorella, J. Soltwisch and K. Dreisewerd, Infrared MALDI Mass Spectrometry with Laser-Induced Postionization for Imaging of Bacterial Colonies, *J. Am. Soc. Mass Spectrom.*, 2021, **32**(4), 1053–1064.
114. S. Van Nuffel, M. Quatredeniers, A. Pirkl, J. Zakel, J. P. Le Caer and N. Elie, *et al.*, Multimodal Imaging Mass Spectrometry to Identify Markers of Pulmonary Arterial Hypertension in Human Lung Tissue Using MALDI-ToF, ToF-SIMS, and Hybrid SIMS, *Anal. Chem.*, 2020, **92**(17), 12079–12087.
115. T. Alexandrov, Spatial Metabolomics and Imaging Mass Spectrometry in the Age of Artificial Intelligence, *Annu. Rev. Biomed. Data Sci.*, 2020, **3**(1), 61–87.
116. G. Guo, M. Papanicolaou, N. J. Demarais, Z. Wang, K. L. Schey and P. Timpson, *et al.*, Automated annotation and visualisation of high-resolution spatial proteomic mass spectrometry imaging data using HIT-MAP, *Nat. Commun.*, 2021, **12**(1), 1–16.
117. N. Verbeeck, R. M. Caprioli and R. Van de Plas, Unsupervised machine learning for exploratory data analysis in imaging mass spectrometry, *Mass Spectrom. Rev.*, 2020, **39**(3), 245–291.
118. X. Tian, Z. Zou and Z. Yang, Extract Metabolomic Information from Mass Spectrometry Images Using Advanced Data Analysis, *Mass Spectrometry Imaging of Small Molecules*, Springer, 2022, pp. 253–272.

CHAPTER 12

Identification of Microorganisms by Mass Spectrometry in Clinical Microbiology Laboratories

QIN QIN,* YUN LIU, YUXIANG WAN AND HAIFENG QIN

Shanghai Changhai Hospital, Changhai Road 168, Shanghai 200433, China
*Email: qinq78@163.com

Identifying the pathogen of infection is an important part in the accurate diagnosis of infectious diseases, in which clinical microbiology laboratories play an important role. Common diagnostic methods for infectious pathogens include traditional microbial culture (gold standard), microscopy, antigen–antibody testing, and evolving molecular biology techniques. Early and rapid pathogen diagnosis, with timely antibiotic application, can effectively reduce mortality and improve patient prognosis.[1,2] Traditional microbial culture has always been limited by the lengthy turnaround time. Nowadays, most laboratories mainly use biochemical identification methods for colonies, which usually require overnight culture to obtain the results of biochemical reactions. For some special microorganisms, this traditional biochemical identification method cannot obtain accurate results, which need to be confirmed by molecular diagnostic techniques, such as 16S rDNA sequencing. Sequencing technology has the advantages of high efficiency and accuracy for unknown and peculiar pathogens, but it is difficult to

New Developments in Mass Spectrometry No. 13
Detection and Analysis of Microorganisms by Mass Spectrometry
Edited by Liang Qiao and Jia Yi

Published by the Royal Society of Chemistry, www.rsc.org

popularize its high technical requirements in routine clinical microbiology laboratories.

Matrix-assisted laser desorption/ionization time-of-flight mass spectrometry (MALDI-TOF MS) is a rapid microbial diagnosis method developed on the basis of MS technology, which has been widely used in clinical microbiology laboratories all over the world,[3–5] with fast and efficient clinical application value.

12.1 Principles of MALDI-TOF MS for Microbial Identification

The basic principle of MALDI-TOF MS for microbial identification is shown in Figure 12.1. A sample of microorganisms to be identified is

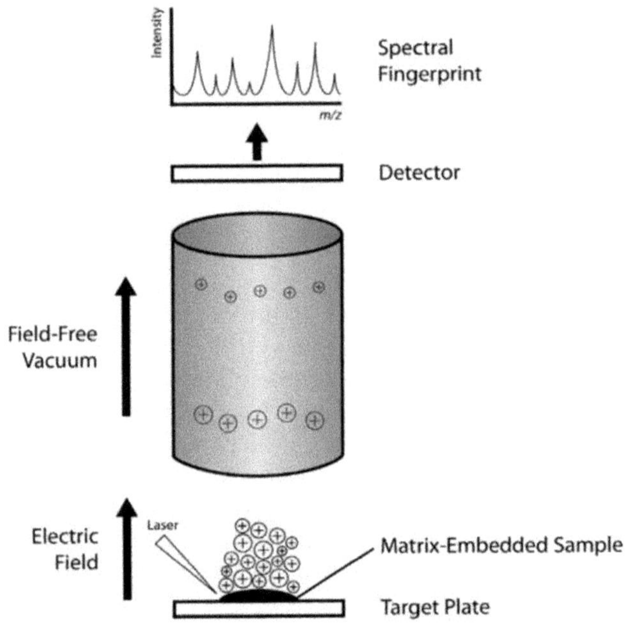

Figure 12.1 Principle of MALDI-TOF MS for microbial identification. Microbial samples are deposited on a conductive metallic target plate and overlaid with a matrix solution composed of an organic acid. Once placed in the instrument, the samples are ionized with short laser pulses followed by acceleration of variably charged particles through an electric field. The amount of time it takes for each particle to pass through a field-free tube, or time of flight (TOF), depends on its mass and charge and is measured using a detector. An organism-specific spectral fingerprint is generated and compared with thousands of spectral profiles in a database to obtain an identification of the organism (Courtesy of Wigington K., Calgary, AB). Reproduced from ref. 6 with permission from Elsevier, Copyright 2013.

deposited on the target plate, and a matrix solution is added to co-crystallize with the sample and lyse the organism.[7,8] For microorganisms that are difficult-to-lyse, pretreatment with strong organic acids or mechanical lysis methods are usually necessary.[9–11] The target plate with the prepared sample is placed in the instrument, and the sample is ionized with short laser pulses to form gas-phase ions with minimal fragmentation (*i.e.*, soft ionization), and the particles are then accelerated through a vacuum tube by an electric field. Particles with different masses and charges take different times to reach the detector, and the detection of these protein signals yields a characteristic spectral fingerprint. This technique can detect proteins ranging from 2 kDa to 20 kDa, or even larger proteins when specific matrices are applied. The main proteins detected by MALDI-TOF MS are ribosomal proteins, as well as some highly abundant cytoplasmic proteins (such as DNA binding proteins and cold shock proteins).[12–14]

Finally, the spectrum of the unknown microorganism is compared with the reference database through the software system, so as to obtain the identification result of the detected microorganism. Since the spectrum obtained by MALDI-TOF MS usually does not completely coincide with the existing spectrum in the database, the software will match a 'value' for each identification result according to its similarity, and judge the reliability of the identification according to the score.

12.2 Applications of MALDI-TOF MS in Clinical Microbiology Laboratories

As shown in Figure 12.2, the application process of MALDI-TOF MS in clinical microbiology laboratories mainly includes two parts: sample preparation and MS detection and analysis. The sample preparation process includes three methods: direct identification of independent colonies obtained after culture, chemical or mechanical pretreatment of colonies to obtain proteins for identification, and direct identification of clinical samples. At present, MALDI-TOF MS can not only be used for microbial species identification, but also for detection of different antibiotic resistance mechanisms, bacterial strain typing, and virulence marker detection. All of these suggest that in the future MALDI-TOF MS, for use as microbiological diagnostic techniques, will be further revolutionized in many ways.

12.2.1 Identification of Microbial Species

To a large extent, the identification of microorganisms by MALDI-TOF MS depends on two key factors: one is the processing method of microbial samples before analysis; the other is the quality and accuracy of the standard database to be compared. The continuous improvement and perfection of these two factors can promote and enhance the clinical application prospects of MS identification.

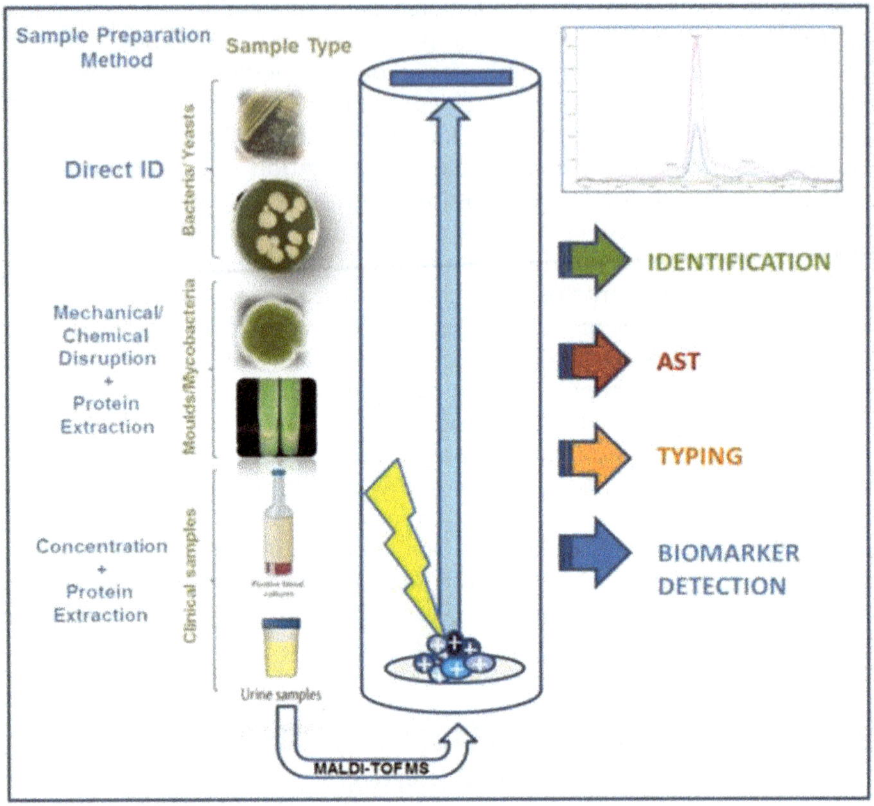

Figure 12.2 Sample types that can be analyzed by MALDI-TOF and the processing method recommended. Reproduced from ref. 56 with permission from Elsevier, Copyright 2021.

12.2.1.1 *Identification of Bacteria*

The routine identification of bacterial isolates by MALDI-TOF MS has been widely used and optimized until now; with fast identification and high accuracy, it can reliably identify most bacterial species commonly found in clinical microbiology laboratories.[9]

Currently, the use of MALDI-TOF for the identification of Gram-negative bacteria is considered to be very accurate. The most common genera in the *Enterobacteriaceae* (*Citrobacter*, *Enterobacter*, *Escherichia*, *Klebsiella*, *Morganella*, *Proteus*, *Salmonella*, and *Serratia*) have been extensively demonstrated to be well diagnosed by MALDI-TOF.[5,15] However, MALDI-TOF can mistake *Shigella* for *Escherichia coli*, and it has a limited ability to further differentiate between *Citrobacter freundii*, *Enterobacter cloacae*, *Klebsiella pneumoniae*, *Klebsiella oxytoca* complex, and *Salmonella enterica* serotypes.[5,16]

MALDI-TOF can accurately identify common non-fermentative bacteria in clinical practice (*Pseudomonas aeruginosa*, *Stenotrophomonas maltophilia*, *Moraxella catarrhalis*), and studies have shown that its accuracy rate can reach 93.6–96.6%.[17] However, due to the high homology displayed by species within the same complex and genus, some other genera of non-fermenting bacteria are identified only at the complex level (such as *Burkholderia cepacia* and *Acinetobacter baumannii* complex) or at the genus level (*Achromobacter* spp., *Chryseobacterium* spp. or *Ralstonia* spp).

With the continuous improvement and update of the database, some other fastidious Gram-negative bacteria have also been successfully identified, such as the HACEK group, whose species-level accuracy can reach 100%.[18] The identification accuracy of *Neisseria* at the genus level based on commercial databases can reach more than 90%, but the identification accuracy at the species level needs to be improved.[18]

The identification of Gram-positive bacteria presents a challenge to MALDI-TOF, due to their complex cell wall composition and the close affinities of major species within the genus. So far, only some species have been reliably resolved to the complex level, such as *Streptococcus pneumoniae* and *Streptococcus mitis* complex, and the identification of *Streptococcus* at the species level is a challenging task. In a MALDI-TOF MS analysis of 659 Gram-positive bacteria (including *Nocardia*, *Listeria*, *Actinomyces*, and *Bacillus*), which were difficult to identify by conventional identification methods,[19] 98.5% of those isolates were correctly identified to the species level (excluding *Listeria*). Most *Listeria* (93%) can only be identified at the genus level. In addition, MALDI-TOF MS was able to distinguish *Corynebacterium* species (*C. diphtheriae*, *C. pseudotuberculosis*, *C. ulcerans*), which appear to be closely related. As the library is continuously updated, MALDI-TOF has been able to provide species-level accuracy, which is similar to molecular techniques for anaerobic bacteria, in a faster and more cost-effective way.[20]

Due to the mycoacid-rich cell wall and infectivity of *Mycobacterium tuberculosis* (MTB), the pretreatment of MTB usually involves heat killing and protein extraction steps. Commercial instrument manufacturers recommend a combo solution of heating, ethanol treatment, and mechanical cracking. Although MALDI-TOF MS can recognize members of the MTB complex, it cannot identify them to the species level. The discrimination of species within the MTB complex remains one of the main challenges to overcome for MALDI-TOF. As for the identification of non-tuberculous mycobacteria, standardized pretreatment methods significantly improved the reliability of species-level identification. For example, the "*in vitro* diagnostics" (IVD) program developed by Bruker Daltonics is based on the mechanical destruction of bacterial bodies using crystal beads, followed by the extraction of proteins with formic acid and acetonitrile. Heat-inactivated mycobacterial cells can also be treated with ultrasonic or high-efficiency beating procedures, all of which provide highly reliable results at the species level. In addition, the identification of species belonging to complex groups, such as the *Mycobacterium avium* complex, has been successfully achieved by detection of species-specific marker peaks.[21]

12.2.1.2 Identification of Yeast and Filamentous Fungi

Routine identification of yeast by MALDI-TOF is well established in clinical laboratories.[22] Although many clinical laboratories still retain the step of formic acid treatment to extract proteins in the identification process of yeast, reliable results can be obtained directly without pretreatment for common genera *Candida* and *Cryptococcus*. Also, MALDI-TOF MS is superior to traditional methods in identifying *Candida krusei* and *Candida dubliniensis*.[23]

Filamentous fungi, similar to mycobacteria, require pre-disruption of the cell wall, and because filamentous fungi are more complex than bacteria, the pattern of expressed proteins changes over time. In addition, the lack of a comprehensive commercial database also contributes to the difficulty of identification. To date, although similar combinations of mechanical and chemical wall breaking have been applied successfully,[24] no standardized method has been established for identifying filamentous fungi. Recently, a database of more than 11 000 reference spectra of 938 clinical fungi has been established for online comparison of fungal strain spectra, which have been shown to correctly identify *Aspergillus*, *Mucor*, *Microsporum*, and other clinical important species,[25] to make it more feasible for clinical microbiology laboratories to achieve routine identification of filamentous fungi.

12.2.1.3 Direct Identification of Microorganisms in Clinical Specimens

The direct identification of microorganisms from clinical samples can greatly shorten the TAT time with great clinical significance. MALDI-TOF has realized the direct identification of microorganisms from blood culture samples. In addition, two sample types (urine and cerebrospinal fluid) have also been tried.

Pretreatment of clinical samples is an important part of MS identification to concentrate pathogens and remove other interfering cells and interfering substances. Table 12.1 lists pretreatment methods for blood culture and other body fluid samples. There are three main methods. The first method involves differential centrifugation, which removes possible interfering substances in blood culture by low-speed (1000 rpm) centrifugation, and then high-speed (14 000 rpm) centrifugation to concentrate the microorganisms to be detected. Microbial pellets can be added directly to MALDI target plates or subjected to protein extraction followed by MALDI-TOF analysis. The second method involves inoculation of blood culture on Columbia agar medium, incubation for 3–4 hours, and transfer of the

Table 12.1 Different methods applied for the direct identification of microorganisms from blood cultures and other clinical samples. Reproduced from ref. 56 with permission from Elsevier, Copyright 2021.

Direct identification of microorganisms from clinical samples using MALDI-TOF		
Differential centrifugation	Short-incubation	Commercial kits
– 10 min centrifugation at 1000 rpm	– Add some drops from the blood culture broth to a Columbia agar plate	MBT SEPSITYPER, Bruker Daltonics
– Recover the supernatant	– Do not spread	VITEK® MS Blood Culture
– 2 min centrifugation at 14 000 rpm	– Incubate during 3–4 h at 36 °C	Kit, bioMérieux
– Transfer the pellet to the MALDI target plate or subject it to protein extraction	– Recover the microorganism from the plate and transfer it to the MALDI target plate	Rapid BACpro® II Nittobo, Ltd

colonies for MS detection.[26] The third method is the direct use of special commercial kits, which can easily standardize and increase the identification power at the species and genus levels.[19] A limitation of these methods is their inability to reliably identify multi-bacterial co-infections.[27,28] Although clinical samples such as urine and cerebrospinal fluid (CSF) samples have been successfully identified by MALDI-TOF, they still face many difficulties. The main reasons are the low concentrations of pathogens ($<1.5 \times 10^5$–10^6) in such samples, insufficient sample size, the presence of other substances that may interfere with the identification, and the mixing of multiple bacteria.[28]

12.2.2 Drug Resistance/Resistance Mechanism Detection

12.2.2.1 Antibiotic Sensitivity Test Based on MALDI-TOF MS

Some researchers have developed a method to determine the resistance/susceptibility of *Bacillus anthracis* to streptomycin.[29] In this method, bacteria are cultured in two stages (without antibiotics and isotopes, with antibiotics and isotopes), and drug-resistant strains can continue to grow in medium containing antibiotics and isotopes. The mass spectrum of those drug-resistant strains contains shifted peaks resulting from the addition of stable isotopes to bacterial proteins. The same principle was used to detect the resistance of strains by identifying the peak shift between two phases (with and without antibiotics). The resistance of *P. aeruginosa* to meropenem, tobramycin, and ciprofloxacin was also validated using a medium supplemented with isotopically labeled lysine.[30] The method could theoretically be standardized for other antibiotics and could potentially lead to the development of an algorithm to correlate drug catalytic rates with minimum inhibitory concentrations (MICs). In addition, some researchers tried another method:[31] bacteria were grown with and without antibiotics, and after short-term incubation (1 h), MALDI-TOF MS spectra were obtained using an internal standard (RNase B). If the antibiotic concentration is set to the first concentration above the clinical breakpoint MIC, this semi-quantitative analysis can determine bacterial growth in the presence or absence of antibiotics, and thus antibiotic resistance. This method reduces the treatment time from 6 hours to 3 hours, which is very important for the clinical management of critically ill patients infected with carbapenem-resistant strains. On this basis, some researchers modified the method and developed a "microdroplet growth assay",[32] in which microorganisms (6 μL) were directly dropped onto the target plate and incubated for 4–5 hours. After removing the excess liquid, the spots were covered with matrix and then analyzed using MALDI-TOF MS. Low peak intensities were observed in the spectra obtained due to the inhibition of bacterial growth, which indicated that the strains were drug resistant. This method can detect the resistance of all drug-resistant strains, but how to standardize the antibiotic incubation time is a problem that must be solved.

All of the assays mentioned earlier, however, are in the proof-of-concept stage. If they are to be widely promoted and applied in clinical laboratories, further improvement and standardization are required.

12.2.2.2 Detection of Resistance Mechanisms

One of the most wide applications of MALDI-TOF MS for the detection of resistance mechanisms is the detection of β-lactamases and carbapenemases by the degree of hydrolysis of bacteria resistant to different antibiotics.[33] If the antibiotic is hydrolyzed, the peak pattern of the antibiotic changes relative to the natural antibiotic; in this way the detected spectral results can be compared to positive and negative controls. If the spectrum has the same peak shape as that of the positive control, this indicates that the antibiotic is hydrolyzed and the bacteria tested carry β-lactamases and/or carbapenemases. If the spectrum has the same peak shape as the negative control, this indicates that the antibiotic is not hydrolyzed and the bacteria do not carry β-lactamases and/or carbapenemases. The basic operation method is as follows: mix the bacteria to be tested with the antibiotic buffer, shake and culture at 37 °C, after incubation (30 min–3 h), centrifuge the mixture, and take the supernatant and place it on the target plate. A matrix can be added after the drying process for MALDI-TOF analysis. The detection effects obtained by this method using different antibiotics are inconsistent. For example, in positive hydrolysis experiments with bacteria producing extended-spectrum β-lactamase (ESBL) or AmpC-type β-lactamases, cefotaxime, ceftazidime, ceftriaxone, cefpodoxime, and cefepime showed differences.[34] Cefepime has low specificity and is prone to false positives; ceftriaxone has the highest sensitivity and specificity and the shortest detection time; for cefotaxime and cefpodoxime, the sensitivity is above 90%, but the recommended incubation time is 1 h; for ceftazidime the sensitivity is less than 90%, and the recommended incubation time is 3 hours. For carbapenems, the sensitivity of ertapenem and meropenem is about 90–95%, and the recommended incubation time is about 2 hours. For imipenem, the reduction in its mass peak relative to the internal standard can be semi-quantitatively determined by the internal standard in the matrix.[19]

In addition, the resistance mechanism can also be judged by the detection of specific peaks in MS. Some studies have distinguished between MSSA and MRSA strains by detecting a small molecule polypeptide, phenol-soluble regulatory protein (PSM), encoded on some staphylococcal chromosome mec (SCCmec).[35] *K. pneumoniae* carbapenemase (KPC)-producing *K. pneumoniae* was detected, which was indicated by the peak observed at 11 109 Da associated with protein p019 expression.[36] By using MALDI-TOF to detect lipid A, as well as its modified form with an added phosphoethanolamine group, the resistance of bacteria to colistin can be detected, wherein the resistance was mediated by the mcr-1 gene or not.[37] Other researchers have been able to identify specific mass spectrometry peaks from *Bacteroides fragilis* carrying the cfiA resistance gene, as well as from *Enterococcus faecium* carrying the vanA resistance gene.[38] Altered or lost expression of porins is associated with high levels of bacterial resistance. We can detect this resistance mechanism through special extraction and analysis of porins.[39] This method is suitable not only for bacteria, but also for mycobacteria and fungi.[40,41] It has high sensitivity and specificity, but its cumbersome operation hinders clinical application.

12.2.3 Microbiological Typing

Microbial typing has important epidemiological significance for nosocomial infection control, and using MALDI-TOF as a typing tool provides a new detection method for clinical practice.

Recently, MALDI-TOF, in comparison with pulsed-field gel electrophoresis (PFGE), has correctly achieved the typing of *E. coli*.[42] In addition, MALDI-TOF performed accurate and rapid identification of 5 high-risk clones of *Pseudomonas aeruginosa*: ST111, ST175, ST235, ST253, and ST395 serial types.[43] Typing of *C. difficile* was achieved by the MALDI-TOF method, based on whole cell high molecular weight (HMW) protein profiles (mass range 30–50 kDa).[44] Kumar *et al.* and Lartigue *et al.* used MALDI-TOF MS fingerprinting to quickly correctly type A, B, C, and G streptococci and subspecies A and B. At the same time, the technology could also subtype *S. pneumoniae* and even distinguish whether there is a capsule or not.[45,46] MS also typed methicillin-resistant *Staphylococcus aureus*[47–50] and *L. monocytogenes*,[51] to determine the epidemiological relevance of vancomycin-resistant *E. faecium* strains.[52] Additionally, rapid subspecies identification of *Salmonella* and *Yersinia enterocolitica* isolates was achieved.[53–55] In addition, MALDI-TOF MS can also be used for the typing of fungi and viruses.

Specialized software for spectral analysis is now available on the market, such as BioNumerics (Applied Maths, bioMérieux), MASS-UP, and Clover MS Data Analysis Software (http://www.clovermsdataanalysis.com/).[56]

12.2.4 Virulence Marker Detection

Many pathogenic factors in bacteria are proteins, and MALDI-TOF is theoretically a viable detection tool. Recently, MALDI-TOF has realized the detection of *Clostridium difficile* toxins,[57] agr-positive methicillin-resistant *Staphylococcus aureus*,[58] and *Clostridium botulinum* neurotoxin,[59] and the detection of anthrax lethal factor (LF), the sensitivity to which was 100%.[60]

12.2.5 Potential Applications of MALDI-TOF MS in Microbiology

The combination of immunization and MALDI-TOF MS (immune–MALDI) may be more specific. Antigen capture using antibodies, coupled with MALDI-TOF-specific mass identification, can significantly improve the specificity of the assay.[61] This method has advantages for small proteins and peptides. The application of this technology, to solve microorganism-related problems such as toxin quantification, requires further research.

Another MALDI-TOF MS-based technique, imaging mass spectrometry (IMS), enables molecular mapping of different kinds of biomolecules in natural environments.[62–64] IMS can analyze a wide variety of analytes without any labeling, from proteins, peptides, and protein modifications, to small molecules, drugs and their metabolites as well as pharmaceutical components, endogenous cell metabolites, lipids, and other analytes *in situ*. The direct molecular analysis of tissue samples by MALDI-TOF MS provides a means to study pathogen–host interactions and to discover potential markers of infection.[65,66] In addition, IMS enables the study of metabolic exchange factors in intraspecific, interspecific, and poly-microbial interactions.[67] While the role of IMS in microbiological research and even diagnostics remains to be researched and developed, it may bring fundamentally new insights into the study of microbial communities.[68]

12.3 Current Situation and Development Trends

In the past ten years, MALDI-TOF MS has been continuously improved and developed. As a fast and reliable method for microbial identification, MALDI-TOF MS is being more and more widely used in routine work of clinical microbiology laboratories worldwide. Although the MS equipment is not cheap yet, its high accuracy, low cost, user-friendliness, and ability to provide reliable results in less than 5 minutes are all conducive to its clinical application and further expansion.

At present, the lack of standardized solutions and the need for continuous improvement of reference databases are the main problems faced by

MALDI-TOF in practical applications. In terms of microbial identification, although MALDI-TOF MS is generally better than traditional methods in accuracy, it cannot distinguish between highly related microorganisms whose ribosomal proteins are very conserved. When identifying some related pathogens, it can also generate some errors. All of these aforementioned weaknesses require further optimization and improvement. The application prospects of MALDI-TOF MS in other aspects of clinical microbiology are also promising: for example, the identification of microorganisms directly from positive blood cultures and clinical specimens, the further improvement of bacterial typing or virulence marker detection, and the combination with other emerging technologies. In addition, high-performance lasers equipped with new-generation MALDI-TOF MS systems (currently the fastest commercial microbial ID MALDI-TOF MS system operates at only 200 Hz, while the new-generation systems can emit up to 10 000 laser pulses per second) demonstrate huge clinical application prospects in MALDI-TOF application.

In addition, with the development of laboratory automation, it is expected to establish and realize a fully automatic assembly line, instead of manual MS, which is of great significance to improve the diagnostic efficiency of clinical laboratories.

References

1. M. E. Mikkelsen, D. F. Gaieski, R. A. Band, J. M. Pines, R. Massone, F. F. Furia, F. S. Shofer and M. Goyal, *Crit. Care Med.*, 2010, **38**, 1045–1053.
2. D. R. Anand Kumar, K. E. Wood, B. Light, J. E. Parrillo, S. Sharma, R. Suppes, D. Feinstein, S. Zanotti, L. Taiberg, D. Gurka, A. Kumar and M. Cheang, *Crit. Care Med.*, 2006, **34**, 1589–1596.
3. A. Bizzini and G. Greub, *Clin. Microbiol. Infect.*, 2010, **16**, 1614–1619.
4. C. Mesquita, E. Carbonnelle, E. Bille, N. Day, B. Dauphin, J.-L. Beretti, A. Ferroni, L. Gutmann and X. Nassif, *Clin. Biochem.*, 2011, **44**, 104–109.
5. M. Drancourt, P. Seng, F. Gouriet, B. La Scola, P. E. Fournier, J. M. Rolain and D. Raoult, *Clin. Infect. Dis.*, 2009, **49**, 543–551.
6. T. C. Dingle and S. M. Butler-Wu, *Clin. Lab. Med.*, 2013, **33**, 589–609.
7. H. N. Shah, S. Emonet, A. Cherkaoui and J. Schrenzel, *Clin. Microbiol. Infect.*, 2010, **16**, 1604–1613.
8. C. Fenselau and P. A. Demirev, *Mass Spectrom. Rev.*, 2001, **20**, 157–171.
9. A. Bizzini, C. Durussel, J. Bille, G. Greub and G. Prod'hom, *J. Clin. Microbiol.*, 2010, **48**, 1549–1554.
10. S. C. Smole, L. A. King, P. E. Leopold and R. D. Arbeit, *J. Microbiol. Methods*, 2002, **48**, 107–115.
11. S. Q. van Veen, E. C. Claas and E. J. Kuijper, *J. Clin. Microbiol.*, 2010, **48**, 900–907.
12. R. D. Holland, C. R. Duffy, F. Rafii, J. B. Sutherland, T. M. Heinze, C. L. Holder, K. J. Voorhees and J. O. Lay Jr., *Anal. Chem.*, 1999, **71**, 3226–3230.
13. V. Ryzhov and C. Fenselau, *Anal. Chem.*, 2001, **73**, 746–750.

14. K. T. Liwei Sun, H. Sato, M. Torimura, H. Tao and T. Shintani, *Rapid Commun. Mass Spectrom.*, 2006, **20**, 3789–3798.
15. D. Martiny, L. Busson, I. Wybo, R. A. El Haj, A. Dediste and O. Vandenberg, *J. Clin. Microbiol.*, 2012, **50**, 1313–1325.
16. C. Rodrigues, V. Passet, A. Rakotondrasoa and S. Brisse, *Front. Microbiol.*, 2018, **9**, 3000.
17. K. Alby, P. H. Gilligan and M. B. Miller, *J. Clin. Microbiol.*, 2013, **51**, 3852–3854.
18. M. R. Couturier, E. Mehinovic, A. C. Croft and M. A. Fisher, *J. Clin. Microbiol.*, 2011, **49**, 1104–1106.
19. M. Oviano, K. Sparbier, M. J. Barba, M. Kostrzewa and G. Bou, *Int. J. Antimicrob. Agents*, 2016, **48**, 655–660.
20. B. Rodriguez-Sanchez, L. Alcala, M. Marin, A. Ruiz, E. Alonso and E. Bouza, *Anaerobe*, 2016, **42**, 101–107.
21. A. B. Pranada, E. Witt, M. Bienia, M. Kostrzewa and M. Timke, *J. Med. Microbiol.*, 2017, **66**, 670–677.
22. M. Welker, A. Van Belkum, V. Girard, J. P. Charrier and D. Pincus, *Expert Rev. Proteomics*, 2019, **16**, 695–710.
23. N. Dhiman, L. Hall, S. L. Wohlfiel, S. P. Buckwalter and N. L. Wengenack, *J. Clin. Microbiol.*, 2011, **49**, 1614–1616.
24. P. Escribano, M. E. Zvezdanova, A. Ruiz, M. C. Martínez-Jiménez, T. Peláez, A. Collazos, J. Guinea, E. Bouza and B. Rodríguez-Sánchez, *Med. Mycol.*, 2019, **57**, 63–70.
25. P. Becker, A. C. Normand, F. Gabriel, C. Cassagne, I. Accoceberry, M. Gari-Toussaint, L. Hasseine, D. De Geyter, D. Pierard, I. Surmont, F. Djenad, J. L. Donnadieu, M. Piarroux, S. Ranque, M. Hendrickx and R. Piarroux, *J. Clin. Microbiol.*, 2017, **55**, 2661–2670.
26. A. Curtoni, R. Cipriani, E. S. Marra, A. M. Barbui, R. Cavallo and C. Costa, *Curr. Microbiol.*, 2017, **74**, 97–102.
27. R. Patel, *Clin. Chem.*, 2015, **61**, 100–111.
28. D. M. Wolk and A. E. Clark, *Clin. Lab. Med.*, 2018, **38**, 471–486.
29. P. A. Demirev, N. S. Hagan, M. D. Antoine, J. S. Lin and A. B. Feldman, *J. Am. Soc. Mass Spectrom.*, 2013, **24**, 1194–1201.
30. J. S. Jung, T. Eberl, K. Sparbier, C. Lange, M. Kostrzewa, S. Schubert and A. Wieser, *Eur. J. Clin. Microbiol. Infect. Dis.*, 2014, **33**, 949–955.
31. C. Lange, S. Schubert, J. Jung, M. Kostrzewa and K. Sparbier, *J. Clin. Microbiol.*, 2014, **52**, 4155–4162.
32. K. Sparbier, E. A. Idelevich, M. Kostrzewa and K. Becker, *Clin. Microbiol. Infect.*, 2018, **24**, 738–743.
33. M. Oviano and G. Bou, *Clin. Microbiol. Rev.*, 2019, **32**, e00037.
34. M. Oviano, M. Gomara, M. J. Barba, M. J. Revillo, L. P. Barbeyto and G. Bou, *J. Antimicrob. Chemother.*, 2017, **72**, 2259–2262.
35. D. D. Rhoads, H. Wang, J. Karichu and S. S. Richter, *Diagn. Microbiol. Infect. Dis.*, 2016, **86**, 257–261.
36. A. F. Lau, H. Wang, R. A. Weingarten, S. K. Drake, A. F. Suffredini, M. K. Garfield, Y. Chen, M. Gucek, J. H. Youn, F. Stock, H. Tso, J. DeLeo,

J. J. Cimino, K. M. Frank and J. P. Dekker, *J. Clin. Microbiol.*, 2014, **52**, 2804–2812.
37. L. Dortet, R. A. Bonnin, I. Pennisi, L. Gauthier, A. B. Jousset, L. Dabos, R. C. D. Furniss, D. A. I. Mavridou, P. Bogaerts, Y. Glupczynski, A. Potron, P. Plesiat, R. Beyrouthy, F. Robin, R. Bonnet, T. Naas, A. Filloux and G. Larrouy-Maumus, *J. Antimicrob. Chemother.*, 2018, **73**, 3359–3367.
38. S. Nakano, Y. Matsumura, K. Kato, T. Yunoki, G. Hotta, T. Noguchi, M. Yamamoto, M. Nagao, Y. Ito, S. Takakura and S. Ichiyama, *Int. J. Antimicrob. Agents*, 2014, **44**, 256–259.
39. J. C. Cai, Y. Y. Hu, R. Zhang, H. W. Zhou and G. X. Chen, *J. Clin. Microbiol.*, 2012, **50**, 2179–2182.
40. P. J. Ceyssens, K. Soetaert, M. Timke, A. Van den Bossche, K. Sparbier, K. De Cremer, M. Kostrzewa, M. Hendrickx and V. Mathys, *J. Clin. Microbiol.*, 2017, **55**, 624–634.
41. A. W. Fothergill, M. A. Saracli, D. A. Sutton and N. P. Wiederhold, *Med. Mycol.*, 2015, **53**, 736–742.
42. S. T.-S. Adrian Egli, M. Oberle, D. Goldenberger, R. Frei and A. F. Widmer, *PLoS One*, 2015, **10**, e0120624.
43. N. Cabrolier, M. Sauget, X. Bertrand and D. Hocquet, *J. Clin. Microbiol.*, 2015, **53**, 1395–1398.
44. K. Rizzardi and T. Akerlund, *PLoS One*, 2015, **10**, e0122457.
45. M. Vairamani, M. Palani Kumar, R. Prasada Raju, C. Lobo, N. Anbumani, C. P. Girish Kumar, T. Menon and S. Shanmugasundaram, *Indian J. Med. Res.*, 2004, **119**, 283–288.
46. G. Héry-Arnaud, M.-F. Lartigue, E. Haguenoer, A.-S. Domelier, P.-O. Schmit, N. van der Mee-Marquet, P. Lanotte, L. Mereghetti, M. Kostrzewa and R. Quentin, *J. Clin. Microbiol.*, 2009, **47**, 2284–2287.
47. N. Pakulat, K. Bernardo, M. Macht, O. Krut, H. Seifert, S. Fleer, F. Hünger and M. Krönke, *Proteomics*, 2002, **2**, 747–753.
48. M. Wolters, H. Rohde, T. Maier, C. Belmar-Campos, G. Franke, S. Scherpe, M. Aepfelbacher and M. Christner, *Int. J. Med. Microbiol.*, 2011, **301**, 64–68.
49. K. A. Jackson, V. Edwards-Jones, C. W. Sutton and A. J. Fox, *J. Microbiol. Methods*, 2005, **62**, 273–284.
50. J. Walker, A. J. Fox, V. Edwards-Jones and D. B. Gordon, *J. Microbiol. Methods*, 2002, **48**, 117–126.
51. S. B. Barbuddhe, T. Maier, G. Schwarz, M. Kostrzewa, H. Hof, E. Domann, T. Chakraborty and T. Hain, *Appl. Environ. Microbiol.*, 2008, **74**, 5402–5407.
52. G. R. Price, P. M. Griffin, J. M. Schooneveldt, S. Schlebusch, M. H. Tilse, T. Urbanski, B. Hamilton and D. Venter, *J. Clin. Microbiol.*, 2012, **50**, 2918–2931.
53. R. Helmuth, R. Dieckmann, M. Erhard and B. Malorny, *Appl. Environ. Microbiol.*, 2008, **74**, 7767–7778.
54. C. Marinach, A. Alanio, M. Palous, S. Kwasek, A. Fekkar, J. Y. Brossas, S. Brun, G. Snounou, C. Hennequin, D. Sanglard, A. Datry, J. L. Golmard and D. Mazier, *Proteomics*, 2009, **9**, 4627–4631.

55. R. Stephan, N. Cernela, D. Ziegler, V. Pfluger, M. Tonolla, D. Ravasi, M. Fredriksson-Ahomaa and H. Hachler, *J. Microbiol. Methods*, 2011, **87**, 150–153.
56. M. Oviano and B. Rodriguez-Sanchez, *Enferm. Infecc. Microbiol. Clin.*, 2021, **39**, 192–200.
57. S. F. Kuo, T. L. Wu, H. L. You, C. C. Chien, J. H. Chia and C. H. Lee, *Diagn. Microbiol. Infect. Dis.*, 2015, **83**, 229–231.
58. M. Josten, J. Dischinger, C. Szekat, M. Reif, N. Al-Sabti, H. G. Sahl, M. Parcina, I. Bekeredjian-Ding and G. Bierbaum, *Int. J. Med. Microbiol.*, 2014, **304**, 1018–1023.
59. M. J. Perry, D. A. Centurioni, S. W. Davis, G. E. Hannett, K. A. Musser and C. T. Egan, *Toxins*, 2017, **9**, 94.
60. M. Gallegos-Candela, A. E. Boyer, A. R. Woolfitt, J. Brumlow, R. C. Lins, C. P. Quinn, A. R. Hoffmaster, G. Meister and J. R. Barr, *Anal. Biochem.*, 2018, **543**, 97–107.
61. K. Sparbier, T. Wenzel, H. Dihazi, S. Blaschke, G. A. Muller, A. Deelder, T. Flad and M. Kostrzewa, *Proteomics*, 2009, **9**, 1442–1450.
62. A. Walch, S. Rauser, S. O. Deininger and H. Hofler, *Histochem. Cell Biol.*, 2008, **130**, 421–434.
63. P. Neubert and A. Walch, *Expert Rev. Proteomics*, 2013, **10**, 259–273.
64. M. Aichler and A. Walch, *Lab. Invest.*, 2015, **95**, 422–431.
65. J. L. Moore, K. W. Becker, J. J. Nicklay, K. L. Boyd, E. P. Skaar and R. M. Caprioli, *Proteomics*, 2014, **14**, 820–828.
66. J. L. Moore, R. M. Caprioli and E. P. Skaar, *Curr. Opin. Microbiol.*, 2014, **19**, 45–51.
67. J. Y. Yang, V. V. Phelan, R. Simkovsky, J. D. Watrous, R. M. Trial, T. C. Fleming, R. Wenter, B. S. Moore, S. S. Golden, K. Pogliano and P. C. Dorrestein, *J. Bacteriol.*, 2012, **194**, 6023–6028.
68. S. Schubert and M. Kostrzewa, *Curr. Issues Mol. Biol.*, 2017, **23**, 17–20.

CHAPTER 13

Mass Spectrometry-based Microbial Identification and Profiling for Environmental Science

SIMON J. S. CAMERON

School of Biological Sciences, Institute for Global Food Security, Queen's University Belfast, UK
Email: s.cameron@qub.ac.uk

13.1 Requirement for Microbial Identification in Environmental Science

Environmental science brings together a wide range of disciplines across the domains of physics, chemistry, biology, and geography to understand the environment as individual processes and as an interconnected complex system. These disciplines include ecology, plant science, zoology, oceanography, limnology, soil science, geology, atmospheric science, and mineralogy. Typically, environmental science is concerned with the study of the natural environment, but with the increasing complexity of human interactions with the environment, understanding and providing solutions to environmental problems is important.[1] There is also crossover with elements of built environments, and how these interact with their occupants and the natural environment in which they are based. Microorganisms can be found across all the different environments on Earth and are central to many of the planet's natural nutrient systems and cycles. They also play an ever

New Developments in Mass Spectrometry No. 13
Detection and Analysis of Microorganisms by Mass Spectrometry
Edited by Liang Qiao and Jia Yi

Published by the Royal Society of Chemistry, www.rsc.org

more central role in many of the human–environment interactions and the potential solutions to associated problems. This can include areas such as bioremediation,[2] microbial interactions with agricultural plants and crops,[3] and associations with, or in response to, mechanisms of climate change.[4]

13.1.1 Importance of Microbial Identification

The identification of microorganisms in environmental science is a key component in the process of understanding community structure, function, and interactions. Indeed, many of the early pioneers in microscopy (Robert Hooke and Antonie van Leeuwenhoek) used environmental samples to observe microscopic organisms.[5] However, as the field of microbiology grew, it became more focused on the isolation and identification of pathogenic microorganisms. This is arguably due to the difficulty of culturing the diverse range of environmental microorganisms on agar-based culture media.[6] This is a required step in the traditional identification workflow that relied on the phenotypic and morphological characteristics of isolated colonies and their biochemical properties. Over the last few decades, our understanding of the microbial communities involved in environmental science has been greatly expanded. This has primarily been driven, though, by advances in DNA and RNA sequencing for metataxonomics/metagenomics and transcriptomics, respectively.[7] These methods are considered to be culture-independent as they are not reliant on the isolation of microbial isolates on agar culture plates. Although they considerably broadened the taxonomic scope of microbial species which could be captured, and discovered many new clades, there are considerable gaps in the information provided.[8]

13.1.2 Utilisation of Molecular Methods for Microbial Identification

Culture-independent molecular profiling has primarily used PCR amplification of microbial marker genes, typically the 16S rRNA gene for bacteria and archaea, and the internal transcribed spacer (ITS) region between the small and large subunits of the rRNA genes for eukaryotic microbes such as fungi. These regions are present in all bacterial and fungal microorganisms and therefore provide a universal target for molecular community profiling. When sequenced, they can be constructed into a phylogenetic tree and taxonomy assigned under the investigatory approach termed metataxonomics.[9] However, current limitations associated with available sequencing platforms prevent the entire length of microbial marker genes from being sequenced, and thus, taxonomic information is typically limited to genus-level identification. It is further disadvantaged by the lack of functional information that can be obtained. Although some approaches can map predicted functional capabilities to metataxonomic data, such as through the Tax4Fun[10] and PICRUSt[11] pipelines, these are reliant on the quality of reference databases. They are further disadvantaged by the innate differences within functional capabilities of different species within genera, which would be overlooked due to limitations associated with sequencing read lengths.

As the cost of DNA sequencing has reduced considerably in the last decade, microbial community profiling has increasingly utilised shotgun metagenomic sequencing. This technique overcomes the restrictions associated with the targeted amplicon sequencing used in metataxonomics by sequencing the double-stranded DNA within a sample. This means that sequencing coverage extends beyond just marker genes, such as 16S rRNA gene and ITS sequences, to cover the broader genome. This gives information on the broader functional capacity of microbial communities, such as genes involved in nutrient cycling, bioremediation, or stress response.[12] Although this provides a greater level of detail, shotgun metagenomic sequencing still has the drawback of being unable to sequence the entirety of double-stranded DNA within a sample. This is combined with limitations for bioinformatically connecting genes from the same microorganism's genome together, and lack of database resources for assigning roles to genes, leaving many with unknown functions.

Metagenomics provides the functional capacity of a microbial community within an environmental sample. This does not, however, necessarily translate to the community's activity which may respond differently based on temporal, spatial, or environmental conditions such as temperature, moisture, and salinity. Transcriptomics uses much of the same analytical technologies as shotgun metagenomics but instead gives information on the messenger RNA (mRNA) that is present within a sample and thus a snapshot of the genes that are transcribed into protein products or involved in expression control. This can give much greater information on the activity of microorganisms but is still limited by many of the same factors as metagenomics.[13] It is also complicated by the requirement to rapidly quench the activity of the sampled microbial community to prevent an alteration in transcriptomic profile from a shock response. This is usually by means of freezing in liquid nitrogen, which carries impracticalities and hazards in environmental science research.[14] As with PCR amplification-based metataxonomics of the 16S rRNA gene or ITS region, the effectiveness of both metagenomics and transcriptomics approaches is limited by the breadth, depth, and accuracy of the databases they draw on. This is something particularly stark in the environmental sciences where it is notably difficult to culture members of microbial communities.

This leaves a large gap in our understanding of microbial communities at a functional level which is increasingly important for understanding their structure, function, and interactions between microorganisms and the environmental niche they inhabit. Although phenotypic and biochemical characteristics have long been used to classify microorganisms, these are usually suited to a narrow range of microbial species or only allow higher-level taxonomic classification for microorganisms typically associated with

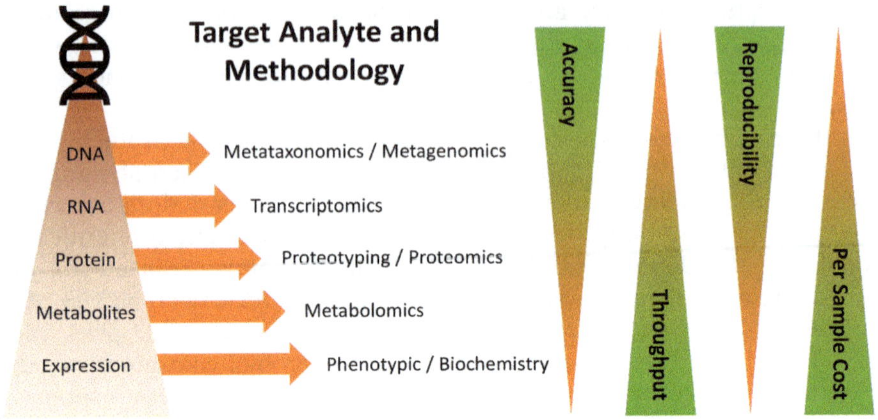

Figure 13.1 Target analytes and associated methodologies for the environmental sciences. Methods of microbial identification based on the systems biology hierarchy of analytes and associated methodologies. Each level has different benefits and drawbacks in terms of accuracy, throughput, reproducibility, and per sample cost. To date, DNA and expression methods have been the mainstay of microbial identification in the environmental sciences. However, MS has increasing utility for protein and metabolite analysis for both identification and functional profiling.

environmental samples.[15] As outlined in Figure 13.1, proteins and metabolites occupy the space in the systems biology hierarchy between RNA transcriptomics and phenotypic and biochemical expression. In the last decade, this space has been illuminated using mass spectrometry (MS) as an analytical tool in the environmental sciences. This has been through revolutionising methods for both microbial identification and functional characterisation.

13.1.3 The Role That MS Can Play

MS is an analytical method that measures the mass-to-charge ratios of ionised compounds. It is a broad field that draws on multiple disciplines within physics, chemistry, and biology to process, analyse, and interpret samples and results. As outlined in Figure 13.2, MS can be combined with multiple pre- and post-ionisation methodologies to increase or improve the level of analytical information obtained from a sample. As a conventional workflow, samples are collected and then processed to extract the desired compounds (such as metabolites, proteins, and toxins) and minimise the presence of non-target compounds. This is usually by physical and chemical disruption and extraction into an organic solvent. Extracted samples then undergo a chromatographic process that separates compounds based on their physical or chemical properties prior to their ionisation. Compounds

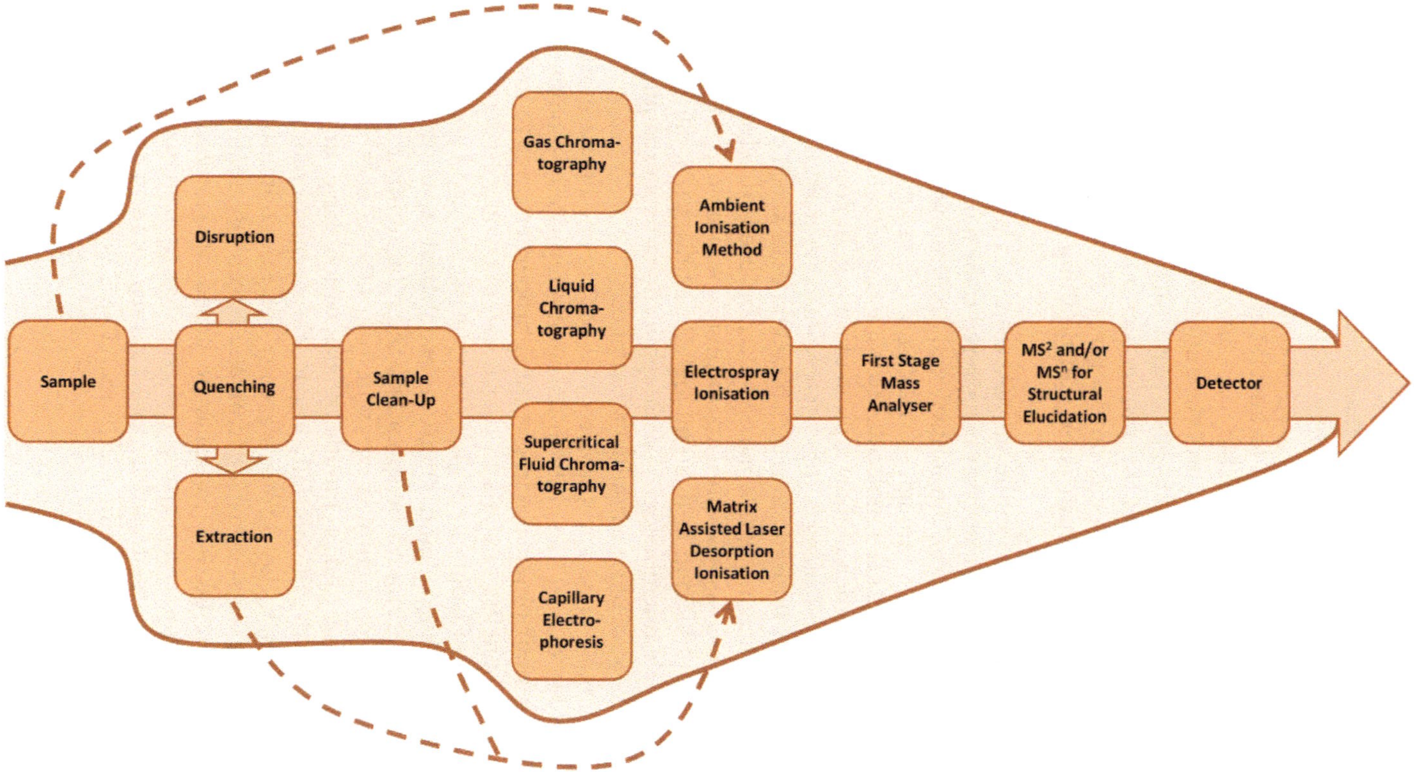

Figure 13.2 Schematic of analytical choices available in MS analysis. MS offers a broad range of analytical choices with different combinations offering various benefits and drawbacks. The schematic shows processes from sample collection to metabolic quenching and mechanical or chemical extraction. Following this, sample clean-up is completed to remove non-target analytes and then chromatography is typically used to separate analytes based on physical and/or chemical properties. Analyte ionisation follows depending on the method of analysis with the resulting ions separated based on mass-to-charge ratios and/or fragmentation through MS^2 or MS^n to elucidate structural information. Various methods such as ambient ionisation and MALDI MS can skip certain steps because of their specific analytical approach. These are indicated by dashed lines.

then undergo the ionisation process which endows them with a charged state that allows them to be manipulated within electrical and magnetic fields. This can be used to measure their mass-to-charge ratios and thus potential elemental compositions. Chromatographic separation and/or structural information obtained through fragmentation of ions can give additional information on the functional arrangement of these elements.[16] In each step, multiple options are available that differ in their speed, efficacy, and target compound and overall provide hundreds of different combinations of analytical workflows in MS. These are summarised in Figure 13.2, but the total complexity of this process lies outside of the scope of this chapter, which will instead focus on the areas of MS that have the most applicability to microbial identification and profiling in the environmental sciences.

Matrix-assisted laser desorption/ionisation (MALDI) MS is typically linked to a time-of-flight (ToF) mass analyser. It is considered to be a direct method of MS as samples, or their extracts, are ionised and analysed without chromatographic separation. It works through the mixing of the sample or extract with a matrix containing crystallised molecules that are capable of absorbing laser energy. This triggers ablation and desorption of ionised molecules into a plume which is then analysed using a mass spectrometer – typically a ToF mass analyser that separates ionised molecules based on their speed of travel through a vacuum chamber. Different matrices can be used to selectively ionise molecules of interest, such as proteins, peptides, and lipids. MALDI-ToF MS has seen broad utilisation in microbiological identification, particularly in clinical diagnostic microbiology,[17] with applications now expanding beyond this into fields including environmental science.[18]

Liquid chromatography–mass spectrometry (LC–MS) is one of the most widely used tools in MS due to its ability to chromatographically separate compounds based on their physical and chemical properties prior to ionisation. This typically benefits analysis in that it can overcome issues associated with separation of isomers – which cannot be separated based on mass-to-charge ratios alone – and improves analytical sensitivity by avoiding ion generation saturation by spreading the total number of analytes over the time range of a chromatographic run. LC–MS uses a combination of a solid phase and a mobile phase which have different properties and consequently differentially impact the speed of movement of different analytes through the chromatography column containing the stationary phase. These interactions are typically based on polar interactions but separation for LC–MS can also be based on size-exclusion chromatography using molecular sizes, ion-exchange chromatography using interactions of charged sites, and affinity chromatography using analyte interactions and specific receptors. LC–MS is typically linked to triple quadrupole mass spectrometers for targeted analysis or ToF or Orbitrap instruments for untargeted profiling analysis.

Gas chromatography–mass spectrometry (GC–MS) is used primarily for the study of volatile analytes or analytes which can be volatilised through addition of derivatising agents that introduce functional groups through alkylation, acylation, sialylation, or oximation. Importantly, these derivatising agents are very sensitive to the presence of water, and thus, samples must be thoroughly dried before analysis, which introduces potential sources of bias. The introduction of these groups allows the analyte to move through a gaseous mobile phase, normally an inert gas such as helium, and through a stationary solid phase where the movement of analytes is determined by polar interactions. This chromatographic separation can take up to 60 minutes with ions analysed typically with either single or triple quadrupole instruments for targeted analysis, or ToF or Orbitrap instruments for untargeted analysis. Compared to LC–MS, GC–MS has a narrower analytical range as metabolites must be either already volatile or capable of being volatilised. It has found many utilities in terms of microbial identification through profiling of fatty acid composition, which can be indicative of taxonomy.

Ambient ionisation mass spectrometry (AIMS) is a collection of ionisation methods that allow for the formation of ions in a normal atmosphere. These methods typically do not require extensive sample preparation, pre-treatment, or extraction prior to analysis. The AIMS field started with the introduction of the direct analysis in real time (DART) and desorption electrospray ionisation (DESI) techniques in 2004. These methods are characterised by their high-throughput nature and the ability to analyse samples in their near-native state. They are, however, limited by the lack of chromatographic separation prior to ionisation which reduces potential sensitivity and analytical resolution. In terms of microbial identification, many methods have shown utility as a means of taxonomic identification of microorganisms and their further functional profiling. These are not, however, considered to be means for routine analysis in the environmental sciences but show promise, which will be discussed in more detail later in this chapter.

13.2 Culture-dependent Identification in the Environmental Sciences

Culture-dependent methods revolve around the isolation of microbial isolates onto, typically, a solid agar growth medium. Although liquid growth media can be used for initial broad or selective enrichments, the physical isolation of individual colony-forming units (CFU) on solid agar is necessary to ensure purity. Culture-dependent methods have long been hampered by the issue of approximately 99% of microbial species not being culturable.[19] This should, however, be strongly caveated by being applicable to currently used methods. There is considerable scope to expand the range of culture conditions that can be used in the environmental sciences, but this will be driven largely by the environmental source of samples.

13.2.1 The Increasing Importance of Culturomics

The isolation of microorganisms onto solid agar culture was the mainstay of microbiological experimentation for nearly a century. This was because isolation of microorganisms was needed to perform phenotypic and biochemical tests to taxonomically identify them. Arguably, this resulted in the field being biased towards microorganisms that were in high abundance, non-fastidious, or fast growing. Thus, culture-dependent analysis of environmental samples may give a biased viewpoint of the microbial community which is present. The advent of genomic sequencing allowed the utilisation of the ubiquitous 16S rRNA gene and ITS regions for taxonomic analysis of environmental samples. This gave a much broader insight into microbial communities, but as discussed previously in this chapter, there are limitations in terms of identifying the functional capability of isolates, and also the extent to which viable microorganisms are detected. As the limitations of genomic sequencing approaches are increasingly understood, there has been a field-wide realisation that the isolation of viable microorganisms in culture is essential for a fuller understanding of microbial communities in environmental samples. This is by no means a novel concept, as one of the first examples of the valuable insight that can be obtained by culturing microorganisms in a realistic environment is a Winogradsky column.[20] It uses materials from the environment – typically the water and sediment from a pond or stream – to provide a sealed system in which nutrient and oxygen gradients form over time and provide environmental niches for microbial communities to occupy.

There are two main approaches to culturomics in the environmental sciences: the modification of artificial media to meet growth requirements of microorganisms or the incubation of microbial samples in their naturally occurring environment. The former has been shown to have marginal and incremental improvements in the growth of previously 'unculturable' microorganisms. This is largely due to the difficulty in understanding the chemical composition of an environment, how this is related to microbial growth and viability, and the isolation, purification, or synthetic manufacturing of compounds. It does, however, have an arguable benefit for the selective isolation of target microorganisms about which sufficient information is known.[21] The alternative approach of isolating microorganisms by culturing them under the conditions of their environmental source has seen substantial advances in the past two decades. Although the Winogradsky column format allows the culture of microbial communities, it does not allow the physical isolation of single species. This is a necessary step for additional work on the functional characterisation of microorganisms and mechanistic studies. Thus, the area of culturomics that has seen the most substantial advances in the past two decades has been where traditional agar-based methods have been combined with incubation of samples in their naturally occurring environments.

One of the early examples of this utilised a diffusion chamber in which microbial cells obtained from an environmental sample are placed between membranes with pores, typically on a nanometre scale, and returned to the sample's environmental origin. These pores allow the diffusion of nutrients, metabolites, proteins, and other naturally occurring factors important for growth into the chamber, but no movement of microbial cells out from the chamber. In 2002, this approach was shown to result in a 30 000% increase in the recovery of microbial species compared to traditional culture techniques.[22] However, this approach was fraught with the difficulty of isolation of colonies requiring microscopic dissection from solid growth media. This is likely reflective of the natural behaviour of microbial cells within their environmental niche that is a polar opposite to that which is provided under laboratory conditions. The diffusion chamber technique was also considered to exhibit somewhat low throughput.[23] To overcome this problem, Nichols *et al.* modified the aforementioned approach to include a much greater number of individual chambers that would allow the growth of single species. This device with improved throughput was termed the isolation chip, or the 'Ichip'.[24] The improved culturability of the Ichip device has been used in a broad range of microbial isolation studies using environmental samples, including the discovery of an antibiotic called teixobactin, which is the first of a novel class of antibiotics.[25] With improvements to our ability to culture microorganisms from environmental samples reducing the range of 'unculturable' microorganisms, the bottleneck that is increasing in prominence is that of microbial identification. Taxonomic identification is an essential component of microbiological research as it forms the basis of understanding evolutionary relationships, functional capabilities, and, importantly, whether the isolated colony is a known or novel species. Traditionally reliant on slow and sometimes inaccurate biochemical and phenotypic tests, or slow and costly sequencing, microbial identification has seen a transformation through the widespread adoption of MS-based methods.

13.2.2 MALDI-ToF MS

As a methodology, the identification of microorganisms has largely been led by the requirements of the clinical microbiology field. This is arguably because of the importance of accurate and timely identification of disease-causing pathogens to guide appropriate treatment. This is made somewhat more straightforward by the relatively narrow range, compared to the environmental microbiology diversity, of disease-causing microorganisms in humans, animals, and plants. Although there is crossover between opportunistic pathogens that typically have an environmental origin, such as *Pseudomonas aeruginosa* and *Bacillus anthracis*, microorganisms with an exclusively environmental niche have been historically understudied. The introduction of MALDI-ToF MS into clinical microbiology diagnostic laboratories has seen a step-change in the routine identification of microorganisms. This has been particularly pronounced in the last decade with the introduction of commercial instruments, but the use of MALDI-ToF MS to identify bacteria based on their protein was first proposed in 1996.[26,27] This field grew over the years to include a broader range of bacteria and then yeasts and filamentous fungi. The introduction of the bioMérieux VITEK MS and Bruker Biotyper systems in 2009 saw substantial uptake in the clinical microbiology field, and which over time diffused into related fields where microbial identification is important. This included, for example, food microbiology,[28] the pharmaceutical industry,[29] and public health microbiology.[30]

MALDI-ToF MS is a culture-dependent MS method for the identification of microorganisms. This is because it requires the direct analysis of a single-species isolated colony transferred from an agar culture plate. As taxonomic identification is determined by matches against a reference database, the algorithms that are used are unable to separate mass spectra from multiple sources. The effectiveness of MALDI-ToF MS in microbial identification is determined largely by the diversity of the reference database that mass spectral fingerprints are compared against.[31] This was the key enabler of MALDI-ToF MS adoption as a routine method of identification in clinical microbiology laboratories, and one of the key limitations in the broader utility in fields where microbial diversity is considerably greater, such as in the environmental sciences. Nevertheless, as the technology has become more broadly adopted outside of the clinical microbiology field, microorganisms of exclusively environmental sources are now included in many vendor databases. Laboratory users are also able to develop in-house databases using the software options of many vendors based on already isolated and identified microorganisms.

The main benefit that has been brought about by the widespread adoption of MALDI-ToF MS is the reduction in both time and resources that are required for microbial identification. In this respect, it can be seen as a key enabler of the increasing focus on culturomics in the environmental sciences. A reliance on traditional biochemical and phenotypic tests and/or more modern but costly and time-consuming genetic speciation would not be practical considering the large number of isolated single-species colonies associated with effective culturomics workflows. At the genus level of classification, MALDI-ToF MS typically has an accuracy rate of more than 99%.[32] At the species level, the accuracy level is lower but typically exceeds 95% for most bacterial and yeast species contained within the database,[33,34] with varying performance for filamentous fungi below this.[35] Strain-level identification using MALDI-ToF has been indicated as a possibility for some species but is largely considered to be non-viable as a general approach.[36] Specific examples of how MALDI-ToF MS has been used in microbial identification in the environmental sciences will be discussed in Section 13.5.

13.2.3 AIMS

MALDI-ToF MS has many benefits in terms of microbial identification, particularly when compared to conventional and traditional methods reliant on biochemical and phenotypic tests. Nevertheless, its main limitations are that it is a largely manual method that requires movement of microbial biomass from an agar culture plate to a MALDI target plate for analysis, and that it has difficulty with strain-level identification. This is arguably because MALDI-ToF MS is reliant on the detection of ribosomal proteins that are of high abundance in microbial cells. The field of AIMS, however, typically involves methods which ionise smaller compounds such as fatty acids, complex lipids, and metabolites – all typically under 2 kDa.[37] Interestingly, fatty acids analysis using MS for the identification of bacteria was proposed decades before the use of proteins.[38] Furthermore, as many techniques are characterised by a lack of requirements for sample preparation or extraction, they are more suited to high-throughput and automated methods of analysis. This broadens the potential application areas of AIMS. The field of AIMS was founded by the invention of direct analysis in real time (DART) MS[39] and desorption electrospray ionisation (DESI) MS.[40] Soon after these methods were published in 2004 and 2005, their application to microbial identification was explored.

DESI MS has been shown to have both species-level and potential strain-level identification resolution, albeit using a small number of biological replicates,[41] and DART MS has been shown to have similar capabilities. However, based on the instrumental setup of these approaches, they require microbial biomass to be removed from an agar culture plate prior to analysis. As such, they do not offer the potential for higher-throughput analysis than MALDI-ToF MS. The two AIMS methods that have shown an ability to complete direct-from-culture microbial identification are liquid extraction surface analysis (LESA) MS[42] and rapid evaporative ionisation mass spectrometry (REIMS).[43] These two technologies differ in their mobilisation of ionised molecules for analysis. LESA MS involves the deposition of a solvent droplet onto the area for analysis with compounds extracted and then analysed through a connected mass spectrometer using electrospray ionisation. LESA MS has been shown to be capable of bacterial analysis through both protein[44] and lipid[45] profiles, although methods for identification have not been clearly established. Arguably the method that has seen most development as a competitor to MALDI-ToF MS is REIMS. First developed as a tool for intraoperative classification of human tissue,[46] it was adapted for direct-from-culture identification of bacteria and yeasts. REIMS works through the heating of biomass, using electrical,[47] contact,[48] or radiative[49] energy, which causes it to produce analyte-containing smoke that is then directly analysed with a mass spectrometer. Focusing on clinically-relevant species, a taxonomic classification model containing 25 species was shown to have a species-level accuracy of 97.2%.[50] The REIMS approach has also been shown to be capable of strain-level identification within *P. aeruginosa*.[51]

AIMS approaches show clear advantages compared to what has now become the central technology of MALDI-ToF MS. However, their development has focused on clinical microbiology, and as such, there are no examples of where these techniques have been more widely used in the environmental sciences. Broader adoption of AIMS methods will require a more extensive reference database and harmonisation of methods. Importantly, as with MALDI-ToF MS, AIMS will require a commercial instrument manufacturer to take forward a technology as a competitive alternative to the currently established technological approach.

13.3 Culture-independent Identification in the Environmental Sciences

MALDI-ToF MS has become the central method for culture-dependent identification of microbial isolates from environmental samples. It has also been developed for culture-independent identification from some types of clinical specimens. For urine, the method was shown to have a sensitivity of approximately 67% but with a detection limit of 10^5–10^6 CFU per mL and only where a single species was dominant.[52] However, there is little potential crossover with typical environmental samples that are usually of much lower biomass and much higher diversity. This means that the culture-independent method of microbial species in environmental samples is metataxonomic sequencing based on marker genes – such as the 16S rRNA or ITS gene regions. As already discussed, this approach has several drawbacks. Alternatives based on using MS include metaproteomics and metabolomics. Of these two, metaproteomics shows the most promise for culture-independent identification of microorganisms from environmental samples. Metabolomics may have future potential but is currently severely restricted by the abundance of unknown compounds within samples, with insufficient knowledge about how detectable metabolites can be tracked to their species of origin. This would be essential for the approach to be used in culture-independent identification.[53] Although complex lipids have been suggested as potential metabolomic-based markers for microbial identification, there is currently an insufficient understanding of underlying biochemical pathways for this to have a high level of confidence in assignment.[54]

13.3.1 Metaproteomics Using MS

Metaproteomics is the analytical approach that uses MS to define and characterise the entire protein content of a given sample at a given spatial and temporal point. It was first applied to a microbial community within an environmental sample in 2004.[55] Typically, a metaproteomic analysis involves four steps. Firstly, the protein component of a collected sample is extracted, purified to remove contaminants, and concentrated. This is followed by denaturation and reduction of the proteins. These proteins are then enzymatically digested and separated using nanoflow LC chromatography and MS. Usually, a quadrupole ToF instrument or quadrupole Orbitrap instrument is used, with ion mobility separation sometimes also combined to further separate compounds. Protein identification is then completed using collected mass spectra.[56] As a field, metaproteomics has been advanced by improvements in sample extraction and preparation, chromatographic separation, the sensitivity of mass spectrometers, and their ability to conduct rapid MS/MS fragmentation and analysis to determine protein sequence and structure.[57] New methods of metaproteomic data collection on mass spectrometers have also substantially advanced the field. These have enabled an extension of the number of proteins that can be detected within a sample and the dynamic range over which they can be detected and quantified. These methods are termed data-independent acquisition (DIA) and are characterised by the collection of tandem mass spectrometry data that is independent of the composition of precursor ions. Alternative methods termed data-dependent acquisition (DDA) analyse selected precursor ions which can be a source of bias and undersampling of the protein component of a sample.[58]

Metaproteomic analysis arguably combines the benefits of genomic and transcriptomic profiling as it gives information on the protein complement that is within a sample. As such, it details not only what proteins are held within the genomic potential of a sample or microbial community, but also which ones are transcribed and translated at the temporal and spatial point of sampling. This approach has been used in the study of microbial communities from a broad range of environmental samples. For example, the analytical approach was used to decipher the metabolic pathway for naphthalene degradation within the microbial community of Spanish soil contaminated with polycyclic aromatic hydrocarbons. This identified the gentisate pathway as the most active in response to the dynamics of the associated microbial community.[59] A similar approach for identifying microbial biochemical pathways associated with bioremediation of 2,4,5-trichlorophenoxyacetic acid (2,4,5-T) was used in a controlled enrichment culture within a Southern Vietnamese facility. This identified 60 protein groups which had a significant fold-change in the presence of 2,4,5-T and could therefore be used to select microbial species, or communities, with genes for these specific proteins to improve the bioremediation of the contaminant.[60]

13.3.2 Leveraging Genomic Information with Proteogenomics

Metaproteomics is a powerful culture-independent technique that can be used to profile the entire protein component of a sample. This gives valuable information but is limited in terms of broader understanding by not elucidating the microbial source of proteins at the species- and/or strain-level of identification. This connection is important for both an improved fundamental understanding of microbial communities within environmental samples and improvement in applications in research areas such as bioremediation, wastewater management, and nutrient cycling. Proteogenomics is a fusion of the fields of proteomics and genomics and leverages the ability to translate genomic information into predicted protein sequences. Based on the exponentially increasing volume of genomic and transcriptomic sequencing data from culture-dependent and culture-independent analysis, proteogenomics can fill in missing information in proteomic databases due to the lack of experimental coverage. Crucially, it also allows the species- and/or strain-level identification of the microbial source of a protein in a sample to be predicted.[61]

Metaproteomics and proteogenomics use the same tools and techniques for sample acquisition, processing, extraction, and chromatographic and MS analysis. Their difference is largely in the computational strategies used for data interpretation. While metaproteomics is concerned with the assembly of peptide fragments into protein sequences and their structural identification, the proteogenomic approach requires these sequences to be traced back to their genomic source through differences in amino acid and codon sequences.[62] This can be fraught with difficulties as there is a reliance on the accuracy of genomic sequences deposited in databases, which can sometimes be lacking.[63] Proteogenomics has been used successfully in several environmental microbiology studies to elucidate the mechanisms and the species-level source of important proteins. For example, the importance of cytochrome Cyt_{579} was identified for microbial community stability in a biofilm community of an acid mine that was then isolated and characterised.[64,65] Recent advances in quantitative proteogenomics have also enabled improved system models to be created in areas such as biotransformation of pharmaceuticals and pollutants.[66]

13.4 Using MS for Functional Profiling of Microbial Communities

Metaproteomics and proteogenomics are powerful tools for both culture-independent identification of microorganisms within an environmental sample, through identification of ribosomal proteins, and broader functional profiling at the protein level. Within the systems biology hierarchy, metabolomics lies below proteomics, but as previously discussed, it is limited in its ability to provide taxonomic identification of microorganisms. Nevertheless, many of these small molecules act as important intermediaries in intra- and inter-species communications within microbial communities and at the host–microbiome interface. For environmental samples, small molecules identified through metabolomic techniques are important for understanding the broader functional profiles of microbial communities.

13.4.1 Methods for Functional Profiling and Analytical Considerations

Functional profiling and metabolic profiling are terms typically used interchangeably. Although nuclear magnetic resonance (NMR) spectroscopy is used in these fields, analytical techniques typically employ variations of MS coupled to a method for chromatographic separation. The two main methods for functional profiling in metabolomics are LC–MS and GC–MS. Although other methods exist that offer unique balances of benefits and drawbacks, they have not been broadly adopted and there remain technical specialisations and niches. LC–MS is a broadly adaptable technique that offers flexibility in terms of target analytes through variations in the chemistries of mobile and stationary phases.[67] GC–MS typically has a narrower range of detectable metabolites, usually restricted to those which are volatile, or which can be easily made volatile through chemical derivatisation.[68] Of particular importance when completing functional profiling of an environmental sample, especially when microbial communities are a driving force in biochemical processes, are considerations around sample collection. This is to help ensure an accurate reflection of the *in situ* microbial community is captured, rather than a reflection of the stress response of the community to an altered environment. This may be particularly pronounced for samples from anaerobic environments, such as deep soils and wastewater sludge. To minimise this, samples should be quenched of activity rapidly either using ultralow temperature freezing or *via* treatment with chemicals, such as sodium azide.[69] After collection, considerations are then needed with regard to removal of water content from the sample to concentrate or allow dry-weight normalisation – although this can cause the loss of some volatile metabolites.[70] Mechanical disruption might be required to homogenise samples or allow release of intra-cellular metabolites of microorganisms. Again, however, this may be a source of bias as metabolites may be differentially affected by heat. Solvent extraction of metabolites also offers considerable opportunity for variation and introduction of bias, which may be somewhat intended if analysis of particular chemical classes of metabolites is desired.[71]

Once MS analysis has been completed, data analysis becomes the most challenging component of the process. The reasons for this are two-fold. Firstly, the mass-to-charge information gives a prediction of the elemental composition of a detected ion. This is not equivalent to structural information, and thus, the exact chemical nature of the detected compound is not discernible without additional fragmentation analysis and comparison to standards – which are not available for all compounds.[72] Secondly, as with our current inability to culture the vast majority of microorganisms under laboratory conditions, the field of metabolomics is confronted with the difficulty that no single method, nor a combination of methods, is able to detect, quantify, and resolve the chemical structure of all compounds within a sample. This is termed the 'dark matter' of metabolomics.[73] Considering the complexity of microbial communities in many environmental samples, and the relative lack of experimental exploration compared to human and animal systems, the metabolomic 'dark matter' is even more pronounced.

13.4.2 Multiomic Analysis for Sequencing and Metabolomics

As there is a current limitation in the use of metabolomics for culture-independent taxonomic identification of microorganisms, metabolomic analysis is typically paired with a genetic sequencing approach, such as ribosomal-based metataxonomics or shotgun metagenomics. The use of multiple 'omic' technologies, and their integration, is termed multiomic analysis. It has powerful potential to show relationships between microbial taxons and metabolites. There are difficulties associated with experimental analysis as there are different, and sometimes conflicting, requirements in sample collection. For example, a quenching agent that stabilises DNA or RNA within a sample for sequencing analysis may cause interferences in MS analysis. However, with careful planning and awareness of limitations, valuable insights can be obtained. Nevertheless, the observational nature of typical studies will always limit conclusions to correlations rather than elucidating causative mechanisms.[74] Many methods for multiomic data integration focus on the use of correlations to establish relationships between two variables, such as a sequence assigned to a specific microbial taxon and a metabolite feature, or across multiple variables in the form of co-occurrence networks. This approach has proved powerful in the study of several microbial communities in environmental samples. For example, correlations between the soil microbiome using 16S rRNA gene sequencing and the sugarcane (*Saccharum officinarum*) plant metabolome using LC–MS metabolomics were able to elucidate relationships that are responsible for sugarcane bitterness. This offers the potential for modification of the soil microbiome to improve the taste properties of sugarcane.[75] Further work in soil, but using metatranscriptomics and GC–MS metabolomics with integration through a network modelling approach, was able to identify how the microbial community was able to respond to environmental stressors. Through the use of the network modelling approach, it was possible to determine that genes and metabolites associated with drought response were actually more broadly utilised in other abiotic stress responses.[76]

13.5 Case Studies in the Environmental Sciences

The field of microbiome research has been heavily dominated by investigations into human and animal systems. This has largely been driven by the potential for translational impact on health and disease outcomes. Nevertheless, the importance of environmental microbiomes in factors important to environmental sustainability, climate change, and agriculture should not be overlooked. MALDI-ToF MS for identification of microorganisms from environmental samples is becoming increasingly common as the cost of the equipment decreases, and the coverage of environmental isolates in reference databases is improved.[77] There have also been numerous studies that have shown the power of MS-based methods for microbial identification, using both culture-dependent and culture-independent methods, and broader functional profiling in environmental microbiomes.

13.5.1 Understanding and Improving Bioremediation Using Microorganisms

Hydrocarbons derived from petroleum and associated refined products are some of the main environmental pollutants globally. These can be released into the environment by accident, such as through oil or sewage spills, through intentional and controlled release, such as from industrial or agricultural effluents, or through application of petroleum-derived products such as herbicides and pesticides. These can have considerable impacts on microorganisms and the homeostasis of their environments, such as soil. Hydrocarbons can be broken down by a variety of microorganisms, but their susceptibility varies depending on their chemical structures; meaning that many can remain abundant in an environment without intervention. This intervention is termed bioremediation. MS-based methods are highly suited for understanding natural microbial processes for the breakdown of hydrocarbons. This is because they can detect hydrocarbon pollutants, the microorganisms present within a community that can break down hydrocarbons, and the biochemical processes involved. Approaches employing MS to improve the process of bioremediation of hydrocarbons have seen considerable success. As previously discussed however, the complexity of such systems means that an approach using a single MS technique will be unable to cover all aspects of an environmental sample and its microbial community. Metaproteomics using MS has seen several instances of successful use in improving the understanding of bioremediation processes associated with hydrocarbon pollution. When also combined with LC–MS metabolomics, a fuller picture of the bioremediation processes of three separate Mediterranean Sea crude oil spillages could be gained. By comparing the proteomic and metabolomic profiles from the microbial communities associated with each spillage, considerable heterogeneity could be seen. This suggests that different communities adopt different processes to bioremediation. Interestingly, both proteomes and metabolomes showed a high prevalence of anaerobic biochemical processes that are not linked to petroleum degradation and likely associated with survival in anaerobic conditions. This means that bioremediation occurs in anaerobic environments, such as sediments, and suggests that different microbial communities would be suited to bioremediation of petroleum in environments with differing levels of oxygen.[78]

13.5.2 MALDI-ToF MS for Water Quality Testing and Processing

Water systems are a common interface in the human–environment interaction axis, with microorganisms playing a central role. This is because water safety is typically associated with the abundance of pathogenic microorganisms and the treatment of sewage and effluents involves microbial communities in wastewater treatment plants. MS is an important means for monitoring and understanding these systems. Environmental monitoring of water bodies is typically centred on the enumeration of microorganisms associated with faecal contamination, such as the Enterobacteriaceae family. Traditionally reliant on traditional phenotypic and biochemical identification, public health laboratories responsible for testing water quality now rely on MALDI-ToF MS for identification.[77] This extends to not only sampling of seawaters, such as those used for public bathing, but also testing of groundwater for drinking water. Several studies using MALDI-ToF for microbial identification have been completed to provide near-real-time monitoring of the microbiome of groundwater sites in response to anthropogenic activity. In addition to identifying the potential risks of opportunistic pathogens present in groundwater samples, the finer-level taxonomic identification achieved using MALDI-ToF MS, as compared to traditional approaches, means that there is a greater ability to understand contamination sources and mitigation and future prevention.[79,80] Besides its uses in water quality testing, MALDI-ToF MS has also seen deployment as a research tool for understanding the microorganisms associated with the environment of desalination water plants. MALDI-ToF MS could be used not only as a tool for taxonomic identification but also as a tool for its ability to utilise antiscalants found in plants as carbon and energy sources for growth. This method could also be used as a tool for community profiling of environmental water samples to inform the location of desalination water plants and the development of novel materials for resistance to biological degradation and increase the overall operation efficiency.[81]

13.5.3 Improving Plant–Microbiome Interactions for Agricultural Efficiency

The human gastrointestinal tract is one of the most highly studied environments in terms of microbiomes. However, compared to the plant microbiome it is a somewhat simplified system in that its environmental conditions are largely maintained at a constant by the host. The plant microbiome is subjected to a much broader range of both abiotic and biotic stressors. It is essential to the correct growth of the plant and has been identified as a potential means of manipulation to improve agricultural efficiency. This does, however, require a broad understanding of the microbial community in terms of both its taxonomic composition and functional operation. Although genomic and transcriptomic approaches have made some progress in this area, the complexity of both plant and microbial genomes involved means that there is substantial unknown material and function. Further information on this 'dark matter' can, however, be accessed through the utilisation of MS techniques.[82] Volatile organic compounds (VOCs) that are produced by plants have been identified as highly important in interactions with the plant microbiome. In controlled environments, which allow tighter control of extraneous variables, microbial production of VOCs has been positively correlated with both pathogen suppression and plant growth promotion. This suggests that the plant microbiome could be manipulated not only to suppress pathogens which threaten crop production, but also to maximise plant growth.[83] However, how these findings translate to 'real-life' environmental conditions is unknown and represents a real field-wide challenge.

13.6 Concluding Remarks

The microbial communities within environmental samples are complex but hold huge importance in addressing global challenges associated with biodiversity, climate change and resilience, agriculture and food security, and environmental pollution. Due to the difficulty with culturing environmental microorganisms, the field has long been reliant on culture-independent genomic and transcriptomic sequencing approaches. As MS instrumentation and approaches have improved, there is an increasing adoption of these techniques not only in taxonomic identification of microorganisms but also in probing broader functional relationships between microbiomes and the environment. This will likely increase as there are further improvements to instrument sensitivity, ease of use, and supporting databases for taxonomic and compound identification. Nevertheless, a major barrier to the greater impact of these methods is how they are currently implemented in the analysis of environmental samples and associated microbial communities. As evident through the examples of research discussed in this chapter, many instances where MS approaches have been used are focused on observational studies. Although they are powerful methods for understanding aspects of microbial communities, they are limited in their ability to move beyond correlations and towards mechanistic understanding at the level of biochemical pathways. This level of understanding will be required for manipulation of microbial communities to overcome many of the challenges that are faced at the human–environment interface. To begin to address this, controlled mechanistic experiments, typically under highly controlled conditions, will be required. Although the broader translation of these findings to 'real-life' environmental conditions will be challenging, this can be overcome through interdisciplinary research collaborations across not only the environmental sciences but also related bioanalytical chemistry and microbiology disciplines. We have come a long way since the day of van Leeuwenhoek being only able to visualise microorganisms using low-magnification microscopes, but it is only recently that we have had the tools to move away from only asking 'who is there?'. Now we can begin to ask: 'what are they doing?', 'how are they doing it?', and 'how can we harness these processes?'. These are questions that MS has a central role in answering.

Abbreviations

2,4,5-T	2,4,5-Trichlorophenoxyacetic acid
AIMS	Ambient ionisation mass spectrometry
CFU	Colony-forming unit

DART	Direct analysis in real time
DDA	Data-dependent acquisition
DESI	Desorption electrospray ionisation
DIA	Data-independent acquisition
DNA	Deoxyribonucleic acid
GC–MS	Gas chromatography–mass spectrometry
Ichip	Isolation chip
ITS	Internal transcribed spacer
LC–MS	Liquid chromatography–mass spectrometry
LESA	Liquid extraction surface analysis
MALDI	Matrix-assisted laser desorption/ionisation
mRNA	Messenger ribonucleic acid
MS	Mass spectrometry or mass spectra
NMR	Nuclear magnetic resonance
PCR	Polymerase chain reaction
REIMS	Rapid evaporative ionisation mass spectrometry
RNA	Ribonucleic acid
rRNA	Ribosomal ribonucleic acid
ToF	Time of flight
VOCs	Volatile organic compounds

References

1. P. C. Stern, A second environmental science: human-environment interactions, *Science*, 1993, **260**(5116), 1897–1899.
2. R. Singh, Microorganism as a tool of bioremediation technology for cleaning environment: a review, *Proc. Int. Acad. Ecol. Environ. Sci.*, 2014, **4**(1), 1.
3. N. S. Raymond, B. Gómez-Muñoz, F. J. van der Bom, O. Nybroe, L. S. Jensen, D. S. Müller-Stöver, A. Oberson and A. E. Richardson, Phosphate-solubilising microorganisms for improved crop productivity: a critical assessment, *New Phytol.*, 2021, **229**(3), 1268–1277.
4. R. Cavicchioli, W. J. Ripple, K. N. Timmis, F. Azam, L. R. Bakken, M. Baylis, M. J. Behrenfeld, A. Boetius, P. W. Boyd and A. T. Classen, Scientists' warning to humanity: microorganisms and climate change, *Nat. Rev. Microbiol.*, 2019, **17**(9), 569–586.
5. H. Gest, The discovery of microorganisms by Robert Hooke and Antoni Van Leeuwenhoek, fellows of the Royal Society, *Notes Rec. R. Soc. London*, 2004, **58**(2), 187–201.
6. E. J. Stewart, Growing unculturable bacteria, *J. Bacteriol.*, 2012, **194**(16), 4151–4160.
7. I. Laudadio, V. Fulci, L. Stronati and C. Carissimi, Next-generation metagenomics: methodological challenges and opportunities, *OMICS: J. Integr. Biol.*, 2019, **23**(7), 327–333.
8. N. Lombard, E. Prestat, J. D. van Elsas and P. Simonet, Soil-specific limitations for access and analysis of soil microbial communities by metagenomics, *FEMS Microbiol. Ecol.*, 2011, **78**(1), 31–49.

9. J. R. Marchesi and J. Ravel, The vocabulary of microbiome research: a proposal, *Microbiome*, 2015, **3**(1), 31.
10. K. P. Aßhauer, B. Wemheuer, R. Daniel and P. Meinicke, Tax4Fun: predicting functional profiles from metagenomic 16S rRNA data, *Bioinformatics*, 2015, **31**(17), 2882–2884.
11. M. G. Langille, J. Zaneveld, J. G. Caporaso, D. McDonald, D. Knights, J. A. Reyes, J. C. Clemente, D. E. Burkepile, R. L. V. Thurber and R. Knight, Predictive functional profiling of microbial communities using 16S rRNA marker gene sequences, *Nat. Biotechnol.*, 2013, **31**(9), 814–821.
12. C. Quince, A. W. Walker, J. T. Simpson, N. J. Loman and N. Segata, Shotgun metagenomics, from sampling to analysis, *Nat. Biotechnol.*, 2017, **35**(9), 833–844.
13. P. M. Schenk, L. C. Carvalhais and K. Kazan, Unraveling plant–microbe interactions: can multi-species transcriptomics help?, *Trends Biotechnol.*, 2012, **30**(3), 177–184.
14. R. Eisenhofer, J. J. Minich, C. Marotz, A. Cooper, R. Knight and L. S. Weyrich, Contamination in low microbial biomass microbiome studies: issues and recommendations, *Trends Microbiol.*, 2019, **27**(2), 105–117.
15. J. M. Janda and S. L. Abbott, Bacterial identification for publication: when is enough enough?, *J. Clin. Microbiol.*, 2002, **40**(6), 1887–1891.
16. K. Dettmer, P. A. Aronov and B. D. Hammock, Mass Spectrometry-Based Metabolomics, *Mass Spectrom. Rev.*, 2007, **26**(1), 51–78.
17. E. Carbonnelle, C. Mesquita, E. Bille, N. Day, B. Dauphin, J.-L. Beretti, A. Ferroni, L. Gutmann and X. Nassif, MALDI-TOF mass spectrometry tools for bacterial identification in clinical microbiology laboratory, *Clin. Biochem.*, 2011, **44**(1), 104–109.
18. I. C. Santos, Z. L. Hildenbrand and K. A. Schug, Applications of MALDI-TOF MS in environmental microbiology, *Analyst*, 2016, **141**(10), 2827–2837.
19. A. Bodor, N. Bounedjoum, G. E. Vincze, Á. Erdeiné Kis, K. Laczi, G. Bende, Á. Szilágyi, T. Kovács, K. Perei and G. Rákhely, Challenges of unculturable bacteria: environmental perspectives, *Rev. Environ. Sci. Bio/Technol.*, 2020, **19**(1), 1–22.
20. M. Grote, Petri dish versus Winogradsky column: a longue durée perspective on purity and diversity in microbiology, 1880s–1980s, *Hist. Philos. Life Sci.*, 2018, **40**(1), 1–30.
21. M. Nowrotek, Ł. Jałowiecki, M. Harnisz and G. A. Płaza, Culturomics and metagenomics: In understanding of environmental resistome, *Front. Environm. Sci. Eng.*, 2019, **13**(3), 1–12.
22. T. Kaeberlein, K. Lewis and S. S. Epstein, Isolating "uncultivable" microorganisms in pure culture in a simulated natural environment, *Science*, 2002, **296**(5570), 1127–1129.
23. B. Berdy, A. L. Spoering, L. L. Ling and S. S. Epstein, In situ cultivation of previously uncultivable microorganisms using the ichip, *Nat. Protoc.*, 2017, **12**(10), 2232–2242.

24. D. Nichols, N. Cahoon, E. Trakhtenberg, L. Pham, A. Mehta, A. Belanger, T. Kanigan, K. Lewis and S. Epstein, Use of ichip for high-throughput in situ cultivation of "uncultivable" microbial species, *Appl. Environ. Microbiol.*, 2010, **76**(8), 2445–2450.
25. L. J. Piddock, Teixobactin, the first of a new class of antibiotics discovered by iChip technology?, *J. Antimicrob. Chemother.*, 2015, **70**(10), 2679–2680.
26. T. Krishnamurthy, U. Rajamani and P. Ross, Detection of pathogenic and non-pathogenic bacteria by matrix-assisted laser desorption/ionization time-of-flight mass spectrometry, *Rapid Commun. Mass Spectrom.*, 1996, **10**(8), 883–888.
27. R. Holland, J. Wilkes, F. Rafii, J. Sutherland, C. Persons, K. Voorhees and J. LayJr, Rapid identification of intact whole bacteria based on spectral patterns using matrix-assisted laser desorption/ionization with time-of-flight mass spectrometry, *Rapid Commun. Mass Spectrom.*, 1996, **10**(10), 1227–1232.
28. M. Pavlovic, I. Huber, R. Konrad and U. Busch, Application of MALDI-TOF MS for the identification of food borne bacteria, *Open Microbiol. J.*, 2013, 7, 135.
29. S. Schubert and M. Kostrzewa, MALDI-TOF MS in the microbiology laboratory: current trends, *Curr. Issues Mol. Biol.*, 2017, **23**(1), 17–20.
30. B. Rodríguez-Sánchez, E. Cercenado, A. T. Coste and G. Greub, Review of the impact of MALDI-TOF MS in public health and hospital hygiene, 2018, *Eurosurveillance*, 2019, **24**(4), 1800193.
31. A. Wieser, L. Schneider, J. Jung and S. Schubert, MALDI-TOF MS in microbiological diagnostics—identification of microorganisms and beyond (mini review), *Appl. Microbiol. Biotechnol.*, 2012, **93**(3), 965–974.
32. T. C. Dingle and S. M. Butler-Wu, MALDI-TOF mass spectrometry for microorganism identification, *Clin. Lab. Med.*, 2013, **33**(3), 589–609.
33. A. Croxatto, G. Prod'hom and G. Greub, Applications of MALDI-TOF mass spectrometry in clinical diagnostic microbiology, *FEMS Microbiol. Rev.*, 2012, **36**(2), 380–407.
34. M. Fraser, Z. Brown, M. Houldsworth, A. M. Borman and E. M. Johnson, Rapid identification of 6328 isolates of pathogenic yeasts using MALDI-ToF MS and a simplified, rapid extraction procedure that is compatible with the Bruker Biotyper platform and database, *Med. Mycol.*, 2016, **54**(1), 80–88.
35. S. Ranque, A. C. Normand, C. Cassagne, J. B. Murat, N. Bourgeois, F. Dalle, M. Gari-Toussaint, P. Fourquet, M. Hendrickx and R. Piarroux, MALDI-TOF mass spectrometry identification of filamentous fungi in the clinical laboratory, *Mycoses*, 2014, **57**(3), 135–140.
36. T. R. Sandrin, J. E. Goldstein and S. Schumaker, MALDI TOF MS profiling of bacteria at the strain level: a review, *Mass Spectrom. Rev.*, 2013, **32**(3), 188–217.
37. C. L. Feider, A. Krieger, R. J. DeHoog and L. S. Eberlin, Ambient ionization mass spectrometry: recent developments and applications, *Anal. Chem.*, 2019, **91**(7), 4266–4290.

38. J. P. Anhalt and C. Fenselau, Identification of bacteria using mass spectrometry, *Anal. Chem.*, 1975, **47**, 219–225.
39. R. B. Cody, J. A. Laramée and H. D. Durst, Versatile new ion source for the analysis of materials in open air under ambient conditions, *Anal. Chem.*, 2005, 77(8), 2297–2302.
40. Z. Takats, J. M. Wiseman, B. Gologan and R. G. Cooks, Mass spectrometry sampling under ambient conditions with desorption electrospray ionization, *Science*, 2004, **306**(5695), 471–473.
41. Y. Song, N. Talaty, W. A. Tao, Z. Pan and R. G. Cooks, Rapid ambient mass spectrometric profiling of intact, untreated bacteria using desorption electrospray ionization, *Chem. Commun.*, 2007, 61–63.
42. V. Kertesz and G. J. Van Berkel, Fully automated liquid extraction-based surface sampling and ionization using a chip-based robotic nanoelectrospray platform, *J. Mass Spectrom.*, 2010, **45**(3), 252–260.
43. K. C. Schäfer, J. Dénes, K. Albrecht, T. Szaniszló, J. Balog, R. Skoumal, M. Katona, M. Tóth, L. Balogh and Z. Takáts, In vivo, in situ tissue analysis using rapid evaporative ionization mass spectrometry, *Angew. Chem., Int. Ed.*, 2009, **48**(44), 8240–8242.
44. E. C. Randall, J. Bunch and H. J. Cooper, Direct analysis of intact proteins from Escherichia coli colonies by liquid extraction surface analysis mass spectrometry, *Anal. Chem.*, 2014, **86**(21), 10504–10510.
45. A. N. Conner, J. R. Jarvis, L. J. Alderwick and R. L. Griffiths, Direct liquid extraction surface analysis mass spectrometry of cell wall lipids from mycobacteria: Salt additives for decreased spectral complexity, *Rapid Commun. Mass Spectrom.*, 2021, **35**, e8523.
46. J. Balog, L. Sasi-Szabó, J. Kinross, M. R. Lewis, L. J. Muirhead, K. Veselkov, R. Mirnezami, B. Dezső, L. Damjanovich, A. Darzi, J. K. Nicholson and Z. Takats, Intraoperative tissue identification using rapid evaporative ionization mass spectrometry, *Sci. Transl. Med.*, 2013, **5**, 194ra93.
47. N. Strittmatter, E. A. Jones, K. A. Veselkov, M. Rebec, J. G. Bundy and Z. Takats, Analysis of Intact Bacteria using Rapid Evaporative Ionisation Mass Spectrometry, *Chem. Commun.*, 2013, **49**, 6188–6190.
48. K. Verplanken, S. Stead, R. Jandova, C. Van Poucke, J. Claereboudt, J. V. Bussche, S. De Saeger, Z. Takats, J. Wauters and L. Vanhaecke, Rapid evaporative ionization mass spectrometry for high-throughput screening in food analysis: The case of boar taint, *Talanta*, 2017, **169**, 30–36.
49. S. J. Cameron, Z. Bodai, B. Temelkuran, A. Perdones-Montero, F. Bolt, A. Burke, K. Alexander-Hardiman, M. Salzet, I. Fournier, M. Rebec and Z. Takats, Utilisation of Ambient Laser Desorption Ionisation Mass Spectrometry (ALDI-MS) Improves Lipid-Based Microbial Species Level Identification, *Sci. Rep.*, 2019, **9**(1), 3006.
50. S. J. Cameron, A. Perdones-Montero, L. Van Meulebroek, A. Burke, K. Alexander-Hardiman, D. Simon, R. Schaffer, J. Balog, T. Karancsi, T. Rickards, M. Rebec, S. Stead, L. Vanhaecke and Z. Takats, Sample

Preparation Free Mass Spectrometry Using Laser-Assisted Rapid Evaporative Ionization Mass Spectrometry: Applications to Microbiology, Metabolic Biofluid Phenotyping, and Food Authenticity, *J. Am. Soc. Mass Spectrom.*, 2021, **32**(6), 1393–1401.

51. E. E. Bardin, S. J. Cameron, A. Perdones-Montero, K. Hardiman, F. Bolt, E. Alton, A. Bush, J. C. Davies and Z. Takáts, Metabolic Phenotyping and Strain Characterisation of Pseudomonas aeruginosa Isolates from Cystic Fibrosis Patients Using Rapid Evaporative Ionisation Mass Spectrometry, *Sci. Rep.*, 2018, **8**(1), 10952.
52. M. L. DeMarco and C. A. D. Burnham, Diafiltration MALDI-TOF mass spectrometry method for culture-independent detection and identification of pathogens directly from urine specimens, *Am. J. Clin. Pathol.*, 2014, **141**(2), 204–212.
53. S. J. Cameron and Z. Takáts, Mass Spectrometry Approaches to Metabolic Profiling of Microbial Communities within the Human Gastrointestinal Tract, *Methods*, 2018, **149**, 13–24.
54. S. Lamichhane, P. Sen, M. A. Alves, H. C. Ribeiro, P. Raunioniemi, T. Hyötyläinen and M. Orešič, Linking gut microbiome and lipid metabolism: moving beyond associations, *Metabolites*, 2021, **11**(1), 55.
55. P. Wilmes and P. L. Bond, The application of two-dimensional polyacrylamide gel electrophoresis and downstream analyses to a mixed community of prokaryotic microorganisms, *Environ. Microbiol.*, 2004, **6**(9), 911–920.
56. C. Su, L. Lei, Y. Duan, K.-Q. Zhang and J. Yang, Culture-independent methods for studying environmental microorganisms: methods, application, and perspective, *Appl. Microbiol. Biotechnol.*, 2012, **93**(3), 993–1003.
57. T. Schneider and K. Riedel, Environmental proteomics: analysis of structure and function of microbial communities, *Proteomics*, 2010, **10**(4), 785–798.
58. F. Zhang, W. Ge, G. Ruan, X. Cai and T. Guo, Data-independent acquisition mass spectrometry-based proteomics and software tools: a glimpse in 2020, *Proteomics*, 2020, **20**(17–18), 1900276.
59. M.-E. Guazzaroni, F.-A. Herbst, I. Lores, J. Tamames, A. I. Peláez, N. Lopez-Cortés, M. Alcaide, M. V. Del Pozo, J. M. Vieites and M. Von Bergen, Metaproteogenomic insights beyond bacterial response to naphthalene exposure and bio-stimulation, *ISME J.*, 2013, 7(1), 122–136.
60. U. Lechner, D. Türkowsky, T. T. H. Dinh, H. Al-Fathi, S. Schwoch, S. Franke, M. S. Gerlach, M. Koch, M. von Bergen and N. Jehmlich, Desulfitobacterium contributes to the microbial transformation of 2, 4, 5-T by methanogenic enrichment cultures from a Vietnamese active landfill, *Microb. Biotechnol.*, 2018, **11**(6), 1137–1156.
61. A. I. Nesvizhskii, Proteogenomics: concepts, applications and computational strategies, *Nat. Methods*, 2014, **11**(11), 1114–1125.
62. K. V. Ruggles, K. Krug, X. Wang, K. R. Clauser, J. Wang, S. H. Payne, D. Fenyö, B. Zhang and D. Mani, Methods, tools and current perspectives in proteogenomics, *Mol. Cell. Proteomics*, 2017, **16**(6), 959–981.

63. J. Armengaud, E. Marie Hartmann and C. Bland, Proteogenomics for environmental microbiology, *Proteomics*, 2013, **13**(18–19), 2731–2742.
64. R. J. Ram, N. C. VerBerkmoes, M. P. Thelen, G. W. Tyson, B. J. Baker, R. C. Blake, M. Shah, R. L. Hettich and J. F. Banfield, Community proteomics of a natural microbial biofilm, *Science*, 2005, **308**(5730), 1915–1920.
65. S. W. Singer, C. S. Chan, A. Zemla, N. C. VerBerkmoes, M. Hwang, R. L. Hettich, J. F. Banfield and M. P. Thelen, Characterization of cytochrome 579, an unusual cytochrome isolated from an iron-oxidizing microbial community, *Appl. Environ. Microbiol.*, 2008, **74**(14), 4454–4462.
66. J. Heidler and R. U. Halden, Fate of organohalogens in US wastewater treatment plants and estimated chemical releases to soils nationwide from biosolids recycling, *J. Environ. Monit.*, 2009, **11**(12), 2207–2215.
67. B. Zhou, J. F. Xiao, L. Tuli and H. W. Ressom, LC-MS-based metabolomics, *Mol. BioSyst.*, 2012, **8**(2), 470–481.
68. D. J. Beale, F. R. Pinu, K. A. Kouremenos, M. M. Poojary, V. K. Narayana, B. A. Boughton, K. Kanojia, S. Dayalan, O. A. Jones and D. A. Dias, Review of recent developments in GC–MS approaches to metabolomics-based research, *Metabolomics*, 2018, **14**(11), 1–31.
69. W. M. Van Gulik, Fast sampling for quantitative microbial metabolomics, *Curr. Opin. Biotechnol.*, 2010, **21**(1), 27–34.
70. J. Saric, Y. Wang, J. Li, M. Coen, J. Utzinger, J. R. Marchesi, J. Keiser, K. Veselkov, J. C. Lindon and J. K. Nicholson, Species variation in the fecal metabolome gives insight into differential gastrointestinal function, *J. Proteome Res.*, 2007, 7(01), 352–360.
71. S. Moosmang, M. Pitscheider, S. Sturm, C. Seger, H. Tilg, M. Halabalaki and H. Stuppner, Metabolomic analysis—Addressing NMR and LC-MS related problems in human feces sample preparation, *Clin. Chim. Acta*, 2019, **489**, 169–176.
72. V. Aguiar-Pulido, W. Huang, V. Suarez-Ulloa, T. Cickovski, K. Mathee and G. Narasimhan, Metagenomics, metatranscriptomics, and metabolomics approaches for microbiome analysis: supplementary issue: bioinformatics methods and applications for big metagenomics data, *Evol. Bioinf.*, 2016, **12**, EBO.S36436.
73. L. Peisl, E. L. Schymanski and P. Wilmes, Dark matter in host-microbiome metabolomics: tackling the unknowns–a review, *Anal. Chim. Acta*, 2018, **1037**, 13–27.
74. J. K. Jansson and E. S. Baker, A multi-omic future for microbiome studies, *Nat. Microbiol.*, 2016, **1**(5), 1–3.
75. W. Huang, D. Sun, L. Chen and Y. An, Integrative analysis of the microbiome and metabolome in understanding the causes of sugarcane bitterness, *Sci. Rep.*, 2021, **11**(1), 1–11.
76. R. S. McClure, J.-Y. Lee, T. R. Chowdhury, E. M. Bottos, R. A. White, Y.-M. Kim, C. D. Nicora, T. O. Metz, K. S. Hofmockel and J. K. Jansson, Integrated network modeling approach defines key metabolic responses of soil microbiomes to perturbations, *Sci. Rep.*, 2020, **10**(1), 1–9.

77. M. Y. Ashfaq, D. A. Da'na and M. A. Al-Ghouti, Application of MALDI-TOF MS for identification of environmental bacteria: A review, *J. Environ. Manage.*, 2022, **305**, 114359.
78. R. Bargiela, F. A. Herbst, M. Martínez-Martínez, J. Seifert, D. Rojo, S. Cappello, M. Genovese, F. Crisafi, R. Denaro and T. N. Chernikova, Metaproteomics and metabolomics analyses of chronically petroleum-polluted sites reveal the importance of general anaerobic processes uncoupled with degradation, *Proteomics*, 2015, **15**(20), 3508–3520.
79. P. Jančová, V. Pachlová, E. Čechová, K. Cedidlová, J. Šerá, H. Pištěková, F. Buňka and L. Buňková, Occurrence of biogenic amines producers in the wastewater of the dairy industry, *Molecules*, 2020, **25**(21), 5143.
80. Y. Suzuki, K. Niina, T. Matsuwaki, K. Nukazawa and A. Iguchi, Bacterial flora analysis of coliforms in sewage, river water, and ground water using MALDI-TOF mass spectrometry, *J. Environ. Sci. Health, Part A*, 2018, **53**(2), 160–173.
81. M. Y. Ashfaq, M. A. Al-Ghouti, H. Qiblawey, D. F. Rodrigues, Y. Hu and N. Zouari, Isolation, identification and biodiversity of antiscalant degrading seawater bacteria using MALDI-TOF-MS and multivariate analysis, *Sci. Total Environ.*, 2019, **656**, 910–920.
82. L. Xu, G. Pierroz, H. M.-L. Wipf, C. Gao, J. W. Taylor, P. G. Lemaux and D. Coleman-Derr, Holo-omics for deciphering plant-microbiome interactions, *Microbiome*, 2021, **9**(1), 1–11.
83. W. Raza, J. Wang, A. Jousset, V.-P. Friman, X. Mei, S. Wang, Z. Wei and Q. Shen, Bacterial community richness shifts the balance between volatile organic compound-mediated microbe–pathogen and microbe–plant interactions, *Proc. R. Soc. B*, 2020, **287**(1925), 20200403.

CHAPTER 14

Mass Spectrometry (MS)-based Bacterial Identification in the Food Industry

HONGYAN BI

College of Food Science and Engineering, Shanghai Ocean University, Hucheng Ring Road 999, Pudong New District, 201306 Shanghai, China
Email: hybi@shou.edu.cn

14.1 Introduction

Food can provide nutritional support for the human body. Food can include any processed, semi-processed, or raw product that is destined for human consumption.

Food safety is crucial to protect consumers from health risks. These health risks may be related to common allergens and/or foodborne illnesses which are usually caused by bacteria, such as *Salmonella*, *Listeria*, *Vibrio cholerae*, and pathogenic *Escherichia coli*, and/or by viruses, such as the Norovirus and Hepatitis A virus. These pathogenic microbes may enter the body through contaminated food or water.

Food bacteria play important roles in food safety as well as in food processing. Pathogenic bacteria may cause foodborne illnesses that can threaten consumers' lives. On the other hand, beneficial bacteria (probiotics) are used to improve the shelf-life and characteristics of foods.

Detection of pathogenic microbes as well as antibiotic residues in animal-based foods is a worldwide food security issue. Traditional methods for identification of bacteria, mainly relying on culturing methods, enzyme-linked immunosorbent assays (ELISA), or DNA-based assays, are usually

New Developments in Mass Spectrometry No. 13
Detection and Analysis of Microorganisms by Mass Spectrometry
Edited by Liang Qiao and Jia Yi

Published by the Royal Society of Chemistry, www.rsc.org

laborious and time-consuming with poor diagnosis competences and/or require costly and bulky equipment though of high sensitivity. The food industry has sought to develop efficient microbial detection strategies.

Recently, techniques like polymerase chain reaction (PCR)-based methods, nucleic acid-based methods, ELISA-based methods, aptamer-based methods, biosensor-based methods, matrix-assisted laser desorption/ionization time-of-flight mass spectrometry (MALDI-TOF MS)-based methods, and electronic nose-based methods have been developed for detecting the presence of bacterial contaminants and antibiotic residues.

MS, as a powerful analytical technique, has shown its strong capability and promise in structure elucidation of molecules. MS has been successfully applied in food microbiology for the supervision of food safety. The goal of the present chapter is to comprehensively cover MS-based analysis of different aspects of bacteria-related food safety in the food industry.

14.2 Bacterial Detection in Foods by MS

Foodborne pathogens can be associated with a wide variety of food products. It is very important to identify foodborne pathogens to supply safe food and prevent foodborne infections. Performing microbial identification can thus assist in ensuring safe products and identifying microbial contaminations.

MALDI-TOF MS is a promising technique for fast and accurate identification of foodborne pathogens. MALDI-TOF MS can be a tool for rapid and specific identification of different strains and species of bacteria in foods. It can be found that MALDI-TOF MS-based protocols have been broadly developed for efficient quality control against microbial agents. Based on the previous reports, MALDI-TOF MS can identify bacteria with accuracies from 70%[1–4] to 100%.[5] The presence of less than 10 CFU per mL *E. coli* is still detectable when utilizing anti-*E. coli* functionalized magnetic beads and MALDI-TOF MS for capture and identification of *E. coli*, used as a model bacterium.[6,7]

Direct analysis of bacteria can greatly simplify the identification of bacteria. Consuming food contaminated with the bacterium *Listeria monocytogenes* usually causes listeriosis, which is a serious infection. MALDI-TOF MS has been utilized to directly detect *L. monocytogenes* from liquid culture and enrichment broths containing foods.[8]

Early detection of pathogenic strains and strains implicated in food poisoning or food spoilage has gained attention. *Citrobacter* is a gastrointestinal pathogen. *Citrobacter* spp. cause urinary tract infections (UTIs), some of the most common bacterial infections, in less than 3% of adults and children. Qin *et al.* investigated the contamination of *Citrobacter* spp. in food and sales environment samples and the antibiotic resistance profiles of the isolates,[9] where the strains of the analyzed bacteria were identified by MALDI-TOF MS.

Enterococcus is a genus with great variability, including pathogenic strains, spoilage strains, and apparently safe strains, including some probiotic strains. Quintela-Baluja *et al.* characterized different food-isolated *Enterococcus* strains by MALDI-TOF MS, paving a way towards the identification of different strains and species of *Enterococcus* in food based on molecular protein markers.[10]

Furthermore, other MS techniques have also been utilized for bacterial analysis in foods. Li *et al.* applied high-resolution pyrolysis gas chromatography–mass spectrometry (HRPGC–MS) for detecting *L. monocytogenes* in food products such as beef and milk.[11]

14.3 MS Applied in Pathogen Detection in Animal-based Foods

14.3.1 Identification of Pathogens in Poultry Meat by MALDI-TOF MS

Manipulation, preparation, and consumption of poultry meat may cause most of the reactive arthritis and irritable bowel syndrome cases due to the intake of contaminated food and/or water. Kumar *et al.* summarized different techniques used for the detection of bacterial pathogens and antibiotic residues in chicken meat.[12]

Arcobacter butzleri is an enteric pathogen. *Campylobacter* and *Salmonella* are gastroenteric pathogens. *Campylobacteriosis*, a common cause of diarrhea in humans, is an infection caused by one of the several species of *Campylobacter* bacteria, and is also associated with Guillain–Barré syndrome. *Salmonella* may cause gastrointestinal illness and fever called salmonellosis.

MALDI-TOF MS, sometimes assisted by biochemical methods or ISO methods, has been utilized to characterize *Arcobacter*, *Campylobacter*, and *Salmonella* in chicken meat. Dekker *et al.* analyzed the frequency of *Arcobacter*, *Campylobacter*, and *Salmonella* antibiotic resistance and the underlying mechanisms of resistance to fluoroquinolones in locally produced and imported poultry sold in urban Ghana,[13] where biochemical methods and MALDI-TOF MS were utilized to identify bacterial strains. Rodrigues *et al.* estimated the prevalence of *Campylobacter* in chicken carcasses from slaughterhouses located in the states of Paraná, Santa Catarina, and Rio Grande do Sul in the South of Brazil[14] and analyzed the samples for *Campylobacter* enumeration using the ISO method 10272-2; the species *C. jejuni* and *C. coli*, important for public health, were identified through MALDI-TOF MS.

Broilers are not equally exposed to bacterial contamination during rearing and processing. This leads to areas with different bacterial communities on carcasses at retail.

MALDI-TOF MS has been utilized for bacterial analysis in broilers. *Enterococcus faecalis* causes approximately 85% to 90% of enterococcal infections. Stepien-Pysniak *et al.* characterized the phenotype and genotype of Enterococci isolated from yolk sac infections (YSI) of broiler chickens from Poland and the Netherlands by MALDI-TOF MS,[15] and found seven peaks in Polish and Dutch isolates, which may be labeled as species-specific biomarkers in *E. faecalis* from YSI. Mangmee *et al.* developed a MALDI-TOF MS-based method to improve the simultaneous identification of the species, subspecies, and serovars of non-typhoidal *Salmonella* (NTS) isolated from broiler samples in a Thai slaughtering and processing factory.[16]

MALDI-TOF MS identification, 16S rRNA amplicon sequencing, or other genetic methods, together with cultures, have been used to assess food microbiological indicators applied on poultry carcasses.[17] Jribi *et al.* isolated *Arcobacter* species (*A. butzleri* and *A. cryaerophilus*) by the culture method from different species of raw poultry meat, and verified the isolates by multiplex PCR (m-PCR) assay and MALDI-TOF MS for determination of the antibiotic resistance profiles of the isolates.[18] Woźniak-Biel *et al.* identified *Campylobacter* strains, isolated from turkeys and broilers, using both PCR and MALDI-TOF MS to reveal the accuracy of identification, as well as to evaluate the antimicrobial and genetic resistance of the investigated strains.[19] Dec *et al.* evaluated the potential use of Amplified 16S Ribosomal DNA Restriction Analysis (16S-ARDRA) and MALDI-TOF MS as methods for species identification of *Lactobacillus* strains in poultry,[20] and a total of 80 *Lactobacillus* strains isolated from the cloacae of chicken, geese, and turkeys were identified to the species level by MALDI-TOF MS (on-plate extraction method) and 16S-ARDRA. MALDI-TOF MS and rRNA sequence analysis were utilized for species identification of *C. jejuni* and *C. coli* isolates from raw poultry products.[21] MALDI-TOF MS profiling and biochemical and serological tests were utilized to isolate and characterize *Salmonella Enteritidis* and *Salmonella Typhimurium* from chicken meat in Egypt.[22]

Combination of MALDI-TOF MS and 16S rDNA analysis has been utilized to discover bacteria in turkeys. Dec *et al.* identified *Lactobacillus* isolates derived from turkeys from six Polish farms and characterized their phenotypic and genotypic antibiotic resistance profiles by MALDI-TOF MS and 16S rDNA analysis,[23] and found that the dominant species were *L. salivarius* (35%), followed by *L. crispatus* (21%), *L. ingluviei* (14.5%), and *L. johnsonii* (10%). A high prevalence of resistance to tetracycline (68% resistant isolates), lincomycin (64.5%), and enrofloxacin (60%) was observed among the lactobacilli tested. Fewer than 50% isolates were resistant to ampicillin (47%), erythromycin (45%), streptomycin (31%), chloramphenicol (29%), and gentamicin (10%).

The distribution of bacteria on different areas of broiler carcasses was investigated by MALDI-TOF MS and 16S amplicon sequencing. Yu *et al.* assessed the bacterial communities on the neck, breast, and back skin on broiler carcasses at retail through classical International Organization of Standardization-based isolation methods combined with identification by MALDI-TOF MS and 16S amplicon sequencing,[24] and found that the distribution of bacteria on the different areas could be related to the contamination during slaughter as well as bird-rearing methods.

Integration of MALDI-TOF and PCR has been utilized for the differentiation of bacterial species in poultry meat. Hilgarth *et al.* analyzed

Photobacterium carnosum sp. nov., isolated from spoiled modified atmosphere packaged poultry meat by PCR and MALDI-TOF MS,[25] and fingerprinting based on M13-RAPD-PCR (randomly amplified polymorphic DNA-polymerase chain reaction) band patterns and MALDI-TOF MS profiles not only allowed intraspecies differentiation among the analyzed isolates but also supported their distinct lineage to a novel species.

14.3.2 Detection of Bacteria in Aquatic Animal-based Foods by MALDI-TOF MS

Identification and detection of pathogenic bacteria in aquatic animal-based foods are imperative for the safe consumption of these foods.

MS-based strategies have been developed for the analysis of histamine-producing bacteria (HPB), *Vibrio* species, and other pathogenic bacteria in fish or other aquatic food products.

14.3.2.1 Detection of HPB in Fish

HPB produce histamine from histidine contained in food through the action of histidine decarboxylase.

MALDI-TOF MS, sometimes with 16S rDNA sequence analysis or other methods, has been utilized to identify and analyze HPB from fish. Uehara *et al.* developed a method to identify HPB isolated from fish using MALDI-TOF MS,[26] and the method holds potential to simplify the procedure and reduce the time required to identify HPB. Hu *et al.* investigated the differences in histamine accumulation between blue scad (*Decapterus maruadsi*) and chub mackerel (*Scomber japonicus*), and tested the methods of inhibiting HPB to control histamine accumulation in fish,[27] and found that a potassium sorbate or sodium diacetate dipping treatment can control histamine accumulation in fish. 16S rDNA sequence analysis, the VITEK 2 Compact system, and MALDI-TOF MS were utilized to identify the histamine-forming bacteria isolated from blue scad and chub mackerel, and *Citrobacter freundii*, *Citrobacter braakii*, and *Enterobacter aerogenes* were identified.

14.3.2.2 Detection of Vibrio *Species in Marine Organisms by MALDI-TOF MS and LC–MS/MS*

Vibrio species, generally recognized as pathogens, are predominant in seafood along coastal areas.

MALDI-TOF MS has been utilized for the identification of *Vibrio* species. Cho *et al.* established an identification method for *Vibrio* isolated from Korea, based on MALDI-TOF MS. All *Vibrio* isolates and reference strains prepared by trifluoroacetic acid (TFA) extraction, using bacterial broth cultures, were identified to the species level (17/22, 77.3%) and to the genus level (5/22, 22.7%).[28] Burbick *et al.* utilized MALDI-TOF MS for the identification of pathogenic *Vibrio* in fish.[29] *Vibrio* (*Listonella*) *anguillarum*, a pathogenic bacterium, can cause septicemia in a wide range of marine organisms, inducing severe mortalities. Kazazić *et al.* assessed MALDI-TOF MS as a method of choice for identification of clinical *V. anguillarum* isolates from affected marine fish,[30] and found that MALDI-TOF MS protein fingerprinting with on-target extraction should be performed on strains grown on a NaCl-supplemented medium at temperatures between 15 and 22 °C, incubated for 48–72 hours to achieve reliable identification.

MALDI-TOF MS, sometimes with PCR, has been greatly utilized for the discrimination and characterization of *Vibrio* species. Malainine *et al.* utilized MALDI-TOF MS to discriminate closely related environmental strains of *Vibrio parahaemolyticus* for the typing of isolates from the Khnifiss lagoon, Morocco.[31] Raharjo *et al.* isolated and characterized 283 *Vibrionaceae* from 15 Asian sea bass farms located around the provinces of the Andaman Sea and Gulf of Thailand coasts to uncover the distribution and antimicrobial resistance (AMR) profiles by a combination of analysis of biochemical characteristics, MALDI-TOF MS analysis, and species-specific PCR,[32] and found that the predominant *Vibrionaceae* were *Vibrio harveyi* ($n = 56$), *Photobacterium damselae* ($n = 35$), and *Vibrio vulnificus* ($n = 31$).

Liquid chromatography–tandem mass spectrometry (LC–MS/MS) has been utilized for the analysis of *V. cholerae*, a common waterborne pathogen, causing pandemic cholera in humans. Identification of risk factors of *V. cholerae* in aquatic food products is imperative for assuming food safety. Shan *et al.* determined virulence-associated genes, cross-resistance between antibiotics and heavy metals, and genome fingerprinting profiles of non-O1/O139 *V. cholerae*, where two-dimensional gel electrophoresis (2D-GE) and/or LC–MS/MS techniques were utilized to obtain secretomes and proteomes of *V. cholerae* with distinct genotypes and phenotypes.[33]

14.3.2.3 Detection of Other Pathogenic Bacteria in Seafoods by MALDI-TOF MS

MALDI-TOF MS, combined with genetic analysis, has been utilized for the analysis of other pathogenic bacteria in fish such as *Enterobacterales*, *Photobacterium damselae* subsp. *piscicida*, *Pseudomonas*, and *Streptococcus iniae*.

MALDI-TOF MS was compared with biochemical and sequencing methods for identification of bacteria in fish and shrimps. Zakrzewski *et al.* compared the effectiveness of the available methods of identifying order *Enterobacterales* strains isolated from fresh fish and shrimps including biochemical analysis, sequencing, and identification using MALDI-TOF MS. The genus identification reached 90.3% for the MALDI-TOF technique, while for the EnteroTest 24N set it was nearly 70.0%.[1]

Photobacterium damselae subsp. *piscicida* may cause disease in fish. Kazazić *et al.* analyzed fish photobacteriosis, and found that the incubation time, media, target plate preparation, and replicates can affect the reliability of identification of *P. damselae* subsp. *piscicida* by MALDI-TOF MS.[34]

Pseudomonas spp. are major spoilage bacteria. Kačániová *et al.* performed the comparison of MALDI-TOF MS Biotyper and 16S rDNA sequencing for the identification of *Pseudomonas* species isolated from fish,[35] and revealed a high discriminatory power of the mass spectral investigation and 16S rDNA gene sequencing technology for the identification of *Pseudomonas* spp. associated with freshwater fish.

Streptococcus iniae is a major fish pathogen, causing high losses in farmed marine and freshwater finfish in warmer regions. *Streptococcus iniae* causes severe mortalities among cultured marine species, especially in the olive flounder (*Paralichthys olivaceus*). MALDI-TOF MS was utilized to characterize 89 *S. iniae* isolates, obtained from diseased olive flounders collected from 2003 to 2008 in Jeju Island, South Korea.[36]

MALDI-TOF MS integrated with immunomagnetic separation (IMS) has been utilized for the isolation and identification of bacteria from fish. We recently developed a protocol for specific capture and identification of bacteria from fish muscle based on immunomagnetic beads and MALDI-TOF MS,[6] and performed a microbial safety test on the surface of the fish muscle. We further developed a strategy to integrate a microchip with MALDI-TOF MS for the investigation of bacterial contamination of fish muscle products.[7]

The establishment of databases may assist in the identification of fish pathogenic bacteria. Piamsomboon *et al.* developed a custom Main Spectral Profile (MSP) database and validated the method using MALDI-TOF MS for rapid identification of fish bacterial pathogens,[37] and proposed to develop databases to accommodate identification of fish bacterial pathogens by MALDI-TOF MS.

It can be found that MALDI-TOF MS can offer a comparable power for discrimination of bacteria in fish with 16S rDNA sequencing. The factors that can affect the detection of fish bacteria by MALDI-TOF MS can be better controlled together with the establishment of databases for improved reliability of identification and recognition of bacteria.

14.3.3 Identification of Bacteria in Beef by MALDI-TOF MS

MALDI-TOF MS, together with IMS and database searches, has been utilized for identification of bacteria in beef. Ochoa *et al.* developed a method for the isolation and identification of the enterohemorrhagic *E. coli* (serotype O157:H7) in ground beef by immunomagnetic isolation, MALDI-TOF MS, and database searches,[38] and found that IMS increased the sensitivity of the method and permitted the detection of bacteria in meat, with a detection limit of 2×10^6 cells per mL.

14.3.4 Characterization of Bacteria in Porcine Meat by MALDI-TOF MS

MALDI-TOF MS, together with 16S rRNA amplicon sequencing and ISO methods, has been utilized for characterization of bacteria in boar. Peruzy *et al.*

characterized the initial bacterial contamination on freshly-killed wild boar meat using a culture-dependent approach with ISO methods combined with MALDI-TOF MS identification and 16S rRNA amplicon sequencing.[39]

Lactic acid bacteria (LAB) produce lactic acid as the major end product of the fermentation of carbohydrates, and play a vital role in food production and fermentation and probiotic products. LAB are also major components of the microbial flora of packaged and cured meats and can contribute to spoilage.

Sodium dodecyl sulfonate–polyacrylamide gel electrophoresis (SDS–PAGE), 16S rRNA gene sequencing, and MALDI-TOF MS have been utilized to identify LAB in pork meat and pork meat products for knowing the diversity of LAB across the samples.[40]

14.4 Detection of Pathogens and Toxins in Plant-based Foods

Plant pathogens can reduce crop production by causing plant diseases, leading to massive economic losses. Identification of plant pathogens is important for plant disease management through plant pathogen diagnostics.

MALDI-TOF MS has been utilized to analyze specific peptides or proteins directly desorbed from fungal spores, intact bacteria, nematodes, and other microorganisms. Ahmad *et al.* summarized the potential of MALDI-TOF MS as a rapid detection technique in plant pathology from the viewpoint of identification of plant-associated microorganisms.[41]

14.4.1 Identification of Bacteria in Rice Seeds by MALDI-TOF MS

MALDI-TOF MS has been utilized for the identification of bacteria in rice seeds. Microbial analysis of rice seeds has been performed to discover bacteria or toxins in them. Okumura *et al.* applied whole-cell MALDI-TOF MS analysis to identify methylobacterium isolates collected from the rice seeds of different cultivars harvested in Japan, Thailand, and Kenya.[42]

14.4.2 Identification of Bacteria in Lettuce by NanoLC–MS

NanoLC–MS and selected reaction monitoring (SRM), a targeted MS technique that is emerging in the field of proteomics as a complement to untargeted shotgun methods, have been utilized for bacterial analysis in lettuce. *Staphylococcus aureus*, a major bacterial human pathogen, causes a wide variety of clinical manifestations. Chen *et al.* identified bacteria in lettuce by using magnetic zirconia nanoparticles and SRM MS,[43] where identification of *S. aureus* and *E. coli* was achieved using nanoLC–MS and SRM of bacteria-related peptides.

14.4.3 Metabolite Profiling of Microfungi in Fruit by HPLC–MS

High-performance liquid chromatography–MS (HPLC–MS) has been utilized to study the metabolite profiles of microfungi in fruits. Santini *et al.* assessed the taxonomy of grape berries by studying the secondary metabolite profiles of microfungi of the genus *Penicillium*, isolated from samples of grape berries collected in two different phases during two vegetative seasons in Slovakia, using HPLC–MS.[44]

14.4.4 Study of Emitted VOCs of Potatoes by GC–MS and FT-ICR/MS

Microorganisms can cause rotting of harvested food and induce loss of half of all the world's harvested food despite advances in increased food production.

Plants emit various volatile organic compounds (VOCs) into their surrounding environment. The VOC profiles of healthy crops can be altered upon infection. VOCs emitted by plants can be monitored for early detection of plant disease and prevention of food loss.

GC–MS and Fourier transform ion cyclotron resonance/MS (FT-ICR/MS) have been utilized to study VOCs of potatoes.

Potato tuber soft rot disease can be caused by the pectinolytic bacteria, *Pectobacterium*, in potato tubers. Veltman *et al.* utilized GC–MS to specify the differences between the VOC patterns of the inoculated and non-inoculated potato tubers during early infection.[45]

Rhizoctonia solani is a soilborne and seedborne fungal pathogen of potatoes (*Solanum tuberosum*). Aliferis *et al.* reported an integrated metabolomics networking approach combining GC–MS with FT-ICR/MS and bioinformatics analyses for the study of interactions in the potato sprout–*R. solani* pathosystem and the fluctuations in the global metabolome of sprouts,[46] and provided a snapshot of the sprout's global metabolic network and its perturbations as a result of pathogen invasion.

14.4.5 Monitoring of Microbial Metabolites of Nuts by LC–MS/MS

In nuts, toxic microbial metabolites are detected and monitored to check their safety for consumption. Studies on the metabolite profiles of nuts may provide useful information for policy makers, border agencies, and other health-related institutes.

Mycotoxins are toxic compounds naturally produced by certain types of molds (fungi). Consumption of mycotoxin contaminated foodstuffs may be common in regions where foods are not adequately controlled and routinely monitored, and induce adverse effects on the health of consumers.

LC–MS/MS, coupled with HPLC–fluorescence detection (HPLC–FLD), has been utilized for the analysis of the fungal metabolite profile of cashew nuts. Adetunji *et al.* compared the fungal metabolite profile of cashew nuts from two African countries, Nigeria and South Africa, using HPLC–FLD and LC–MS/MS.[47]

LC–MS/MS was utilized to analyze mycotoxins and other microbial metabolites in roasted nuts vended within two cities of Sierra Leone,[48] and it was found that the aflatoxin contamination of Sierra Leonean peanuts was high.

Several species of *Penicillium* often contaminate nuts during the supply chain, from harvest to storage. *Penicillium* spp. have emerged as producers of mycotoxins and other toxic metabolites in nuts. HPLC–MS/MS has been utilized for the analysis of mycotoxins in nuts. Spadaro *et al.* developed a HPLC–MS/MS method for the detection of selected toxic metabolites produced by *Penicillium* spp. in nuts,[49] and 19 metabolites produced by *Penicillium* spp. in chestnuts, hazelnuts, walnuts, and almonds were detected.

Future studies and legislation should take mycotoxins into account due to the risk of contamination of nuts with molds (fungi).

14.5 Detection of Bacteria in Processed Foods

14.5.1 Identification of Bacteria in Juice by NanoLC–MS and SRM MS

NanoLC–MS and SRM MS have been utilized for bacterial analysis in juice. Chen *et al.* identified bacteria in juice by using magnetic zirconia nanoparticles and SRM MS,[43] where identification of *S. aureus* and *E. coli* was achieved using nanoLC–MS and SRM of bacteria-related peptides.

14.5.2 Detection of Bacterial Grown on Ready-to-eat Meat and Fish by MALDI-TOF/TOF MS and MALDI-TOF MS

MALDI-TOF/TOF MS has been utilized to detect bacterial grown on ready-to-eat (RTE) meat and fish.

Mujahid *et al.* examined protein expression by *L. monocytogenes* grown on RTE sliced turkey meat, using *L. monocytogenes* grown on brain-heart-infusion agar as a control,[50] where 2-DE was utilized to extract and compare the total protein fractions of *L. monocytogenes* from both growth conditions; MALDI-TOF/TOF MS was utilized to identify proteins expressed by turkey meat-grown *L. monocytogenes*, and seventy-seven proteins expressed by turkey meat-grown *L. monocytogenes* were identified.

We previously developed a magnetically-assisted microfluidic chip integrated with MALDI-TOF MS detection for monitoring bacterial growth on salmon sashimi, a delicacy made from raw salmon, during its storage period.[7]

14.5.3 Identification of Bacteria Recovered from Baby Foods by MALDI-TOF MS

MALDI-TOF MS has been utilized to study bacteria recovered from baby foods. *Clostridium* is an anaerobic bacteria, consisting of approximately 100 species.

Some *Clostridium* spp. have been recognized to cause acute food poisoning, botulism, tetanus, and diarrhea in humans. Sulaiman *et al.* utilized MALDI-TOF MS for the identification of foodborne *Clostridium* spp.,[4] and identified 28 of the 33 *Clostridium*-like isolates, recovered from various baby food and surveillance samples, with a confidence value of 99.9% by using a VITEK MS system.

14.5.4 Identification of Bacteria in Fermented Meat and Fish Products by GC–MS, LC–MS/MS, and MALDI-TOF MS

14.5.4.1 MALDI-TOF MS Applied in Marinated Chicken and Turkey Meat

Marinating poultry meat is very important for improving its microbiological quality as well as physical characteristics or technological traits. MALDI-TOF MS has been utilized to identify bacteria of the marinated poultry meat with other ingredients.

Sokołowicz *et al.* determined the effects of marinating with fermented milk products (buttermilk and sour milk) on the physical characteristics, microbiological quality, and sensory acceptability of the meat of Rhode Island Red (RIR) hens after the first year of laying use,[51] where the MS method, MALDI-TOF MS Biotyper, was utilized to identify bacteria.

Augustyńska-Prejsnar *et al.* evaluated the effects of marinating turkey meat with buttermilk and acid whey on the technological traits and microbiological quality of the product,[52] where bacterial identification was performed by MALDI-TOF MS.

14.5.4.2 Detection and Identification of Bacteria in Fermented Fish by GC–MS and MALDI-TOF MS

GC–MS combined with gas chromatography–olfactometry (GC–O) analysis, and MALDI-TOF MS has been utilized to detect bacteria in fermented fish.

Suanyu is one of the traditional fermented fish products in South China. It is mainly consumed in certain provinces of China such as Hunan, Guizhou, and Guangxi. GC–MS combined with GC–O analysis was utilized to identify the characteristic flavor and microorganisms related to flavor formation in the product Suanyu, fermented common carp (*Cyprinus carpio* L.),[53] and 21 characteristic VOCs were found, in which the top 10 bacterial genera associated with flavor formation were *Staphylococcus*, *Enterococcus*, *Vibrio*, *Kocuria*, *Lactobacillus*, *Macrococcus*, *Streptococcus*, *Enterobacter*, *Lactococcus*, and *Pediococcus*.

MALDI-TOF MS has also been utilized to analyze bacteria in fermented fishes.

Gajami-sikhae, a traditional Korean fermented fish food, is commonly served as a side dish, and made by naturally fermenting flatfish (*Glyptocephalus stelleri*) with ingredients such as cooked grains, salt, red pepper powder, white radish, garlic, ginger, and green onion. MALDI-TOF MS was utilized to analyze the diversity of a LAB community during fermentation of Gajami-sikhae,[54] and it was found that *Latilactobacillus*, *Lactiplantibacillus*, *Levilactobacillus*, *Weissella*, and *Leuconostoc* were the dominant genera during fermentation, while the dominant species were *Latilactobacillus sakei*, *Lactiplantibacillus plantarum*, *Levilactobacillus brevis*, *Weissella koreensis*, and *Leuconostoc mesenteroides*.

Jeotgal, with a high concentration of salt, is a Korean traditional fermented seafood. Lee *et al.* isolated LAB from galchi (*Trichiurus lepturus*, hairtail) and myeolchi (*Engraulis japonicus*, anchovy) jeotgal on MRS agar and MRS agar containing 5% NaCl (MRS agar + 5% NaCl), and identified them by using 16S rRNA gene sequencing and MALDI-TOF MS.[55]

14.5.4.3 LC–MS/MS Applied in Fermented Soybean Products

LC–MS/MS has been utilized for the analysis of bacteria in fermented soybean products.

Da-jiang is a typical traditional fermented soybean product in China. Zhang *et al.* analyzed the composition and species of microbial proteins in traditional fermented Da-jiang by metaproteomics based on SDS–PAGE and LC–MS/MS.[56]

14.5.4.4 MALDI-TOF MS Applied in Fermented Beverages

MALDI-TOF MS has been utilized for bacterial identification in fermented beverages.

Kunu-zaki is a cereal-based Nigerian fermented beverage. Ogunremi *et al.* identified a total of 164 out of 220 LAB isolates from Kunu-zaki,[57] and a combination of MALDI-TOF MS profiling and 16S rRNA gene sequencing revealed a total of 15 LAB species, of which the most predominant species were *L. plantarum* (40.46%), *Weissella confusa* (27.27%), and *Pediococcus pentosaceus* (15.00%).

14.6 Detection of Bacteria in Packaged Foods by MALDI-TOF MS and PTR–MS

MALDI-TOF MS and proton transfer reaction–MS (PTR–MS) have been utilized for the analysis of bacteria and emitted VOCs in packaged meat and seafoods for spoiling control.

Photobacterium is a widely studied genus due to its association with and ubiquitous presence in marine environments. However, *P. phosphoreum*, *P. carnosum*, and *P. iliopiscarium* have been recently described to colonize and spoil raw meats without a marine link.

VOCs produced by meat bacteria have been studied to evaluate the spoiling of packaged meat. Mayr *et al.* used PTR–MS to investigate the spoiling of air- and vacuum-packed meat (beef, pork, and poultry) during storage at 4 °C for up to 13 days,[58] and measured and identified partly the emitted VOCs as a function of storage time and found a large increase in these emissions after a few days of storage.

Streptococcus parauberis, belonging to the *Streptococcaceae* family, is known as an etiological agent of mastitis in cows and for producing streptococcosis in farmed fish. Fernández-No *et al.* isolated and characterized *S. parauberis* from vacuum-packaged refrigerated seafood products by MALDI-TOF MS.[59]

14.7 Detection of Probiotics in (Canned) Foods by MALDI-TOF MS

Bacillus species are ubiquitous in nature. *Bacillus coagulans* is a probiotic (beneficial bacterium). Spores of several *Bacillus* species have a long history of consumption, and are safe to use as probiotics. There are a variety of formulations containing these organisms available in the global market.

Bacillus coagulans is a spore-forming probiotic bacterium. Due to the high heat resistance of the spores and the high tolerance of the vegetative cells to catechins and chemicals, the typing of *B. coagulans* strains is important for bacterial control in the manufacture of some canned foods and tea-based beverages.

MALDI-TOF MS has been utilized for strain typing of *Bacillus coagulans*,[60] and it was found that the combination of MALDI-TOF MS and repetitive-PCR analyses was advantageous for the rapid and detailed typing of bacterial strains with respect to both phenotype and genotype.

14.8 Detection of Antibiotic Residues and Drug Resistance of Bacteria by MALDI-TOF MS

The use of antibiotics in food animals is of public health concern, since resistant zoonotic pathogens can be transmitted to humans. The drug resistance of bacteria in poultry has gained attention. MALDI-TOF MS has been utilized to assist the identification of drug-resistant bacteria. Rasmussen *et al.* investigated whether imported chicken meat and meat from locally reared chicken are potential sources for human exposure to multiresistant *E. coli* isolates,[61] and 153 bacterial isolates were cultured and identified as *E. coli* using MALDI-TOF MS. Limayem *et al.* studied the drug resistance profile of *Enterococcus faecium* distributed from poultry retailers to hospitals,[62] and identified presumptive positive *Enterococcus* isolates at the species level by MALDI-TOF MS. Pauly *et al.* detected and quantified methicillin-resistant *S. aureus* (MRSA) in fresh broiler meat at retail in Germany,[63] and MALDI-TOF and PCR were utilized to confirm presumptive MRSA colonies.

14.9 MALDI-TOF MS Applied in Identification of Recombinant Bacteria

The MS technique applied in the analysis of GM foods or gene-engineered foods has drawn increasing attention.

MALDI-TOF MS has been used to not only develop rapid, accurate, and high-throughput methods for the identification of recombinant bacteria, but also provide new ideas for the identification of mutant strains and bio-terrorism pathogens. Xiao *et al.* utilized MALDI-TOF MS for the identification of recombinant bacteria, including 12 recombinant proteins, that is, 10 of *Yersinia pestis*, 1 of *C. jejuni*, and 1 of *Helicobacter pylori*.[64]

On the other hand, unexpected contaminations of unauthorized GM microorganisms (GMM) carrying AMR genes were recently reported in microbial fermentation products commercialized on the food and feed chain. Fraiture *et al.* in 2021 developed a DNA walking strategy, based on gene sequencing, to identify unauthorized GM bacteria in microbial fermentation products.[65] MALDI-TOF MS may be a potential tool for the identification of unauthorized GM bacteria in microbial fermentation products.

14.10 MS Applied to Analysis of Bacterially Produced Toxins

14.10.1 Detection of Staphylococcal Toxins in Food Matrices by LC–ESI/MS and LC–QITMS

Staphylococcus species have a ubiquitous habitat in a wide range of foods. *Staphylococcus aureus*, as an important pathogen, has been indicated as the fifth causative agent of food-borne human illnesses throughout the world.

Enterotoxin-producing *S. aureus* is one of the most common causes of food-borne human illnesses worldwide. Staphylococcal enterotoxins (SEs) are toxic compounds excreted mainly by strains of *S. aureus*.

It is well known that MS is a broadly applied tool for proteomic and metabolomic analysis. Sospedra *et al.* summarized the omic sciences applied to the study of staphylococcal toxins analyzed in food matrices.[66]

Reverse phase liquid chromatography coupled to electrospray ionization MS (LC–ESI/MS) and liquid chromatography–quadrupole ion trap MS (LC–QITMS) have been developed for analysis of SEs. Sospedra *et al.* utilized LC–ESI/MS for identification and quantification of SEs A and B.[67] Differential isotopic tags and LC–QITMS were utilized for analysis of *Staphylococcus* enterotoxin B (SEB) in food matrices.[68]

14.10.2 Analysis of Toxic Microbial Metabolites of Nuts by GC–MS and LC–MS/MS

Consumption of mycotoxin-contaminated foodstuffs is common in regions where foods are not adequately controlled and routinely monitored, and this could have adverse effects on the health of consumers.

The toxic microbial metabolites of nuts need to be detected and monitored for their safe consumption.

GC and MS/selected ion monitoring (GC–MS/SIM) and LC–MS/MS have been utilized for the analysis of the metabolite profiles of nuts or other seeds.

Adetunji *et al.* compared the fungal metabolite profiles of cashew nuts from two African countries, Nigeria and South Africa, using HPLC–FLD and LC–MS/MS,[47] and found that a total of 15 unspecific metabolites were recorded in the nuts from both countries.

Gibberellins (GAs) are initially identified as secondary metabolites in the fungus *Gibberella fujikuroi*, with strong implications in plant physiology. Shahzad *et al.* utilized GC–MS/SIM to quantify GAs produced by *Bacillus amyloliquefaciens* RWL-1, an endophyte associated with rice (*Oryza sativa* L. 'Jin so mi') seeds, to see its role in improving host-plant physiology.[69]

14.11 MALDI-TOF MS Applied to Food Product Quality Inspection

MALDI-TOF MS has also been utilized to check whether a food product complies with quality requirements. Celandroni *et al.* identified the *Bacillus* species, and compared the performance of biochemical tests based on the BCL VITEK 2 card and MALDI-TOF MS by using 16S rDNA sequencing as the reference technique where contaminant bacteria (*Lysinibacillus fusiformis*, *Acinetobacter baumannii*, *Bacillus cereus*, *Brevibacillus choshinensis*, *Bacillus licheniformis*, *Bacillus badius*) were detected in some formulations, and it was found that the majority of the analyzed products did not comply with quality requirements.[70]

14.12 MS Applied in Detection of Food Contamination by Analysis of VOCs

14.12.1 Detection of VOCs for Identification of Foodborne Pathogens by GC–MS

VOCs may be associated with a peculiar microbiological contamination of food. GC–MS coupled with solid-phase microextraction has been utilized for the identification of foodborne pathogens by analyzing the emitted VOCs. Wang *et al.* utilized GC–MS coupled with solid-phase microextraction for identification of foodborne pathogens including *Shigella sonnei*, *Staphylococcus aureus*, and *Vibrio parahaemolyticus* in foods,[71] and a total of 17, 13, and 14 VOCs were identified as characteristic VOCs of *S. aureus*, *Vibrio parahaemolyticus*, and *Shigella sonnei*, respectively.

14.12.2 Detection of VOCs for Detection of Microbiological Contamination of Food by GC–MS

Early warning systems, highlighting in real-time bacterial contamination of meat matrices, can be generated to provide information to support companies in accepting or rejecting batches. The identification of unique VOCs' profiles may be taken as markers of meat contamination due to their association with specific pathogens.

HRPGC–MS, solid-phase microextraction (SPME) combined with GC–MS, headspace solid-phase microextraction (HS-SPME), and GC–MS have been utilized for the analysis of VOCs and/or metabolites of microbiological contamination of food. Lee *et al.* analyzed the HRPGC–MS profiles of VOCs produced by microorganisms at refrigerated temperatures,[72] and found twenty different compounds from three different strains of bacteria isolated from spoiled, uncooked chicken, which were grown in pure culture on trypticase soy agar supplemented with yeast extract, where the VOCs produced by each culture were concentrated on a porous polymer precolumn and analyzed by HRPGC–MS. SPME combined with GC–MS was utilized to identify and quantify VOCs as chemical spoilage indices (CSIs) of raw chicken breast stored aerobically at 4, 10, and 21 °C,[73] and it was found that it's possible to analyze the combination of three selected spoilage markers ethanol (EtOH), 1-butanol-3-methyl (1But-3M), and acetic acid (C2) as a rapid evaluation method for poultry quality testing using SPME-GC–MS. Carraturo *et al.* evaluated the ability of the HS-SPME and GC–MS methodologies to detect VOCs,[74] and found that the identification of VOC markers in association with selected bacterial pathogens and their metabolites could support the rapid determination of specific meat samples' contamination.

14.12.3 Detection of VOCs for the Identification of Microorganisms in Harvested Foods by GC–MS and FT-ICR/MS

Rots caused by microorganisms have induced loss of half of the harvested food. Plants can emit VOCs into their surrounding environment. The VOC profiles of healthy crops can be altered upon infection.

GC–MS and FT-ICR/MS have been utilized for the analysis of VOCs or metabolites of infected potatoes. Potato tuber soft rot disease can be caused by the pectinolytic bacteria *Pectobacterium* in potato tubers. Veltman *et al.* utilized GC–MS to specify the differences between the VOC patterns of the inoculated and non-inoculated potato tubers during early infection.[45] *Rhizoctonia solani* is an important soilborne and seedborne fungal pathogen of potatoes (*S. tuberosum*). Aliferis *et al.* reported an integrated metabolomics networking approach combining GC–MS with FT-ICR/MS and bioinformatics analyses for the study of interactions in the potato sprout–*Rhizoctonia solani* pathosystem and the fluctuations in the global metabolome of sprouts,[46] and provided a snapshot of the sprout's global metabolic network and its perturbations as a result of pathogen invasion.

14.12.4 Analysis of VOCs for Evaluation of the Impact of Probiotic Bacteria on Fruit Juice Aroma by GC–MS

GC–MS coupled with SPME has been utilized to analyze VOCs of fruits and probiotic bacteria of fruit juice aroma. Farag *et al.* analyzed the VOCs in 15 Egyptian mango specimens collected from three different regions and of different cultivars, and the impact of probiotic bacteria on mango juice aroma *via* SPME coupled to GC–MS.[75]

14.13 MS Applied in Detection of Spoilage of Animal-based Products

14.13.1 Detection of Spoilage of Poultry Meat During Storage by GC–MS, HPLC–MS/MS, and MALDI-TOF MS

As a result of microbial development and consumption of meat nutrients by spoilage bacteria, spoilage of meat during chilled storage can occur, ultimately resulting in the release of undesired metabolites.

Ultra-high-performance LC–MS/MS (UHPLC–MS/MS), GC–MS, and MALDI-TOF MS have been utilized for analysis of poultry during storage. Zhang *et al.* characterized chilled chicken spoilage using an integrated microbiome and metabolomics analysis,[76] and microbial and metabolic profiles in chicken during chilled storage were determined by 16S rRNA gene sequencing and UHPLC–MS/MS-based untargeted metabolomics analyses.

Besides, principal component analysis (PCA) or a MALDI-TOF MS database was utilized or established for assisting the spoilage control of chicken meat. Tirnsorn *et al.* analyzed fresh and spoiled chicken meats by GC–MS,[77] and used PCA for pattern recognition and classification. A MALDI-TOF MS database was established for meat-borne bacteria to investigate the spoilage of skin-less chicken breast in high O_2 modified atmosphere packaging (MAP) and low O_2 MAP, and the effects of different storage temperatures on the quality of poultry meat.[78]

14.13.2 Detection of Spoilage of Seafoods by MALDI-TOF MS

It is important to ensure food quality and safety by identifying food pathogenic and spoilage bacteria.

MALDI-TOF MS has been utilized for the species identification of seafood spoilage and pathogenic bacteria. Bacterial growth on salmon sashimi during storage was monitored using a magnetically-assisted microfluidic chip integrated with MALDI-TOF MS detection.[7] Böhme *et al.* performed species identification of seafood spoilage and pathogenic Gram-positive bacteria by MALDI-TOF mass fingerprinting, where a collection of the main seafood pathogenic and spoilage Gram-positive bacteria was compiled, including *Bacillus* spp., *Listeria* spp., *Clostridium* spp., *Staphylococcus* spp., and *Carnobacterium* spp. The strains, belonging to 20 different species, were obtained from the culture collections and studied by MALDI-TOF MS.[79]

14.13.3 Detection of Spoilage of Meat by UHPLC–MS/MS

LAB can also contribute to spoilage, and LAB are major components of the microbial flora of packaged and cured meats. Han *et al.* carried out bacteriological analyses of pork meat and pork meat product samples to know the diversity of LAB across the samples.[40]

Leuconostoc mesenteroides are generally recognized as a group of specific spoilage organisms in meat and poultry products, which cause acidification and blown pack spoilage of meat. The spoilage heterogeneity of meat-borne *L. mesenteroides* was evaluated by metabonomics and *in situ* analysis.[80]

Studies have been limited to judging bacteria-derived meat spoilage only by *in vitro* growth or pure standard culture. *In situ* spoilage should be focused on in practical sample analysis. Studies of *in situ* spoilage may provide references for emphasis on future research to reduce the loss of meat in consequence of spoilage. UHPLC–MS/MS has been applied to meat-derived spoilage bacteria. Chen *et al.* analyzed profiles related to growth, acidification, and gas production of *L. mesenteroides in vitro* and *in situ* by UHPLC–MS/MS.[80]

14.13.4 Detection of Spoilage of Packaged Foods by PTR–MS

A PTR–MS-based method was developed to monitor the spoilage of packaged foods by studying the emitted VOCs of meat during its storage. Mayr *et al.* used PTR–MS to investigate the spoiling of air- and vacuum-packed meat (beef, pork, and poultry) during storage at 4 °C for up to 13 days,[58] and measured and identified partly the emitted VOCs as a function of storage time, and found a large increase in these emissions after a few days of storage.

Figure 14.1 compares the MS-based bacterial analysis applied in the food industry for bacteria-related food safety control by the surveyed literature discussed in this chapter. It can be observed that MS is mainly applied in identification and characterization of bacteria in raw food samples. With the emergence of new MS techniques, MS-based strategies will continue to be developed to guarantee food safety.

14.14 Conclusions and Perspectives

Food can provide nutritional support for the human body. Food safety is crucial to protect consumers from health risks. Pathogenic microbes may enter the body through contaminated food or water. Food bacteria play important roles in food safety as well as in food processing. Pathogenic bacteria may cause foodborne illnesses. Beneficial bacteria (probiotics) can

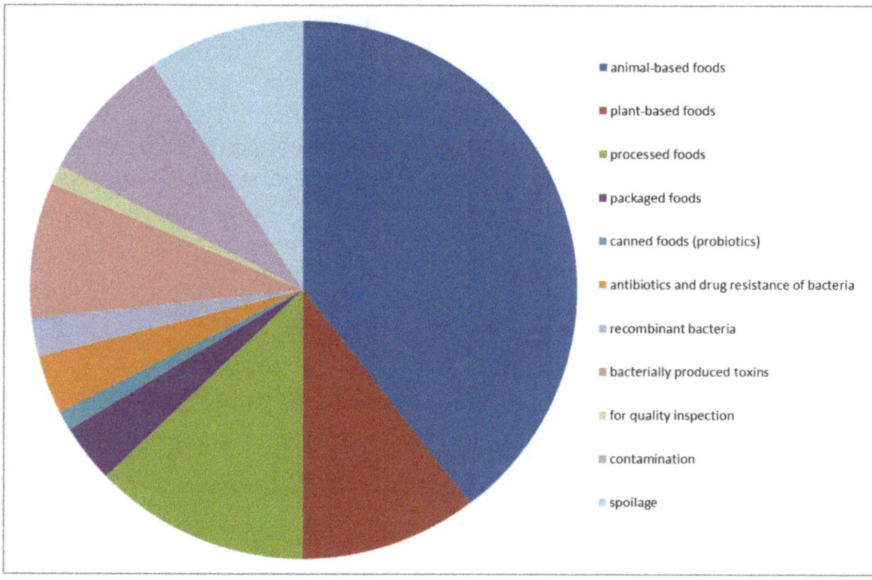

Figure 14.1 MS-based bacterial analysis in the food industry for bacteria-related food safety control.

be used to improve the shelf-life and characteristics of foods. MS-based strategies have been greatly developed for the analysis of bacteria in various food matrices.

The present chapter comprehensively summarized MS-based bacterial identification in different foods. The detection of pathogenic microbes in animal-based foods including poultry, seafoods, beef, and porcine meat as well as plant-based foods including fruits, lettuce, nuts, rice seeds, and potatoes was addressed. The detection of pathogenic bacteria in processed foods including fruit juice, RTE foods, baby foods, and fermented food products, as well as the probiotics used in (canned) foods, was summarized. Identification of recombinant bacteria or unauthorized GM bacteria by MALDI-TOF MS, detection of antibiotic residues and drug resistance of bacteria, and analysis of bacterially produced toxins by MS were discussed. The contamination and spoilage of foods monitored by MS were also summarized. Many other concerns like the factors affecting the reliability of identification of bacteria by MALDI-TOF MS, the accuracy and sensitivity of identification of bacteria by MALDI-TOF MS, the impact of probiotics on fruit juice aroma by GC–MS, the integration of MS techniques with other methods, and MALDI-TOF MS applied to food product quality inspection were also covered. MALDI-TOF MS can offer comparable power for discrimination of bacteria with 16S rDNA sequencing.

MALDI-TOF MS has been broadly utilized in bacteria-related food safety supervision together with GC–MS, LC–MS, nanoLC–MS, HPLC–MS/MS, LC–ESI/MS, UHPLC–MS/MS, FT-ICR/MS, or other emerging MS techniques.

The factors that can affect the detection of bacteria by MALDI-TOF MS may be better controlled together with the establishment of databases for improved reliability of identification and recognition of bacteria. Mycotoxins should be taken into account due to the risk of contamination of nuts with molds (fungi).

It is expected that MS-based technologies will continue to be applied in the food industry for global food safety management and risk analysis.

Abbreviations

AMR	Antimicrobial resistance
BCL	Bacillus identification card
ELISA	Enzyme-linked immunosorbent assay
FT-ICR/MS	Fourier transform ion cyclotron resonance/mass spectrometry
GAs	Gibberellins
GC–MS	Gas chromatography–mass spectrometry
GC–MS/SIM	Gas chromatography and mass spectrometric selected ion monitoring
GM	Genetically modified
GMM	Genetically modified microorganisms
HPB	Histamine-producing bacteria
HPLC–FLD	High-performance liquid chromatography–fluorescence detection
HRPGC–MS	High-resolution pyrolysis gas chromatography–mass spectrometry
HS-SPME	Headspace solid-phase microextraction
IMS	Immunomagnetic separation
LAB	Lactic acid bacteria
LC–ESI/MS	Reverse phase liquid chromatography coupled to electrospray ionization mass spectrometry
LC–MS/MS	Liquid chromatography–tandem mass spectrometry
LC–QITMS	Liquid chromatography–quadrupole ion trap mass spectrometry
MALDI-TOF MS	Matrix-assisted laser desorption/ionization time-of-flight mass spectrometry
MAP	Modified atmosphere
MRSA	Methicillin-resistant *Staphylococcus aureus*
m-PCR	Multiplex polymerase chain reaction
MS	Mass spectrometry
MSP	Main spectral profile
PCA	Principal component analysis
PCR	Polymerase chain reaction
PTR–MS	Proton transfer reaction–mass spectrometry
RAPD-PCR	Randomly amplified polymorphic DNA-polymerase chain reaction
RTE	Ready-to-eat

SDS–PAGE	Sodium dodecyl sulfonate–polyacrylamide gel electrophoresis
SEB	*Staphylococcus* enterotoxin B
SEs	Staphylococcal enterotoxins
SPME	Solid-phase microextraction
SRM	Selected reaction monitoring
2-DE	Two-dimensional [gel] electrophoresis
2D-GE	Two-dimensional gel electrophoresis
TFA	Trifluoroacetic acid
UHPLC–MS/MS	Ultra-high-performance liquid chromatography tandem mass spectrometry
UTI	Urinary tract infection
VOCs	Volatile organic compounds
YSI	Yolk sac infections

Acknowledgements

The author is grateful for the support and contribution of all research assistants of her group. Winnie C. Soko made a proofreading for the whole text. Li Wang helped to update the references. Li Wang, Weijiao Chen, Yunxing Li, Zhaoliang Chai, Xiaoxia Li, Fang Lin, Yuean Chu, and Tongtala Ao assisted in performing literature search. The author would like to thank her family, and her colleagues at Shanghai Ocean University (SHOU) for creating an inspiring environment to work.

References

1. A. J. Zakrzewski, U. Zarzecka, W. Chajęcka-Wierzchowska and A. Zadernowska, *Pathogens*, 2022, **11**, 1099.
2. M. Wenning, F. Breitenwieser, R. Konrad, I. Huber, U. Busch and S. Scherer, *J. Microbiol. Methods*, 2014, **103**, 44–52.
3. S. Ulrich, C. Gottschalk, R. Dietrich, E. Märtlbauer and M. Gareis, *Food Microbiol.*, 2019, **82**, 75–81.
4. I. M. Sulaiman, N. Miranda and S. Simpson, *J. AOAC Int.*, 2021, **104**, 1381–1388.
5. A. Elbehiry, E. Marzouk, M. Hamada, M. Al-Dubaib, E. Alyamani, I. M. Moussa, A. AlRowaidhan and H. A. Hemeg, *New Microbiol.*, 2017, **40**, 269–278.
6. Z. Chai and H. Bi, *Food Chem.: X*, 2022, **13**, 100225.
7. Z. Chai, W. C. Soko, J. Xie and H. Bi, *Food Chem.*, 2022, **396**, 133658.
8. S. Karasu-Yalcin, N. Soylemez-Milli, O. Eren and K. Eryasar-Orer, *J. Food Sci. Technol.*, 2021, **58**, 4102–4109.
9. Q. Liyun, J. Ruiping, G. Weili, L. Guoping, L. Xinchao and S. Xin, *Chin. J. Food Hyg.*, 2021, 131–137.

10. M. Quintela-Baluja, K. Bohme, I. C. Fernandez-No, S. Morandi, M. E. Alnakip, S. Caamano-Antelo, J. Barros-Velazquez and P. Calo-Mata, *Electrophoresis*, 2013, **34**, 2240–2250.
11. X. Li, P. T. Lv, L. Wang, A. L. Guo, M. H. Ma and X. B. Qi, *J. Chromatogr. B: Anal. Technol. Biomed. Life Sci.*, 2014, **971**, 107–111.
12. H. Kumar, K. Bhardwaj, T. Kaur, E. Nepovimova, K. Kuca, V. Kumar, S. K. Bhatia, D. S. Dhanjal, C. Chopra, R. Singh, S. Guleria, T. C. Bhalla, R. Verma and D. Kumar, *Foods*, 2020, **9**, 1504.
13. D. Dekker, D. Eibach, K. G. Boahen, C. W. Akenten, Y. Pfeifer, A. E. Zautner, E. Mertens, R. Krumkamp, A. Jaeger, A. Flieger, E. Owusu-Dabo and J. May, *Foodborne Pathog. Dis.*, 2019, **16**, 352–358.
14. C. S. Rodrigues, P. M. Armendaris, C. de Sa, J. P. A. Haddad and C. B. de Melo, *Curr. Microbiol.*, 2021, **78**, 2242–2250.
15. D. Stepien-Pysniak, T. Hauschild, M. Dec, A. Marek, R. Urban-Chmiel and U. Kosikowska, *Poult. Sci.*, 2021, **100**, 100985.
16. S. Mangmee, O. Reamtong, T. Kalambaheti, S. Roytrakul and P. Sonthayanon, *Food Control*, 2020, **113**, 107188.
17. Z. Yu, M. F. Peruzy, C. Dumolin, M. Joossens and K. Houf, *Food Microbiol.*, 2019, **82**, 53–61.
18. H. Jribi, H. Sellami, S. B. Amor, A. Ducournau, E. SifrÉ, L. Benejat, F. MÉgraud and R. Gdoura, *J. Food Prot.*, 2020, **83**, 2080–2086.
19. A. Woźniak-Biel, G. Bugla-Płoskońska, A. Kielsznia, K. Korzekwa, A. Tobiasz, A. Korzeniowska-Kowal and A. Wieliczko, *Microb. Drug Resist.*, 2018, **24**, 314–322.
20. M. Dec, A. Puchalski, R. Urban-Chmiel and A. Wernicki, *BMC Microbiol.*, 2016, **16**, 105.
21. I. M. Sulaiman, Y. H. Hsieh and S. Simpson, *J. AOAC Int.*, 2020, **103**, 197–204.
22. R. Tarabees, M. S. A. Elsayed, R. Shawish, S. Basiouni and A. A. Shehata, *J. Infect. Dev. Countries*, 2017, **11**, 314–319.
23. M. Dec, A. Nowaczek, D. Stępień-Pyśniak, J. Wawrzykowski and R. Urban-Chmiel, *BMC Microbiol.*, 2018, **18**, 168.
24. Z. J. Yu, M. Joossens and K. Houf, *Front. Microbiol.*, 2020, **11**, 539540.
25. M. Hilgarth, S. Fuertes, M. Ehrmann and R. F. Vogel, *Syst. Appl. Microbiol.*, 2018, **41**, 44–50.
26. S. Uehara, M. Kobayashi, K. Kimura, J. Suzuki and K. Sadamasu, *J. Food Sci. Technol.*, 2021, **58**, 4055–4061.
27. J. W. Hu, M. J. Cao, S. C. Guo, L. J. Zhang, W. J. Su and G. M. Liu, *J. Food Prot.*, 2015, **78**, 383–389.
28. Y. Cho, E. Kim, S. K. Han, S. M. Yang, M. J. Kim, H. J. Kim, C. G. Kim, D. W. Choo, Y. R. Kim and H. Y. Kim, *J. Microbiol. Biotechnol.*, 2017, **27**, 1593–1601.
29. C. R. Burbick, S. D. Nydam, G. K. Hendrix, T. E. Besser, D. Diaz and K. Snekvik, *J. Aquat. Anim. Health*, 2018, **30**, 332–338.
30. S. P. Kazazić, N. Topić Popović, I. Strunjak-Perović, S. Babić, D. Florio, M. Fioravanti, K. Bojanić and R. Čož-Rakovac, *PLoS One*, 2019, **14**, e0225343.

31. S. M. Malainine, W. Moussaoui, G. Prévost, J. M. Scheftel and R. Mimouni, *Lett. Appl. Microbiol.*, 2013, **56**, 379–386.
32. H. M. Raharjo, H. Budiyansah, M. F. Mursalim, P. Chokmangmeepisarn, R. Sakulworakan, S. Madyod, M. Sewaka, M. Sonthi, P. P. Debnath, S. Elayaraja, T. Rung-Ruangkijkrai, H. T. Dong and C. Rodkhum, *J. Fish Dis.*, 2022, **45**, 1355–1371.
33. X. Shan, J. Fu, X. Li, X. Peng and L. Chen, *J. Proteomics*, 2022, **251**, 104408.
34. S. Pečur Kazazić, N. Topić Popović, I. Strunjak-Perović, D. Florio, M. Fioravanti, S. Babić and R. Čož-Rakovac, *J. Fish Dis.*, 2019, **42**, 1201–1209.
35. M. Kačániová, A. Klūga, A. Kántor, J. Medo, J. Žiarovská, C. Puchalski and M. Terentjeva, *Microb. Pathog.*, 2019, **132**, 313–318.
36. S. W. Kim, S. W. Nho, S. P. Im, J. S. Lee, J. W. Jung, J. M. Lazarte, J. Kim, W. J. Lee, J. H. Lee and T. S. Jung, *J. Microbiol.*, 2017, **55**, 260–266.
37. P. Piamsomboon, J. Jaresitthikunchai, T. Q. Hung, S. Roytrakul and J. Wongtavatchai, *BMC Vet. Res.*, 2020, **16**, 52.
38. M. L. Ochoa and P. B. Harrington, *Anal. Chem.*, 2005, **77**, 5258–5267.
39. M. F. Peruzy, N. Murru, Z. Yu, P. J. Kerkhof, B. Neola, M. Joossens, Y. T. R. Proroga and K. Houf, *Int. J. Food Microbiol.*, 2019, **301**, 51–60.
40. S. K. Han, Y. Hong, H. L. Kwak, E. S. Kim, M. J. Kim, A. Shrivastav, M. H. Oh and H. Y. Kim, *J. Food Saf.*, 2014, **34**, 224–232.
41. F. Ahmad, O. O. Babalola and H. I. Tak, *Anal. Bioanal. Chem.*, 2012, **404**, 1247–1255.
42. M. Okumura, Y. Fujitani, M. Maekawa, J. Charoenpanich, H. Murage, K. Kimbara, N. Sahin and A. Tani, *J. Biosci. Bioeng.*, 2017, **123**, 190–196.
43. C.-T. Chen, J.-W. Yu and Y.-P. Ho, *J. Food Drug Anal.*, 2019, **27**, 575–584.
44. A. Santini, P. Mikusova, M. Sulyok, R. Krska, R. Labuda and A. Srobarova, *Mycotoxin Res.*, 2014, **30**, 213–220.
45. B. Veltman, D. Harpaz, S. Melamed, Z. Tietel, L. Tsror and E. Eltzov, *Talanta*, 2022, **247**, 123545.
46. K. A. Aliferis and S. Jabaji, *PLoS One*, 2012, **7**, e42576.
47. M. C. Adetunji, S. O. Aroyeun, M. B. Osho, M. Sulyok, R. Krska and M. Mwanza, *Food Addit. Contam., Part A*, 2019, **36**, 1847–1859.
48. J. I. N. Sombie, C. N. Ezekiel, M. Sulyok, K. I. Ayeni, F. Jonsyn-Ellis and R. Krska, *Food Addit. Contam., Part A*, 2018, **35**, 1573–1580.
49. D. Spadaro, G. R. Meloni, I. Siciliano, S. Prencipe and M. L. Gullino, *Toxins*, 2020, **12**, 307.
50. S. Mujahid, T. Pechan and C. L. Wang, *Int. J. Food Microbiol.*, 2008, **128**, 203–211.
51. Z. Sokołowicz, A. Augustyńska-Prejsnar, J. Krawczyk, M. Kačániová, M. Kluz, P. Hanus and J. Topczewska, *Animals*, 2021, **11**, 3282.
52. A. Augustyńska-Prejsnar, P. Hanus, Z. Sokołowicz and M. Kačániová, *Anim. Biosci.*, 2021, **34**, 2003–2011.
53. J. H. Zang, D. W. Yu, T. R. Li, Y. S. Xu, J. M. Regenstein and W. S. Xia, *Food Res. Int.*, 2022, **155**, 111128.

54. E. Kim, J. E. Won, S. M. Yang, H. J. Kim and H. Y. Kim, *Foods*, 2022, **11**, 1068.
55. Y. Lee, Y. Cho, E. Kim, H. J. Kim and H. Y. Kim, *J. Microbiol. Biotechnol.*, 2018, **28**, 1112–1121.
56. Z. Ping, Z. Pengfei, X. Mengxi, A. Feiyu, Q. Boshu and W. Rina, *J. Food Sci.*, 2018, **83**, 1342–1349.
57. O. R. Ogunremi, S. Freimüller Leischtfeld and S. Miescher Schwenninger, *Int. J. Food Microbiol.*, 2022, **366**, 109563.
58. D. Mayr, R. Margesin, F. Schinner and T. D. Mark, *Int. J. Mass Spectrom.*, 2003, **223**, 229–235.
59. I. C. Fernández-No, K. Böhme, P. Calo-Mata, B. Cañas, J. M. Gallardo and J. Barros-Velázquez, *Food Microbiol.*, 2012, **30**, 91–97.
60. J. Sato, M. Nakayama, A. Tomita, T. Sonoda, M. Hasumi and T. Miyamoto, *PLoS One*, 2017, **12**, e0186327.
61. M. M. Rasmussen, J. A. Opintan, N. Frimodt-Møller and B. Styrishave, *PLoS One*, 2015, **10**, e0139706.
62. A. Limayem, R. S. Donofrio, C. Zhang, E. Haller and M. G. Johnson, *J. Environ. Sci. Health, Part B*, 2015, **50**, 827–832.
63. N. Pauly, H. Wichmann-Schauer, B. Ballhausen, N. Torres Reyes, A. Fetsch and B. A. Tenhagen, *Int. J. Food Microbiol.*, 2019, **292**, 8–12.
64. D. Xiao, X. X. Tao, P. Wang, G. D. Liu, Y. N. Gong, H. F. Zhang, H. B. Wang and J. Z. Zhang, *Biomed. Environ. Sci.*, 2014, **27**, 250–258.
65. M. A. Fraiture, N. Papazova and N. H. C. Roosens, *Int. J. Food Microbiol.*, 2021, **337**, 108913.
66. I. Sospedra, J. M. Soriano and J. Mañes, *Food Res. Int.*, 2013, **54**, 1052–1060.
67. I. Sospedra, C. Soler, J. Mañes and J. M. Soriano, *J. Chromatogr. A*, 2012, **1238**, 54–59.
68. K. D. Bao, A. Letellier and F. Beaudry, *Biomed. Chromatogr.*, 2012, **26**, 1049–1057.
69. R. Shahzad, M. Waqas, A. L. Khan, S. Asaf, M. A. Khan, S. M. Kang, B. W. Yun and I. J. Lee, *Plant Physiol. Biochem.*, 2016, **106**, 236–243.
70. F. Celandroni, A. Vecchione, A. Cara, D. Mazzantini, A. Lupetti and E. Ghelardi, *PLoS One*, 2019, **14**, 13.
71. Y. Wang, S. Liu, Q. Pu, Y. Li, X. Wang, Y. Jiang, D. Yang, Y. Yang, J. Yang and C. Sun, *Food Chem.*, 2018, **262**, 7–13.
72. M. L. Lee, D. L. Smith and L. R. Freeman, *Appl. Environ. Microbiol.*, 1979, **37**, 85–90.
73. M. Mikš-Krajnik, Y. J. Yoon, D. O. Ukuku and H. G. Yuk, *J. Food Sci.*, 2016, **81**, M2006–M2014.
74. F. Carraturo, G. Libralato, R. Esposito, E. Galdiero, F. Aliberti, A. Amoresano, C. Fontanarosa, M. Trifuoggi and M. Guida, *J. Food Sci.*, 2020, **85**, 3467–3477.
75. M. A. Farag, E. U. Dokalahy, T. F. Eissa, I. M. Kamal and A. Zayed, *ACS Omega*, 2022, 7, 2377–2390.

76. T. Zhang, H. Ding, L. Chen, S. S. Zhang, P. F. Wu, K. Z. Xie, Z. M. Pan, G. X. Zhang, G. J. Dai, H. Q. Wu and J. Y. Wang, *Food Res. Int.*, 2021, **144**, 110328.
77. K. Tirnsorn, T. Thoopboochagorn, N. Lertwattanasakul and C. Wongchoosuk, *Biosyst. Eng.*, 2016, **151**, 116–125.
78. L. Höll, J. Behr and R. F. Vogel, *Food Microbiol.*, 2016, **60**, 84–91.
79. K. Böhme, I. C. Fernández-No, J. Barros-Velázquez, J. M. Gallardo, B. Cañas and P. Calo-Mata, *Electrophoresis*, 2011, **32**, 2951–2965.
80. S. Chen, S. Liu, J. Ma, X. Xu and H. Wang, *Food Res. Int.*, 2022, **156**, 111365.

Subject Index